236 Current Topics in Microbiology and Immunology

Springer
Berlin
Heidelberg
New York
Barcelona
Hong Kong
London
Milan
Paris
Singapore
Tokyo

Defense of Mucosal Surfaces: Pathogenesis, Immunity and Vaccines

Edited by J.-P. Kraehenbuhl and M.R. Neutra

With 30 Figures and 10 Tables

Professor Dr. JEAN-PIERRE KRAEHENBUHL
Swiss Institute for Experimental Cancer Research
Institute of Biochemistry
University of Lausanne
Ch. des Boveresses 155
CH-1066 Epalinges
Switzerland

Professor Dr. MARIAN R. NEUTRA
GI Cell Biology Laboratory
Children's Hospital and Harvard Medical School
300 Longwood Ave.
Boston, MA 02115
USA

Cover Illustration: The mouse Peyer's patch is of tissue from a ligated loop that was injected with reovirus Type 1 and collected 1 hour later. An M cell and adjacent enterocyte is shown. The photomicrograph is by Richard Weltzin.

Cover Design: design & production GmbH, Heidelberg

ISSN 0070-217X
ISBN 3-540-64730-9 Springer-Verlag Berlin Heidelberg New York

Library of Congress Catalog Card Number 15-12910
Printed in Germany

Typesetting: Scientific Publishing Services (P) Ltd, Madras

Production Editor: Angélique Gcouta

SPIN: 10643020 27/3020 – 5 4 3 2 1 0 – Printed on acid-free paper

Preface

Mucosal surfaces of the gut, the airways and the urogenital tract are covered by epithelial tissues that form tight barriers between a highly regulated internal compartment and a rapidly changing external environment. To maintain the integrity of these vulnerable cellular barriers mucosal surfaces have acquired specialized innate and adaptive defense mechanisms, including a major branch of the immune system. The mucosal immune system is anatomically and functionally distinct from that found elsewhere in the body, in that it has developed specialized processes for antigen uptake, transport, processing and presentation as well as specialized immune effector mechanisms such as polymeric immunoglobulin secretion. Distinctive immune effector cells which are produced in response to antigens and pathogens sampled in mucosal lymphoid tissues acquire a specific homing program which allows them to return to mucosal sites. In addition, mucosal tissues act as primary lymphoid organs where lymphocytes develop de novo from immature precursors, undergo rearrangement of their antigen receptor genes, and differentiate into effectors involved in the protection of epithelia and modulation of immune functions.

The role of mucosal immunity in the defense against infectious agents has been recognized for decades, but progress in this field has been hampered by the technical difficulty of analyzing immune effectors in mucosal tissues, and by the lack of appropriate *in vitro* systems which include epithelial, lymphoid and antigen-presenting cells. The spatial organization of lymphoid tissue in mucosal surfaces of the gut or the airways is the result of complex interactions between the microorganisms, the mucosal epithelium, and the cells of the immune system. We are only beginning to understand the nature of the cross-talk.

Such information is essential for rational design of vaccines that can be targeted into mucosal tissues. To date, most studies in this area have been restricted to animal models, and the few reported clinical studies indicate that the mouse and human mucosal immune systems differ in their response to both subunit and live mucosal vaccines. One message emerging from the overview of mucosal vaccines in this volume is that much more effort should be devoted to elucidating how antigens and vaccines are processed in mucosal tissues of humans.

Mucosal immunology encompasses a vast field that includes areas of developmental biology, epithelial cell and molecular biology, molecular and cellular immunology, microbiology, virology, and vaccinology. Since this body of information is rapidly expanding, an integration of current knowledge is needed for experts in each field and nonspecialists alike. This volume on mucosal immunity and vaccines presents up-to-date and concise discussions of the key components of the mucosal immune system, mechanisms of microbial-epithelial cell interactions, and basic science relevant to mucosal vaccination. The volume begins with chapters reviewing the organization of mucosal lymphoid tissue, epithelial antigen sampling and antigen presentation in mucosal tissues, induction of immune responses and/or tolerance in mucosal tissues, and the role of immune effectors in protecting mucosal sites. The next chapters review microbial-host interactions, focusing specifically on viruses and bacteria that infect mucosal tissues. Finally the role of mucosal adjuvants in promoting efficient mucosal responses and the properties of subunit and live mucosal vaccines are discussed.

It is our hope that the multidisciplinary effort represented by this volume will be a valuable resource for researchers, clinicians and students who need clear distillation of basic concepts and a guide to the wide-ranging literature in this very active research area.

Jean-Pierre Kraehenbuhl
and Marian Neutra

List of Contents

List of Contributors

(Their addresses can be found at the beginning of their respective chapters.)

Organization of Mucosal Lymphoid Tissue

W.R. Hein

1 Introduction

Mucous membranes form interfaces between internal microenvironments that are subject to homeostatic regulation and external environments that vary widely and are frequently contaminated with microbes and other potentially injurious agents. To offset their inherent vulnerability, mucosal surfaces have therefore been endowed with specialized defense mechanisms, many of which have an immune component.

Mucosal lymphoid tissues perform unique roles in the afferent and efferent limbs of an immune response, including the specialized uptake and processing of antigens and the production of distinctive classes of immunoglobulins and effector cells. These features impart special characteristics to mucosal immunity. In addition, it is becoming increasingly clear that immune events at mucosal surfaces have wider ramifications, since certain mucous membranes have emerged as important sites where lymphoid cells either develop *de novo* from immature precursors or become imprinted and conditioned in ways that regulate their subsequent systemic behavior. The effective expression of these diverse roles hinges on the development of localized lymphoid tissues and the tactical recruitment and deployment of special populations of immune-system cells.

Basel Institute for Immunology, Grenzacherstrasse 487, Postfach CH-4005, Basel, Switzerland
Present Address: Wallaceville Animal Research Centre, P.O. Box 40063, Upper Hutt, New Zealand

The purpose of this introductory chapter is to review, briefly, the overall organization of lymphoid tissues associated with mucosal surfaces. Two levels of organization are considered: a morphological one, dealing with the structure of mucosal lymphoid tissues; and a developmental one, involving those factors that regulate the ontogeny of these tissues. There will be no attempt, here, to address a third and arguably the most important level – that of functional organization – since that topic forms the overall focus of the balance of this book and is covered in detail in the chapters that follow.

2 Morphology of Mucosal Lymphoid Tissue

According to one convention, the property of 'being organized' is itself used as a criterion for classification, and mucosal lymphoid tissues are grouped into those that are organized in some way and those that remain diffuse. While making this distinction is useful in many ways, it nevertheless carries the misleading implication that diffuse lymphoid tissue has no organizational qualities. This is certainly not the case in a functional sense, or probably in a structural one.

2.1 Organized Lymphoid Tissue

All mucosal surfaces potentially contain organized lymphoid tissues and these can be described and categorized at different levels of complexity. At the most basic level, organized mucosal lymphoid tissues are comprised of two structural units – B-cell follicles and interfollicular or parafollicular T-cell sheets. Along the tracheobronchial tree, these structural units are usually solitary in nature, microscopic in size, and they remain sparse in the absence of deliberate antigenic challenge, although there are important differences between species (Pabst and Gehrke 1990). In other cases, follicle morphology has been modified considerably and they occur constitutively. For example, certain regions of the gut mucosa contain dense aggregations of follicles which are easy to detect macroscopically; mucocutaneous junctions are usually invested to some degree with lymphoid follicles, and follicular structures are associated with many secretory glands emptying at mucosal surfaces (Table 1.). The remainder of this section summarizes the structure of these lymphoid elements and highlights the major differences in their prevalence and location among different types of mucosal surface.

Three morphological types of lymphoid follicles can be distinguished – simple follicles, follicle-dome structures and lymphoglandular complexes (Fig. 1). A simple lymphoid follicle consists of a more-or-less spherical core of B lymphocytes assembled within a matrix of loose connective tissue and follicular dendritic cells. Simple follicles may occur singly or in groups, they can be found in or near all mucosal surfaces and usually contain variable numbers of T lymphocytes and

Table 1. Lymphoid tissues found at mucosal surfaces

Type of lymphoid tissue	Distribution	Characteristics
Organized lymphoid tissue:		
Microscopic		
Simple follicles	All mucosal surfaces	Mainly occur singly or in small groups. Dense aggregations found at some body sites, e.g., mucocutaneous junctions
Follicle-dome structures	Intestine	Main lymphoid element of Peyer's patches; comprised of body, corona and dome regions
Lymphoglandular complexes	Intestine	Occur singly or in small aggregates; may be interspersed with simple follicles or follicle-dome structures
Folliculoglandular complexes	Mucocutaneous junctions, pilosebaceous units, palpebral glands	Follicles associated with specialized secretory glands
Cryptopatches	Intestine	Occur as small foci between intestinal crypts; proposed sites for T- and/or B-cell differentiation
Macroscopic		
Tonsil	Base of tongue, pharynx	Dense aggregation of B-cell follicles and interfollicular T-cell sheets
Appendix	Large intestine	Accumulation of B cell follicles in terminal part of cecum
Peyer's patches	Intestine	Sites of antigen sampling from gut lumen and induction of immune responses. Expanded role in production of systemic B cells and primary antibody repertoire in some species.
Bursa of Fabricius	Avian proctodeum	Primary site of B cell differentiation
Diffuse lymphoid tissue:		
Mast cells, intraepithelial T cells, dendritic cells, migratory T cells homing to the lamina propria, IgA plasma cells and their precursors	Occur differentially at all mucosal surfaces, mostly within the lamina propria; may develop *in situ*, be recruited from blood circulation or migrate through lymphatic vessels to the regional lymph nodes	Impart distinctive functional properties to the afferent and efferent arms of mucosal immune responses

macrophages (Fig. 2a,b). In mucosal surfaces with a well-defined muscularis mucosae, they are sometimes situated entirely within the lamina propria, and may then be called proprial follicles. However, simple follicles often assume irregular shapes and many contain a 'tail' of cells that protrudes through a break in either the muscularis mucosae or connective tissue tunics to form a submucosal extension

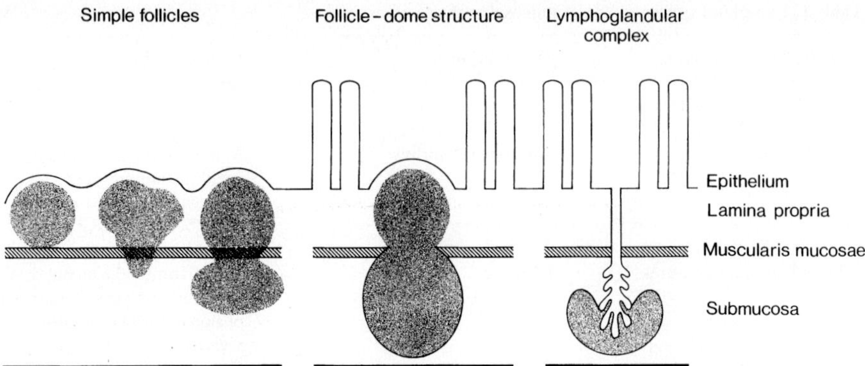

Fig. 1. Schematic diagram showing the main structural characteristics of different lymphoid follicles. The different types of follicles vary in their shape, in their relationship with other tissue layers present at mucosal surfaces and in the degree to which they are encapsulated. Simple follicles may occur at any mucosal surface, whereas follicle-dome structures and lymphoglandular complexes occur in the gut

(Fig. 2c,d). Unlike follicle-dome structures and lymphoglandular complexes, simple follicles are not obviously encapsulated, although they are sometimes partially encircled by a network of small lymphatic vessels. Proprial follicles in the gut are covered by a specialized type of epithelium called follicle-associated epithelium (FAE).

Follicle-dome structures are usually larger and possess extra levels of organization (Fig. 3a,b). The follicle itself consists of identifiable body, corona and dome regions, with the body of the follicle lying submucosally. The apex of the follicle passes through a break in the muscularis mucosae and extends outward to form the corona and dome, which protrude into the lamina propria. Follicle-dome structures form the major lymphoid component of the Peyer's patches in the small intestine and also occur in the large intestine either singly or as aggregates. The follicle body is encapsulated and the dome is covered by FAE which usually contains specialized M cells that play important roles in the transcytosis of particulate material from the gut lumen. These cells form membranous extensions at the luminal surface and interdigitate with groups of lymphocytes, macrophages and dendritic cells at their basal surface.

Lymphoglandular complexes have a more unusual structure (Fig. 3c,d). A central epithelial invagination protrudes vertically through a break in the muscularis mucosae and gives rise to diverticulae which then extend into the submucosal tissues to form lymphoid follicles surrounded by marginal infiltrations of lymphocytes (Lowden and Heath 1995). These structures occur most commonly in the large intestine, less frequently in the small intestine, and the central invagination usually contains ingesta, cells and exudates. The mucosal lining of lymphoglandular complexes does not show the specializations of FAE, but contains columnar or cuboidal cells similar to those found in either villous or glandular epithelium, as well as goblet cells and Paneth cells. An extensive network of lymphatic vessels and

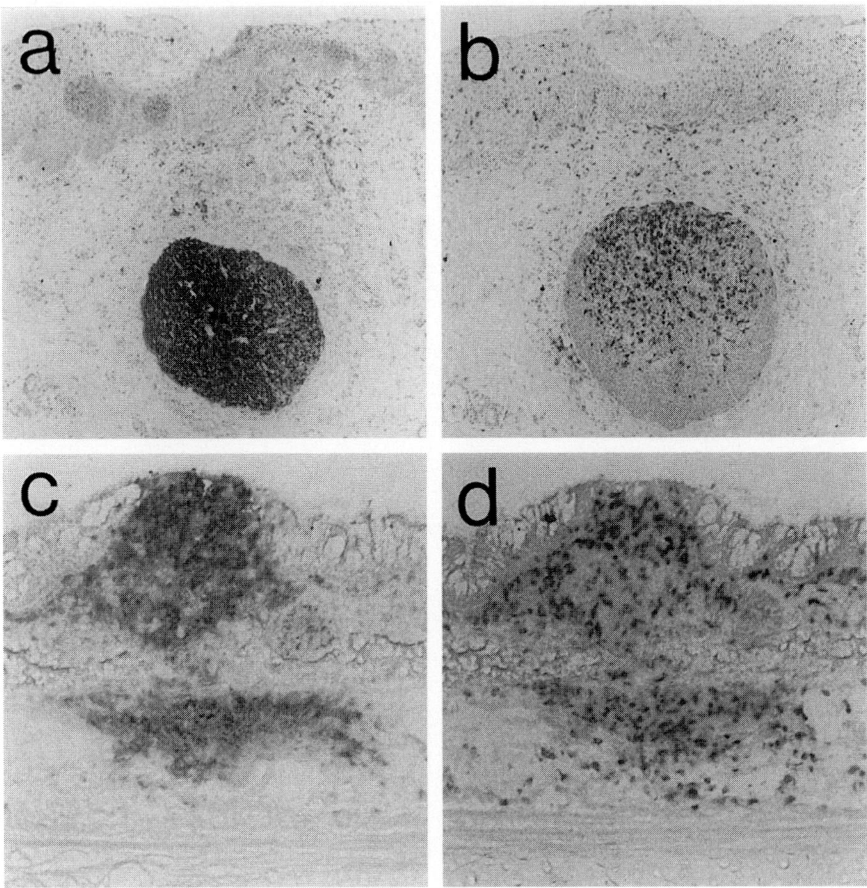

Fig. 2a–d. Simple lymphoid follicles. (**a**) A single spherical follicle lying beneath stratified squamous epithelium containing predominantly $CD72^+$ B cells. (**b**) One pole and the central region of the follicle are infiltrated with $CD4^+$ T cells. A small irregularly shaped lymphoid follicle beneath pseudostratified ciliated epithelium has a similar cellular composition as shown by staining with monoclonal antibodies specific for CD72 (**c**) and CD3 (**d**). In this case, lymphocytes at the apex of the follicle have infiltrated the epithelium and part of the base of the follicle protrudes through a dense layer of elastic connective tissue into the submucosal region. (**a,b**) *Sheep vagina* ×50 (**c,d**) *Sheep trachea* ×80

anastomosing sinuses surrounds both follicle-dome structures and lymphoglandular complexes, although there are no lymph vessels emanating from within the follicle body (Lowden and Heath 1992; 1994; 1995; 1996).

At certain body sites, follicular morphology has become more specialized as a consequence of the aggregation of follicles or their regular association with other tissues; this introduces a second level of morphological complexity. For example, the tonsils, Peyer's patches, the appendix and large-intestinal patches arise as sites where follicles aggregate constitutively, to an extent that these lymphoid structures become visible macroscopically. The aggregated lymphoid tissues have their own morphological characteristics. The tonsils consist of a thickened section of mucosa

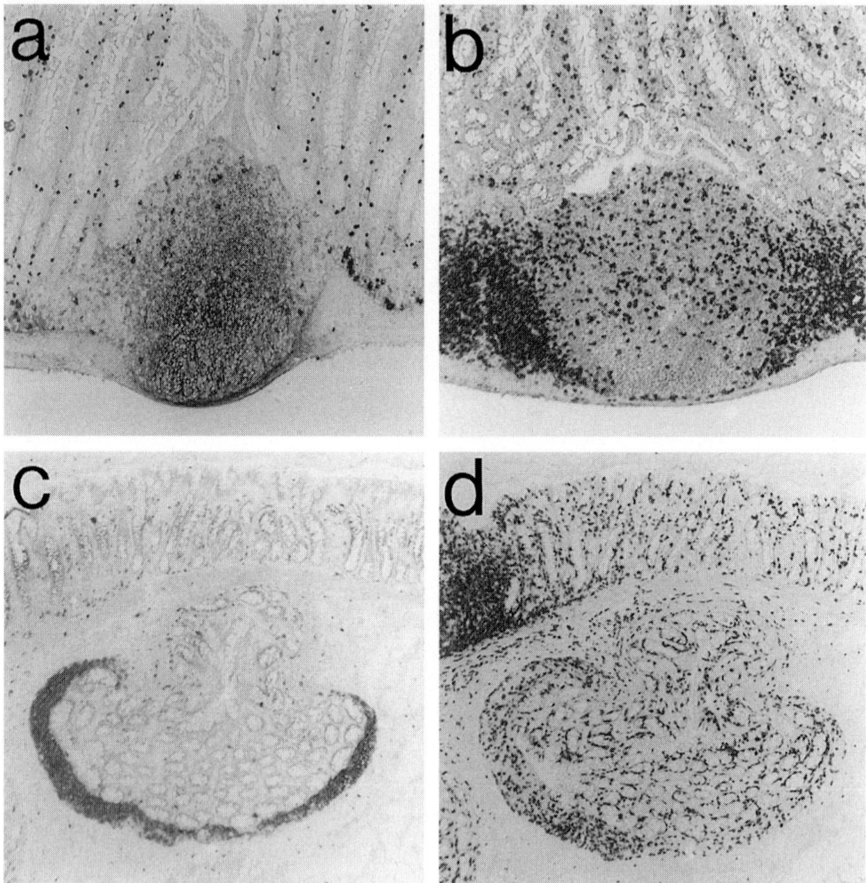

Fig. 3a–d. Follicle-dome structure and lymphoglandular complex. (**a**) An encapsulated follicle-dome structure in the small-intestinal mucosa containing predominantly IgM^+ B cells. (**b**) Some $CD3^+$ cells occur within the follicle body and dome, although T cells are especially concentrated in the interfollicular regions. The follicular dome is covered by distinctive follicle-associated epithelium. (**c**) An inverted lymphoglandular complex containing a prominent peripheral accumulation of $CD72^+$ B cells. (**d**) A significant number of $CD3^+$ cells also occur within the lymphoid and epithelial components of the lymphoglandular complex as well as in adjacent intestinal villi. (**a,b**) *Rat jejunal Peyer's patch* ×50 (**c,d**) *Calf colon* ×35

formed into folds and crypts. Here, B-cell follicles aggregate beneath non-keratinizing stratified squamous epithelium and are separated by dense interfollicular T-cell sheets (Fig. 4).

Peyer's patches occur in the small intestine of all animals, although their number, morphology, relative distribution in different parts of the intestine, life-span, cellular composition and function vary among different species (reviewed by GRIEBEL and HEIN 1996). Peyer's patches situated in the jejunum play important roles in the induction and dissemination of immune responses to gut antigens, as discussed in detail in later chapters. In sheep, and perhaps in other species, the ileal

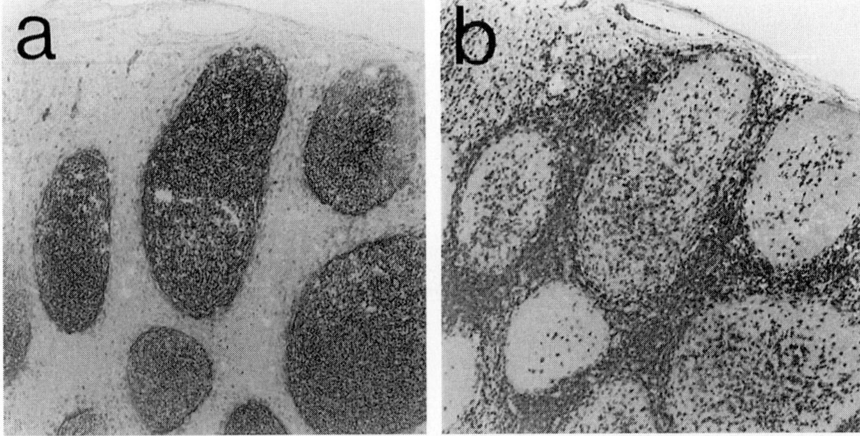

Fig. 4a,b. Palatine tonsil. The sections are stained with monoclonal antibodies specific for (**a**) CD21 and (**b**) CD3 to show the distribution of the B cell-FDC network and T cells, respectively. Tonsillar tissue consists of a dense accumulation of B-cell follicles separated by an interconnected network of T-cell sheets. *Sheep* ×35

Peyer's patches (IPP) have specialized functions. The IPP follicles are a site where B cells undergo intense population expansion, during which rearranged immunoglobulin V-region genes are diversified by hypermutation. The production of B cells and hypermutation appear to be regulated by antigen-independent mechanisms (Reynaud et al. 1995; 1997; Griebel and Hein 1996).

Although by convention the term Peyer's patch is usually applied only to aggregations of follicles that occur in the small intestine, comparable lymphoid patches occur at other intestinal sites. The cecal and colonic mucosae are usually invested to varying degrees with aggregated lymphoid tissues, although their precise location and morphology again vary somewhat among species. Large intestinal lymphoid patches often have a heterogeneous composition and may contain mixtures of two or even three of the follicle structures shown in Fig. 1. In mammals, clusters of follicles are also found characteristically adjacent to the ano-rectal mucocutaneous junction (Fig. 5a,b). The bursa of Fabricius, a specialized aggregation of B-cell follicles lying in the dorsal wall of the proctodeum, plays a key role in the generation of the avian primary-antibody repertoire.

The gut contains the most abundant mucosal lymphoid tissues and includes representatives of all follicle types. However, it remains unclear how the variations in follicle morphology arise and whether they correlate with differences in function. Two general trends can be identified in the progression from simple follicles to follicle-dome structures and then to lymphoglandular complexes (Fig. 1). First, there is an increasing invasion of the follicle body into the submucosal tissues and this is associated with the formation of a distinct capsule and a more intimate relationship to the lymphatic drainage. Secondly, simple follicles may occur anywhere in the gut, but the frequency of the more organized structures increases with progression along the gut (follicle-dome structures, mainly in the jejunum and

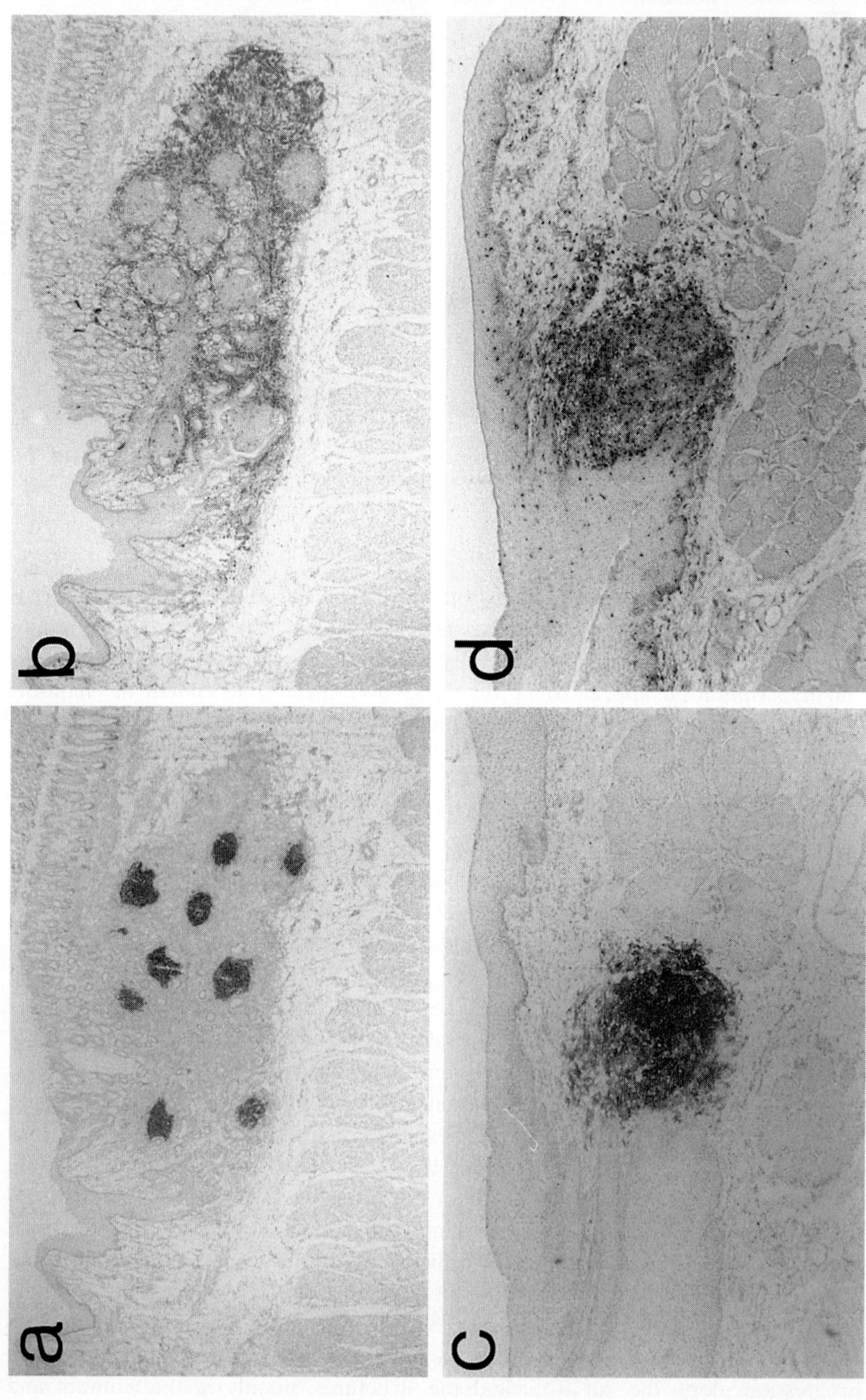
a
b
c
d

ileum; lymphoglandular complexes, mainly in the large intestine). These features may reflect differences between the constitutive ontogeny of each structure and may impart distinctive roles to the different follicle types in terms of the capture, processing and presentation of mucosal antigens.

Lymphoid follicles also occur constitutively near many secretory glands associated with epithelial surfaces and these can be considered as comprising of folliculoglandular complexes (Fig. 5c,d). Examples include the pilosebaceous units associated with some hair follicles, the lymphoid follicles near the palpebral glands and their ducts, and the lymphoid follicles associated with apocrine and other glands emptying onto the moist mucosal surfaces near mucocutaneous junctions (conjunctivae, openings of the urogenital tract, nasal cavities and nostrils). The close association of lymphoid follicles and glands at these locations probably facilitates the attainment of high levels of relevant antibodies in glandular secretions.

Recently, a new type of organized lymphoid tissue termed ‘cryptopatches’ (CP) has been detected in the intestine of mice. These structures are microscopic in size (each CP contains about 1000 lymphocytes), are located in the crypt lamina propria and first become detectable 14–17 days after birth. In terms of their histogenesis, lymphocyte composition and tissue levels of cell division, CPs differ from all other organized lymphoid tissue found in the murine gut. It has been proposed that they represent sites for the early generation of interleukin (IL)-7-dependent progenitors of T and/or B cell lineages (KANAMORI et al. 1996).

2.2 Diffuse Lymphoid Tissue

In addition to the organized elements outlined in section 2.1, mucosal surfaces contain variable numbers of lymphoid cells that do not associate into recognizable higher-order structures. These cells are distributed among the tissue spaces of the mucosa, either interspersed among epithelial cells or in the loose connective tissue of the lamina propria. T cells, B cells and plasma cells all contribute to diffuse lymphoid tissue, although their relative abundance varies considerably between different mucosal surfaces. These cells are rare or absent at most mucosal surfaces during fetal life, and their development and localization after birth is linked to the degree of antigen exposure. The gut is particularly well invested with diffuse lymphoid tissues.

Although they are not lymphoid cells, granulocytes (neutrophils, eosinophils, basophils and mast cells), monocytes and dendritic cells also extravasate into

Fig. 5a–d. Folliculoglandular complexes at or near mucocutaneous junctions. (**a**) A prominent submucosal accumulation of CD21^{+} B-cell follicles and (**b**) CD3^{+} lymphocytes lying at the junction between stratified squamous epithelium of the perianal skin and simple cuboidal epithelium of the rectal mucosa. Crypts of the rectal mucosa are invaginated between B-cell follicles in a part of the complex. (**c**) A secretory gland emptying through a duct onto the moist stratified preputial mucosa is closely associated with a lymphoid follicle containing CD21^{+} B cells and FDCs and (**d**) intrafollicular and parafollicular T cells. (**a,b**) Anorectal junction of 144-day-old fetal lamb × 20 (**c,d**) Sheep prepuce × 35

mucosal surfaces. These cells contribute to mucosal defense by performing effector functions *in situ* or by migrating from the mucosa into organized lymphoid tissues, such as the Peyer's patches or regional lymph nodes. Some of these cells, for example mast cells and dendritic cells, acquire distinctive phenotypic and functional characteristics in mucosal surfaces which distinguish them from corresponding cells in other tissues (Kitamura 1989; Kelsall and Strober 1996). Therefore, even though they show no clear structural organization, the diffuse lymphoid tissues play crucial roles in imparting distinctive properties to the afferent and efferent arms of mucosal immune responses.

3 Development of Mucosal Lymphoid Tissue

A number of genetic and environmental factors regulate the development of mucosal lymphoid tissues. However, because experimental observations have been made in diverse species, kept under a wide range of conditions, no overall consensus has emerged regarding the relative importance or precise roles of these factors. This section will summarize more recent findings relating to the development of gut-associated lymphoid tissues, since these are the best studied experimentally.

The macroscopically organized lymphoid tissues of the gut (tonsil, Peyer's patches, appendix) seem to develop in a constitutive way as part of an ontogenetic plan, in much the same way as animals develop other lymphoid tissues, such as spleen and lymph nodes, at defined and predictable anatomical locations. For example, each species has a characteristic frequency and distribution of Peyer's patches and these structures develop during fetal life in the absence of deliberate antigenic stimulation (Griebel and Hein 1996). This implies that defined sites or regions in the intestine are predestined in some way to become colonized with lymphoid cells, then supporting the growth of follicles and that this is at least partially under genetic control.

Recent studies in mouse embryos have defined three successive stages in Peyer's patch formation (Adachi et al. 1997). The earliest identifiable step involved the expression of vascular cell-adhesion molecule-1 (VCAM-1) on clusters of stromal cells at 15.5 days post coitus (d.p.c.). Round, lymphoid-like cells expressing major histocompatability complex (MHC) class II, IL-7R or CD4 accumulated in the vicinity of the stromal cell clusters by 17.5 d.p.c. Lymphocytes expressing CD3 or B220 were detected only during the final step which started at 18.5 d.p.c. The cell clusters were detected first in the upper jejunum, but later extended to the colon.

Around eight or nine clusters were present by the time of birth, which equates well with the number of Peyer's patches found in normal adult mice. In mutant mice homozygous for the allele "alymphoplasia" (*aly*/*aly*), where no lymph nodes or Peyer's patches are found in adults (see Table 2), none of these three steps was detected. In severe combined immunodeficient (SCID) mice, which lack mature

Table 2. Natural mutations and targeted gene deletions affecting the development of lymphoid tissues and organs.

Mutation/ deletion	Primary B-cell follicles[a]	Secondary B-cell follicles (germinal centers)[a]	Lymph nodes	Peyer's patches	Reference
aly/aly	Absent	Absent	Absent	Absent	Miyawaki et al. (1994), Nanno et al. (1994)
TNFRp55 $^{-/-}$	Absent	Absent	Normal	Present but reduced in number and size	Neumann et al. (1996) Pasparakis et al. (1997)
TNFRp75 $^{-/-}$	Normal	Normal	Normal	Normal	Erickson et al. (1994)
LTα $^{-/-}$	Absent	Absent	Absent	Absent	De Togni et al. (1994), Banks et al. (1995)
LTβ $^{\Delta/\Delta}$	Absent	Absent	Mostly absent[b]	Absent	Alimzhanov et al. (1997)
TNF $^{-/-}$	Absent	Absent	Normal	Present, but reduced in number and size	Pasparakis et al. (1996; 1997)

aly/aly; alymphoplasia; *TNFR* tumor necrosis factor receptor; *LT* lymphotoxin; *TNF* tumor necrosis factor.
[a] Distribution in spleen for those mutations where lymph nodes are absent.
[b] Mesenteric and cervical lymph nodes present; all other nodes absent

lymphocytes, the first and second steps proceeded, but there was a block at stage three. Collectively, these results suggest that a number of separate mechanisms, including the development of receptive local environments and the influx and expansion of lymphocytes are interlinked during Peyer's patch ontogeny and that the initial colonization process is largely completed before birth.

Similar processes probably occur in other species, although they may operate at different times in ontogeny and at different sites in the gut. In chickens, the bursa of Fabricius becomes colonized by B cells over a relatively short developmental window. In fetal lambs (gestation 150 days), lymphoid follicles develop in the colon around day 60, in the jejunum at day 75 and in the ileum at day 110 (Landsverk et al. 1991). The later oligoclonality of Vλ genes in individual sheep IPP follicles suggests that only three to four B cell clones successfully colonize each follicle (Reynaud et al. 1991). The colonizing cells have completed immunoglobulin (Ig)-gene rearrangement and express surface IgM. Within the follicles, they appear to be maintained by a self-replicating process that is dependent on growth factors produced by intrafollicular stromal cells (Griebel and Ferrari 1994). In humans, the colonization and development of Peyer's patches in the small intestine commences at about 11 weeks of gestation (Spencer et al. 1986).

Recent observations in mice containing either spontaneous mutations or targeted gene deletions have begun to identify some of the genes and factors that regulate the development of peripheral lymphoid tissues, including the Peyer's patches (Table 2). The spontaneous autosomal recessive mutation *aly* results in a generalized and complete absence of lymph nodes and Peyer's patches. The thymus and spleen are present, but have abnormal micro-architecture. Mutant homozygotes are deficient in both humoral and cell-mediated immune functions, with severely depressed levels of serum IgG and IgA, although intestinal intraepithelial lymphocytes continue to develop with relatively minor changes in their surface phenotype and number. The nature of the genetic lesion and the factors contributing to the pathogenesis of the phenotype remain unknown, but available evidence points toward a possible disorder in a mesenchymal component of lymph nodes and Peyer's patches (Miyawaki et al. 1994; Nanno et al. 1994).

A number of recent studies indicate that the structurally related cytokines tumor necrosis factor (TNF, also termed TNFα) and lymphotoxin (LT) are critically involved in the organogenesis of lymphoid tissues. The targeted disruption or deletion of genes encoding TNF, LTα, LTβ, and two common receptor proteins TNFRp55 and TNFRp75, has produced a range of phenotypes that sometimes includes differential effects on the development of lymph nodes compared with Peyer's patches (Table 2). The disruption of TNFRp75 has been the least damaging genetic lesion; the mice were viable, contained no overt phenotypic abnormalities in lymphoid organs and showed normal T-cell development, although the resistance of these mice to TNF-induced death was increased (Erickson et al. 1994). At the other end of the phenotypic spectrum, the targeting of LTα and LTβ genes caused severe lesions. The Peyer's patches and all lymph nodes, except mesenteric and cervical, were ablated and there was no development of primary or secondary B-cell follicles in the spleen (De Togni et al. 1994; Banks et al. 1995, Alimzhanov et al. 1997).

The targeting of TNF or TNFRp55 genes produced quite similar and, in a certain sense, more intriguing defects. Neither lesion had an effect on the gross organogenesis of lymph nodes and spleen, which had distinct T- and B-cell areas implying normal B-cell homing, although these organs lacked organized networks of follicular dendritic cells and primary and secondary follicles did not develop (Neumann et al. 1996; Pasparakis et al. 1996; 1997). In contrast, the development of Peyer's patches was more severely affected. In an initial study, targeting of TNFRp55 caused a nearly total ablation of Peyer's patches – only 6 of 18 knockout mice contained a single, small aggregation of follicles in the ileum, whereas 4 control mice contained an average of 6.5 Peyer's patches (Neumann et al. 1996).

In a second study, mice with disrupted TNF or TNFRp55 genes contained an average of two to four Peyer's patches per mouse compared with an average of six to eight patches found in wild-type controls (Pasparakis et al. 1997). Although the severity of the defects differed between these two studies, and this led to some differences between the interpretations placed on the results, the effects were actually quite similar in nature. In both cases the Peyer's patches were markedly reduced in number and they contained small, flat follicles with few B cells, suggesting that TNF and TNFRp55 did not play an essential role in primary organogenesis, but that they were critically involved in the development of mature secondary characteristics (Pasparakis et al. 1997).

The sequence of mechanisms that underlie the development of mucosal lymphoid tissues remain unclear, but it seems plausible that the localized secretion of chemokines, cytokines or other factors by stromal cell foci could establish appropriate concentration gradients to promote the subsequent colonization of lymphoid cells. Once the developmental process has been initiated, inductive events between different participating cell types probably also play important roles in regulating continued morphogenesis. For example, lymphocytes isolated from murine Peyer's patches are able to induce the differentiation of FAE and specialized M cell-like cells from absorptive villous epithelium (Kerneis et al. 1997). These inductive effects were obtained when murine Peyer's patch lymphocytes and the human enterocyte cell line, Caco-2, were co-cultured in Transwell chambers and involved both cell–cell contact and secreted factors. Also, during *in vivo* experiments, the injection of Peyer's patch lymphocytes into sites in the duodenal mucosa that lacked organized lymphoid tissue induced the development of Peyer's patch-like structures which included well-organized follicles and parafollicular regions.

The epithelium overlying the newly developed follicles had properties of FAE and included M-cell like cells. This effect was regularly induced by Peyer's patch lymphocytes, occasionally by splenocytes but never by thymocytes (Kerneis et al. 1997). These results suggest a high degree of specificity of interaction between different populations of lymphoid and epithelial cells and that these have potent regulatory effects on the development and organization of mucosal lymphoid tissues.

Finally, it should always be borne in mind that the immune system interacts in complex ways with other body systems and that these interactions will also influence the development of mucosal lymphoid tissues. Paracrine networks involving the secretion of hormones, cytokines, growth factors and neuropeptides by enterocytes or other cells in the gut mucosa modulate the development of localized lymphoid tissues (see Shanahan 1997; Wang et al. 1997).

4 Conclusions and Perspectives

Although there has been significant recent progress, there is still much to learn about the factors that regulate the development and organization of mucosal lymphoid tissues. The data emerging from gene deletion experiments confirm earlier indications that these processes are multi-factorial and complex and that mucosal lymphoid tissues differ in significant ways from other secondary lymphoid structures in terms of their morphogenesis. In the future, there is likely to be increased focus on distinguishing between those processes that operate constitutively during the initial colonization of mucosal lymphoid tissues and those that operate to amplify an immune response following antigen exposure. It would perhaps not be too surprising if certain molecular mechanisms are common to these two phases but are induced by distinctive stimuli. There is also a need to use the mutant animals that are becoming available to elucidate the factors that regulate lymphoid development at mucosal surfaces other than the gut. All of this points the way towards some exciting experimental challenges for the future.

Acknowledgements. I thank Alan Young for constructive comments on the manuscript. The Basel Institute for Immunology was founded and is supported by F. Hoffmann-La Roche Ltd., Basel, Switzerland.

References

Adachi S, Yoshida H, Kataoka H, Nishikawa S-I (1997) Three distinctive steps in Peyer's patch formation of murine embryo. Int Immunol 9:507–514

Alimzhanov MB, Kuprash DV, Kosco-Vilbois MH, Luz A, Turetskaya RL, Tarakhovsky A, Rajewsky K, Nedospasov SA, Pfeffer K (1997) Abnormal development of secondary lymphoid tissues in lymphotoxin β-deficient mice. Proc Natl Acad Sci USA 94:9302–9307

Banks TA, Rouse BT, Kerley MK, Blair PJ, Godfrey VL, Kuklin NA, Bouley DM, Thomas J, Kanangat S, Mucenski ML (1995) Lymphotoxin-α-deficient mice. Effects on secondary lymphoid organ development and humoral immune responsiveness. J Immunol 155:1685–1693

De Togni P, Goellner J, Ruddle NH, Streeter PR, Fick A, Mariathasan S, Smith SC, Carlson R, Shornick LP, Strauss-Schoenberger J, Russell JH, Karr R, Chaplin DD (1994) Abnormal development of peripheral lymphoid organs in mice deficient in lymphotoxin. Science 264:703–707

Erickson SL, de Sauvage FJ, Kikly K, Carver-Moore K, Pitts-Meek S, Gillett N, Sheehan KC, Schreiber RD, Goeddel DV, Moore MW (1994) Decreased sensitivity to tumour-necrosis factor but normal T-cell development in TNF receptor-2-deficient mice. Nature 372:560–563

Griebel PJ, Ferrari G (1994) Evidence for a stromal cell-dependent, self-renewing B cell population in lymphoid follicles of the ileal Peyer's patch of sheep. Eur J Immunol 24:401–409

Griebel PJ, Hein WR (1996) Expanding the role of Peyer's patches in B-cell ontogeny. Immunol Today 17:30–39

Kanamori Y, Ishimaru K, Nanno M, Maki K, Ikuta K, Nariuchi H, Ishikawa H (1996) Identification of novel lymphoid tissues in murine intestinal mucosa where clusters of c-kit$^+$ IL-7R$^+$ Thy1$^+$ lympho-hemopoietic progenitors develop. J Exp Med 184:1449–1459

Kelsall BL, Strober W (1996) Distinct populations of dendritic cells are present in the subepithelial dome and T cell regions of the murine Peyer's patch. J Exp Med 183:237–247

Kerneis S, Bogdanova A, Kraehenbuhl J-P, Pringault E (1997) Conversion by Peyer's patch lymphocytes of human enterocytes into M cells that transport bacteria. Science 277:949–952

Kitamura Y (1989) Heterogeneity of mast cells and phenotypic change between subpopulations. Ann Rev Immunol 7:59–76

Landsverk T, Halleraker M, Aleksandersen M, McClure S, Hein W, Nicander L (1991) The intestinal habitat for organized lymphoid tissue in ruminants: comparative aspects of structure, function and development. Vet Immunol Immunopathol 28:1–16

Lowden S, Heath T (1992) Lymph pathways associated with Peyer's patches in sheep. J Anat 181:209–217

Lowden S, Heath T (1994) Ileal Peyer's patches in pigs: intercellular and lymphatic pathways. Anat Rec 239:297–305

Lowden S, Heath T (1995) Lymphoid tissues of the ileum in young horses: distribution, structure and epithelium. Anat Embryol 192:171–179

Lowden S, Heath T (1996) Lymph pathways associated with three types of follicle structure found in gut-associated lymphoid tissue of horse ileum. Anat Embryol 193:175–179

Miyawaki S, Nakamura Y, Suzuka H, Koba M, Yasumizu R, Ikehara S, Shibata Y (1994) A new mutation, aly, that induces a generalized lack of lymph nodes accompanied by immunodeficiency in mice. Eur J Immunol 24:429–434

Nanno M, Matsumoto S, Koike R, Miyasaka M, Kawaguchi M, Masuda T, Miyawaki S, Cai Z, Shimamura T, Fujiura Y, Ishikawa H (1994) Development of intestinal intraepithelial T lymphocytes is independent of Peyer's patches and lymph nodes in aly mutant mice. J Immunol 153:2014–2020

Neumann B, Luz A, Pfeffer K, Holzmann B (1996) Defective Peyer's patch organogenesis in mice lacking the 55-kD receptor for tumor necrosis factor. J Exp Med 184:259–264

Pabst R, Gehrke I (1990) Is the bronchus-associated lymphoid tissue (BALT) an integral structure of the lung in normal mammals, including humans? Am J Respir Cell Mol Biol 3:131–135

Pasparakis M, Alexopoulou L, Episkopou V, Kollias G (1996) Immune and inflammatory responses in TNFα-deficient mice: a critical requirement for TNFα in the formation of primary B cell follicles, follicular dendritic cell networks and germinal centers, and in the maturation of the humoral immune response. J Exp Med 184:1397–1411

Pasparakis M, Alexopoulou L, Grell M, Pfizenmaier K, Bluethmann H, Kollias G (1997) Peyer's patch organogenesis is intact yet formation of B lymphocyte follicles is defective in peripheral lymphoid organs of mice deficient for tumor necrosis factor and its 55-kDa receptor. Proc Natl Acad Sci USA 94: 6319–6323 (see also Correction; 94:9510)

Reynaud C-A, Mackay CR, Müller RG, Weill J-C (1991) Somatic generation of diversity in a mammalian primary lymphoid organ: the sheep ileal Peyer's patches. Cell 64:995–1005

Reynaud C-A, Garcia C, Hein WR, Weill J-C (1995) Hypermutation generating the sheep immunoglobulin repertoire is an antigen-independent process. Cell 80:115–125

Reynaud C-A, Dufour V, Weill J-C (1997) Generation of diversity in mammalian gut-associated lymphoid tissues. Restricted V gene usage does not preclude complex V gene organization. J Immunol 159:3093–3095

Shanahan F (1997) A gut reaction: lymphoepithelial communication in the intestine. Science 275: 1897–1898

Spencer J, Macdonald TT, Finn T, Isaacson PG (1986) The development of gut associated lymphoid tissue in the terminal ileum of fetal human intestine. Clin Exp Immunol 64:536–543

Wang J, Whetsell M, Klein JR (1997) Local hormone networks and intestinal T cell homeostasis. Science 275:1937–1939

M Cells in Antigen Sampling in Mucosal Tissues

M.R. Neutra

1 Introduction

The mucosal surfaces of the gastrointestinal and respiratory tracts are lined by epithelial barriers composed of cells joined by tight junctions. These gasket-like junctional structures are generally effective in excluding peptides and macromolecules with antigenic potential (MADARA et al. 1990). Mucosal surfaces are also provided with other defenses such as local secretions containing mucins and secretory immunoglobulin A (IgA) antibodies that tend to prevent antigens and

---

Department of Pediatrics, Harvard Medical School, GI Cell Biology Laboratory, Enders 1220, Children's Hospital, 300 Longwood Ave., Boston, MA 02115, USA

pathogens from contacting the epithelium (Neutra et al. 1994). Nevertheless, to obtain samples from the external environment across this barrier, the mucosal immune system depends on a close collaboration between epithelial cells and antigen-presenting and lymphoid cells. These "lympho-epithelial complexes" allow transport of antigen samples across the mucosal barrier without compromising the integrity and protective functions of the epithelium.

In the gastrointestinal tract, uptake of macromolecules, particulate antigens and microorganisms across the epithelial barrier occurs almost entirely through active transepithelial vesicular transport. Specialized epithelial M cells deliver samples of foreign material by vesicular transport from the lumen directly to intraepithelial lymphoid cells and to organized mucosal lymphoid tissues. Elucidation of the features and selectivity of the M-cell transport system is important to understand the pathogenesis of certain bacterial and viral diseases (Neutra et al. 1995; Siebers and Finlay 1996), to predict the immunological outcomes of antigen or pathogen transport (Neutra et al. 1996a) and to design effective mucosal vaccines (Neutra and Kraehenbuhl 1996).

The vast majority of the cells comprising the intestinal epithelium are absorptive enterocytes that are designed to absorb nutrients, but to exclude foreign macromolecules and microorganisms. Enterocyte apical surfaces are coated by rigid, closely packed microvilli (Mooseker 1985) that are coated by the filamentous brush border glycocalyx, a thick (400–500 nm) layer of membrane-anchored glycoproteins (Ito 1974; Maury et al. 1995). The glycocalyx of enterocytes is an effective diffusion barrier that contains large, negatively charged integral membrane mucin-like molecules (Maury et al. 1995), adsorbed pancreatic enzymes and stalked intramembrane glycoprotein enzymes responsible for terminal digestion (Semenza 1986). This thick, highly glycosylated layer prevents direct contact of most macromolecular aggregates, particles, viruses and bacteria with the microvillus membrane (Frey et al. 1996). The glycocalyx, thus, serves a protective function, preventing the uptake of antigens and pathogens, while providing a highly degradative microenvironment that promotes the digestion and absorption of nutrients. Enterocytes can endocytose small amounts of intact proteins and peptides, and such uptake may be immunologically significant (Mayer et al. 1992). However, most proteins that are taken up by enterocytes are transported to lysosomes and this, along with the digestive activities of the enterocyte surface, tends to discourage transepithelial transport of intact antigens.

2 The Lymphoid Follicle-Associated Epithelium

In contrast to the general villus and surface epithelia of the small intestine and colon that are specialized for absorption of digested nutrients and resorption of fluids, the epithelium overlying mucosal lymphoid follicles (follicle-associated epithelium or FAE) is designed for uptake of macromolecules, particles and microorganisms by transepithelial transport. The FAE contains M cells, a specialized epithelial cell type that occurs only over mucosal follicles (Fig. 1). M cells provide functional openings in the epithelial barrier through vesicular transport activity (Neutra et al. 1996b). Restriction of M cells to these sites seems to be due to the inductive influence of cells and/or secreted factors from the organized lymphoid tissues on epithelial differentiation, as discussed below. This arrangement serves to reduce the inherent risk of transporting foreign material and microbes across the epithelial barrier by assuring immediate exposure to professional phagocytes and antigen-presenting cells.

2.1 Mechanism of Transepithelial Transport

The ability of M cells to conduct transport of intact macromolecules from one side of the barrier to the other involves the directed movement of membrane vesicles. Although the molecular mechanisms of this transport have not been studied in M cells, it is safe to assume that the vectorial membrane traffic conducted by M cells depends on the polarized organization and signaling networks typical of polarized epithelial cells (Drubin and Nelson 1996). Studies in model epithelial systems

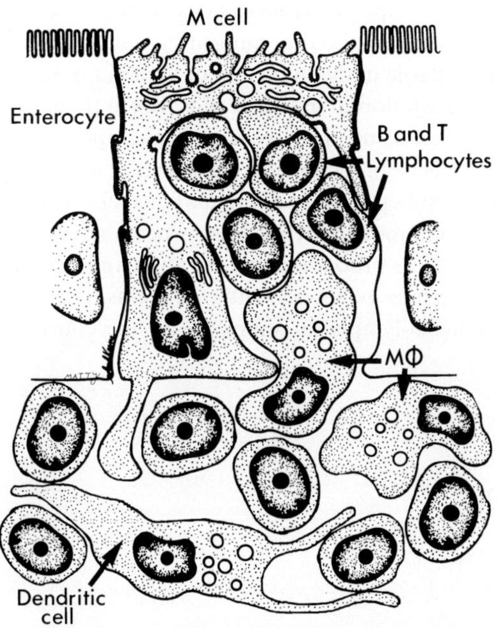

Fig. 1. Diagram of an M cell in the follicle-associated epithelium of the intestine. The M-cell apical membrane contains domains involved in endocytosis. The basolateral surface of the M cell forms an intraepithelial pocket containing B and T lymphocytes and macrophages (*M*Φ). Some M cells send cytoplasmic extensions into the subepithelial tissue. In the dome region under the follicle-associated epithelium, there are lymphocytes, macrophages and a network of dendritic cells (Reproduced with permission from Neutra et al. 1996a)

indicate that transport of proteins across epithelial cells is not accomplished by simple movement of a vesicle derived from one plasma membrane and fusion with the opposite side. Rather, transcytosis probably requires a complex series of events, including formation and fusion of endosomes and polarized recycling of membrane vesicles at both apical and basolateral poles (Apodaca et al. 1994; Mostov and Cardone 1995). The directional information that assures appropriate movement of vesicles is provided both by G proteins (Pimplikar and Simons 1993) and by the highly polarized cytoskeleton of epithelial cells (Matter and Mellman 1994).

2.2 Transepithelial Transport by M Cells

In M cells, unlike most epithelial cells, transepithelial vesicular transport is the major pathway for endocytosed materials. Macromolecules and particles that are endocytosed by M cells can be released into the intraepithelial pocket as rapidly as 10–15 min later. The phenotypes of the cells in the M-cell pocket have been described in rodents (Jarry et al. 1989; Ermak and Owen 1994) rabbits (Ermak et al. 1990) and humans (Farstad et al. 1994), but there is no direct information available concerning the interactions and events that occur in the M-cell pocket. Lectin staining of M-cell basolateral membranes revealed that M cells have basal processes that extend into the underlying lymphoid tissue (Giannasca et al. 1994). If these extensions make direct contact with lymphoid or antigen-presenting cells in the subepithelial tissue, they might play a role in the induction of the unique M-cell phenotype or in the processing and presentation of antigens after M-cell transport. Below the FAE lies an extensive network of dendritic cells and macrophages intermingled with CD4+ T cells and B cells from the underlying follicle (Ermak and Owen 1994; Farstad et al. 1994). These subepithelial cell populations reinforce the idea that M cells serve as gateways to immune inductive sites, where endocytosis and killing of incoming pathogens as well as processing, presentation and perhaps storage of antigens occurs. However, the fates of specific antigens in this tissue and the “natural history” of a given epithelial-lymphoid complex over the course of time are largely unexplored.

3 Unique Features of M Cells

3.1 Membrane Domains

The apical surfaces of M cells are distinguished by morphological criteria: the absence of a typical brush border and the presence of variable microvilli or microfolds (Owen 1977) and inter-microvillar endocytic domains (Neutra et al. 1988). The basolateral surface of the M cell includes the two major subdomains typical of all epithelial cells: the lateral subdomain is involved in cell–cell adhesion and contains Na/K ATPase; the basal subdomain interacts with the extracellular matrix and basal lamina. Unlike other epithelial cells, however, the M-cell basolateral surface has an additional subdomain that lines a deeply invaginated, intra-epithelial pocket. This structural modification, a hallmark of fully-differentiated M cells, shortens the distance that transcytotic vesicles must travel from the apical to basolateral surface and ensures that transcytosis is rapid and efficient. The composition of the membrane that lines the pocket is unknown, but it serves as the major docking site for transcytotic vesicles and interacts closely with the lymphocytes that enter this unique intra-epithelial space. Ultrastructural studies have shown that endocytic vesicles formed at the apical surface of M cells first deliver their cargo to endosomes in the apical cytoplasm (Bockman and Cooper 1973; Owen 1977; Neutra et al. 1987) and that these acidify their content and contain proteases (Allan et al. 1993; Finzi et al. 1993). Whether this intravesicular milieu alters the antigens delivered into the pocket is not known.

3.2 The M-Cell Apical Surface

The ability of M cells to efficiently transport adherent particles and macromolecules to the mucosal immune system is offset by the fact that M cells are relatively rare and represent a very small fraction of the total epithelial surface area. Little is known about the molecular composition of M-cell membranes because of the difficulties in obtaining sufficient numbers of pure M-cell preparations for biochemical or membrane fractionation studies. It is clear, however, that the M-cell apical surface differs from that of intestinal absorptive cells in many respects. First, most M cells in Peyer's patches lack the highly-organized brush border with uniform, closely packed microvilli typical of enterocytes. The actin-associated protein villin, confined to microvilli in enterocytes, is diffusely distributed in M cells (Kernéis et al. 1996), reflecting the modified apical organization and perhaps the ability to rapidly respond to adherence of microorganisms with ruffling and phagocytosis. Second, the apical plasma membrane of M cells contains broad endocytic domains that function in endocytosis and phagocytosis (Neutra et al. 1988). Other differences involve the molecular composition of the membrane itself. While enterocyte brush borders have abundant hydrolytic enzymes, these enzymes are usually reduced or absent on M cells (Owen and Bhalla 1983; Savidge and Smith 1995).

M-cell apical membranes also display glycosylation patterns that distinguish them from their epithelial neighbors. For example, in Peyer's patches of BALB/c mice, lectins that recognize a range of carbohydrate structures containing $\alpha(1\text{–}2)$-fucose selectively stained all M cells in the FAE (Clark et al. 1993; Falk et al. 1994; Giannasca et al. 1994). Giannasca et al. (1994) observed these lectin binding sites not only on M-cell apical membranes, but also on intracellular vesicles and basolateral membranes, including the pocket domain.

Indeed, there are variations in the glycosylation patterns of individual M cells within a single FAE, a phenomenon that would expand possible microbial lectin–

M-cell surface carbohydrate interactions of the local M-cell population and allow the M cells to "sample" a wider variety of microorganisms (GIANNASCA et al. 1994; NEUTRA et al. 1995). Furthermore, glyco-conjugates expressed on M cells in different intestinal regions (cecum, appendix, colon and rectum) are distinct (GEBERT and HACH 1993; FALK et al. 1994; GIANNASCA et al. 1994, JEPSON et al. 1993) and this could explain the regional specificity of certain pathogens that exploit M cells to invade the mucosa.

4 Selective Adherence of Particles and Macromolecules

4.1 Uptake of Particles by M Cells

The ability of M cells to bind and transport particles and microorganisms may not be entirely due to specific molecular features on M cells. Hydrophobic particles and cationic macromolecules adhere much more avidly to M cells than to enterocytes. Several investigators have observed M cell-selective adherence and uptake of polystyrene or latex beads (PAPPO and ERMAK 1989), polylactide/polyglycolide microparticles (ERMAK et al. 1995), liposomes (CHILDERS et al. 1990; ZHOU et al. 1995), cationized ferritin (BYE et al. 1984; NEUTRA et al. 1987) and hydroxyapatite (AMERONGEN and NEUTRA, unpublished observations). The M-cell membrane features that are responsible for adherence of such particles are not understood. There is current interest in improving mucosal vaccine delivery by exploiting the M-cell pathway in humans, but information about the unique molecular components exposed on M-cell surfaces in humans is lacking.

4.2 Interaction of Secretory Immunoglobulin A (IgA) with M Cells

The FAE in both the GI tract and airways is distinct from the rest of the epithelium in that it lacks basolateral polymeric immunoglobulin receptors and, thus, is presumably unable to bind and secrete IgA produced by mucosal plasma cells (PAPPO and OWEN 1988). This is consistent with evidence that mucosal lymphoid follicles contain precursors of IgA B cells but are not sites of terminal plasma-cell differentiation or IgA production.

Secretory (s) IgA is produced by lamina propria plasma cells and is transported into the lumen by crypt epithelial cells throughout the gut. Secreted IgA does not adhere to the apical surfaces of enterocytes. However, secretory IgA has been shown to adhere selectively to the apical membranes of M cells. This was first observed in suckling rabbits as local accumulation of milk sIgA on M cells of Peyer's patches (ROY and VARVAYANIS 1987) and was confirmed by our observations that monoclonal IgA, monoclonal IgA–antigen complexes and polyclonal secretory IgA adhered to apical membranes of M cells and were transported into the intra-epithelial pocket (WELTZIN et al. 1989).

Whether uptake of IgA–antigen complexes by M cells boosts the secretory immune response or has some other modulating effect is not clear. However, there is evidence that in naive mice, IgA can promote uptake of antigens and induction of immune responses. When liposomes containing ferritin as a test antigen and coated with monoclonal IgA were tested by mucosal immunization via the rectum, IgA enhanced the uptake of the liposomes and moderately enhanced the local rectal/colonic secretory immune response to ferritin compared with that seen with liposomes lacking the IgA coat (ZHOU et al. 1995).

A novel set of experiments suggested that secretory IgA itself can serve as a mucosal vaccine carrier: secretory component (SC) was genetically engineered to deliver protective foreign epitopes into mucosal lymphoid tissue. An exposed loop in the first immunoglobulin-like domain of SC was replaced with a 10-amino acid linear epitope from the invasin of *Shigella flexneri*. The "antigenized" SC was recognized by an invasin-specific monoclonal antibody, was able to bind dimeric IgA and evoked immune responses that included antibodies against invasin when administered orally (CORTHESY et al. 1996).

5 Exploitation of the M-Cell Pathway by Pathogens

Sampling of microorganisms by the M-cell transport pathway carries the risk that pathogens may exploit these mechanisms to cross epithelial barriers and invade the body. Epithelial M cells of the intestine are continuously exposed to the lumen of the gut and are relatively accessible to attachment and invasion of pathogens. At mucosal sites containing M cells, the risk of local invasion is high but the occurrence of mucosal disease may be reduced by the close interactions of the FAE with professional antigen processing and presenting cells, and by the organization of mucosal lymphoid tissues immediately under the epithelium. Nevertheless, M-cell adherence and transport of pathogens does result in initiation of mucosal and/or systemic infections. Furthermore, viral and bacterial pathogens that infect M cells can cause selective loss of this cell type (BASS et al. 1988; JONES et al. 1994), and this could compromise the ability of the FAE to transport antigens in a selective, controlled manner. The exploitation of M-cell transport by specific viral and bacterial pathogens, and the use of such microorganisms in attenuated form for vaccine delivery, have been reviewed (NEUTRA et al. 1995, NEUTRA and KRAEHENBUHL 1996; SIEBERS and FINLAY 1996) and are discussed in other chapters of this volume.

5.1 Bacteria

Several gram-negative bacteria that cause disease by colonizing or invading the intestinal mucosa bind either preferentially or exclusively to M cells. These include *Vibrio cholerae* (Owen et al. 1986; Winner et al. 1991), some strains of *E.coli* (Inman and Cantey 1983; Uchida 1987), *Salmonella typhi* (Kohbata 1986), *Salmonella typhimurium* (Jones et al. 1994), *Shigella flexneri* (Wassef et al. 1989), *Yersinia enterocolitica* (Grutzkau et al. 1990), *Yersinia pseudotuberculosis* (Fujimura et al. 1992) and *Campylobacter jejuni* (Walker et al. 1988). The modes of interaction of these organisms with M cells vary widely, reflecting the differing adhesins, surface extensions and secreted products involved in their pathogenic strategies (Finaly and Falkow 1997).

The ultrastructural appearances of bacterial–M-cell interaction sites include tight adherence and phagocytosis (*V. cholerae*), pedestal formation (*E. coli RDEC-1*) and disorganization or ballooning of the M-cell apical surface (*S. typhimurium*), reminiscent of the complex interactions of these bacteria with other cell types (Siebers and Finlay 1996; Neutra et al. 1996b). It is likely that M-cell adherence and uptake involves a sequence of molecular interactions, including initial recognition (perhaps via a lectin–carbohydrate interaction), followed by more intimate associations that require expression of additional bacterial genes, processing of M-cell surface molecules, activation of intracellular signaling pathways and recruitment or reorganization of membrane and cytoskeletal M-cell proteins.

5.2 Viruses

The first virus found to use M cells as a route of entry was the mouse pathogen, reovirus (Wolf et al. 1981; Bass et al. 1988). Reovirus infects mice via the intestine and is processed but not killed by the degradative intestinal environment. Processing of reovirus by proteases in the intestinal lumen increases viral infectivity through cleavage of the major outer capsid protein, sigma 3, and a conformational change resulting in extension of the viral hemagglutinin sigma 1 (Nibert et al. 1991). Studies in our laboratory showed that proteolytic processing of the outer capsid is required for M-cell adherence (Amerongen et al. 1994) and it follows that either the protease-resistant outer capsid protein μ1c or the extended sigma-1 protein is used to bind to M cells. The M-cell surface component that serves as a receptor has not yet been identified, but our recent studies point to a specific sialic acid-containing glyco-conjugate. Reovirus recognizes mouse M cells not only in the Peyer's patches, but also in the colon (Owen et al. 1990) and the airways (Morin et al. 1994), and reovirus binds to rabbit as well as mouse M cells.

Entry of poliovirus via M cells in human intestinal explants has also been demonstrated by means of electron microscopy (EM) (Sicinski et al. 1990). The receptor for poliovirus on neuronal target cell membranes has been identified as a member of the immunoglobulin superfamily, and the cloned gene has been used to create transgenic mice that can be infected by injection of virus (Mendelsohn et al.

1989). However, such mice are not infected when challenged orally, indicating that poliovirus exploits a different, unknown mechanism to adhere to M cells. In addition, autoradiographical and EM studies from this laboratory showed that human immunodeficiency virus (HIV) can be transported by rabbit and mouse M cells, raising the possibility that rectal M cells could provide a portal of entry for HIV in humans (Amerongen et al. 1991).

5.3 Role of the Glycocalyx

What explains the predilection of viral and bacterial pathogens for M cells? One possibility is that the distinctive surface glyco-conjugates or other, as yet unidentified, components on M cells provide unique docking sites that are recognized by viral and bacterial adhesins. Alternatively, common, ubiquitous membrane components are simply more accessible on M cells, because of the lack of a rigid brush border with its thick blanketing glycocalyx. Evidence for the latter mechanism has emerged from studies in this laboratory using microparticles of sizes analagous to bacteria and viruses (Frey et al. 1996). The studies were designed to evaluate the accessibility of glycolipid receptors in intestinal epithelial cell membranes to particulate antigens, by testing the effect of particle size on the ability of a ligand, cholera toxin B subunit (CTB), to bind to a glycolipid receptor on M cells and enterocytes.

Ganglioside G_{M1} has been demonstrated to be the only receptor for cholera toxin in diverse cell types, including enterocytes of rabbit small intestine (Griffiths et al. 1986) and enterocyte-like intestinal cell lines (Orlandi et al. 1994). The carbohydrate head groups of G_{M1} protrude only 2.5 nm above the surface of the membrane lipid bilayer (Mc Daniel et al. 1986) and the G_{M1} binding sites in CTB pentamers are 2.3-nm deep cavities (deduced from its homologue, heat-labile *E. coli* enterotoxin B subunit; Sixma et al. 1991). Thus, to bind to G_{M1}, CTB must come into very close contact with the lipid bilayer of the host cell. While soluble CTB-fluorescein (hydrated diameter 6.4 nm) bound to apical plasma membranes of all cell types in the rabbit small-intestinal epithelium, CTB-colloidal gold (diameter 28 nm) adhered exclusively to Peyer's-patch M cells and 1-μm CTB-latex particles failed to adhere specifically to any epithelial cell surface (Frey et al. 1996). Thus, association of ligand with particles can result in M cell-specific adherence, but only within a restricted size range. Although the relatively thin M-cell glycocalyx has not generally been considered a significant barrier to microorganisms in vivo, the abundant, terminally-glycosylated apical surface glyco-conjugates on M cells was sufficient to prevent access of CTB-coated 1-μm particles to the membrane bilayer.

5.4 Implications for Microbial Pathogenesis

The studies described above illustrate the fact that the presence of an appropriate receptor is necessary, but not sufficient, for microbial attachment; the receptor must also be accessible to the ligands or adhesins on the surfaces of viruses or bacteria. Furthermore, the apical surfaces of intestinal epithelial cells in vivo are highly specialized structures that may not be exactly replicated by cultured cell lines in vitro. For example, on the basis of studies using HT29 and other enterocyte-like cell lines, galactosylceramide has been proposed as the epithelial-cell component that could serve as a receptor for binding of HIV to human rectal epithelial cells (Fantini et al. 1993). However, the human rectal enterocyte glycocalyx is comparable in thickness to that of enterocytes of other regions and species (Neutra 1979), and we have previously observed that HIV failed to penetrate the glycocalyx of rabbit or mouse enterocytes on villi or the FAE (Amerongen et al. 1991). On the other hand, HIV did adhere to rabbit and mouse M cells and was transcytosed. If the relevant glycolipid is present on M cells in human rectum and if HIV enters via this route, receptor accessibility could account for a cell-selective uptake mechanism that would deliver the virus directly to its target cells in mucosal lymphoid tissues.

Accessibility of the components in the lipid bilayer of M cells may be less crucial for bacterial pathogens, since bacteria can produce long surface extensions such as pili to penetrate the membrane glycoprotein barrier of host cells, secrete proteins that penetrate or enzymatically cleave components of the glycocalyx, and reorganize host cell surfaces (Rosenshine et al. 1992; Bliska et al. 1993; Jones and Falkow 1996; Finlay and Falkow 1997). Many enteric bacteria exploit carbohydrate epitopes as receptors for initial adherence (Hultgren et al. 1993), and such sites may be exposed to the lumen in the periphery of the glycocalyx. However, certain epitopes used by bacteria for adherence may be "buried" in the thick glycocalyx of enterocytes, and this could explain the observation that bacteria, such as Salmonella, bind more rapidly to M cells than to enterocytes (Jones et al. 1994; Neutra et al. 1995).

6 Induction and Differentiation of the M-Cell Phenotype

Several lines of evidence indicate that microorganisms play a crucial role in development of organized mucosal lymphoid tissues, and that mucosal follicles, in turn, play an inductive role in the differentiation of the specialized FAE and M cells. This was first demonstrated in studies using germ-free mice. Mice reared in a germ-free environment have a reduced number of Peyer's patches, but lymphoid follicles and M cells rapidly increase in number after transfer to a normal animal-house environment (Smith et al. 1987) or after exposure to a single bacterial species (Savidge et al. 1991). From these observations it could not be determined to what extent the microorganisms acted via lymphoid mediators or directly on epithelial cells. Current evidence now indicates that differentiation of FAE involves both exogenous and endogenous factors in a complex, three-way interplay between epithelial cells, lymphoid cells and the luminal environment.

6.1 Origin of M-Cell Precursors

In normal mice, FAE phenotypes, including M cells, appear in the crypts adjacent to mucosal lymphoid follicles. In adult small intestine, each crypt is a clonal unit that produces multiple cell types, and these migrate upward in columns onto several adjacent villi, such that the epithelium of each villus is derived from several surrounding crypts (Gordon and Hermiston 1994; Schmidt et al. 1985). Similarly, the FAE is derived from the "follicle-associated crypts" surrounding the mucosal lymphoid follicles. These crypts contain two distinct axes of migration from the same ring of crypt stem cells: cells on one wall of the crypt differentiate into absorptive enterocytes, goblet and entero-endocrine cells that migrate onto the villi, while cells on the follicle-facing wall of the same crypt acquire features of M cells and distinct follicle-associated enterocytes (Bye et al. 1984; Gebert et al. 1992; Savidge and Smith 1995). These FAE features include a lack of goblet cells, a lack of polymeric immunoglobulin receptor expression (Pappo and Owen 1988), M cell-like glycosylation patterns (Giannasca et al. 1994) and expression of vimentin (Gebert et al. 1992). As they emerge from the crypt, differentiating M cells begin endocytic activity, fail to assemble brush borders and acquire immune cells in their characteristic intra-epithelial pocket (Bye et al. 1984). These observations suggest that inductive factors may act early in the differentiation pathway, inducing crypt cells to commit to FAE phenotypes.

6.2 Role of Organized Lymphoid Tissue

There is experimental evidence that cell contacts and/or soluble factors from mucosal lymphoid follicles play an important role in the induction of FAE and M cells. Injection of Peyer's-patch lymphocytes into the submucosa of syngeneic mice resulted in local assembly of a new lymphoid follicle and the de novo appearance of FAE with typical M cells (Kernéis et al. 1997). Several lines of evidence indicate that B cells play a crucial role: severe combined immunodeficient (SCID) mice lack mucosal follicles and identifiable M cells, but follicles with FAE appeared after injection of Peyer's-patch cells, and fractions enriched in B cells were most effective in reconstituting these structures (Savidge and Smith 1995). Furthermore, B cell-deficient mice lack mucosal follicles and identifiable M cells (Debard N, Kernéis S, Fischer G, Pringault E and Kraehenbuhl J, unpublished data), whereas T cell-deficient nude mice have small Peyer's patches with FAE and M cells (Ermak and Owen 1987). Induction of new lymphoid follicles also occurs during mucosal inflammation: an inflamed ileal mucosa may contain increased numbers of mucosal follicles and an increased number of FAE and M cells (Cuvelier et al. 1993).

Appearance of FAE over newly-formed follicles could indicate a local inductive influence on undifferentiated cells in adjacent crypts. However, factors or cells from the follicle or the lumen may also act later, to convert some of the FAE enterocyte-like cells to antigen-transporting M cells. Cells with both enterocyte and M-cell features are present in FAE (Gebert et al. 1992). Furthermore, M-cell numbers have been observed to increase rapidly after Salmonella infection (Savidge et al. 1991) and after challenge with a non-intestinal bacterium (Borghesi et al. 1996) on time scales too short to be explained by induction of M cells in crypts.

6.3 Role of Microflora

Bacteria can alter the host epithelial barrier to promote colonization or invasion (as discussed by Finlay and Falkow in this volume), but in doing so they set off epithelial alarms that can mobilize host defenses. In response to adherence or invasion of bacterial or protozoal pathogens, epithelial cells release cytokines and chemokines (Eckmann et al. 1995; Jung et al. 1995) that can attract local inflammatory cells and recruit lymphoid and other cells from blood. For example, locally released signaling molecules could facilitate polynuclear and mononuclear cell extravasation by upregulating endothelial addressins on mucosal venules (Butcher 1991; Wagner et al. 1996). In addition, chemokines could form gradients by binding to extracellular matrix components, thus supporting the directional migration of leukocytes toward the epithelium (Madara 1994). These signaling mechanisms could play an indirect role in FAE differentiation by promoting the assembly of lymphoid follicles.

6.4 Induction of the M-Cell Phenotype in Vitro

Recently, the ability of B cells to convert an enterocyte phenotype to an M-cell phenotype was directly demonstrated using a novel co-culture system (Kernéis

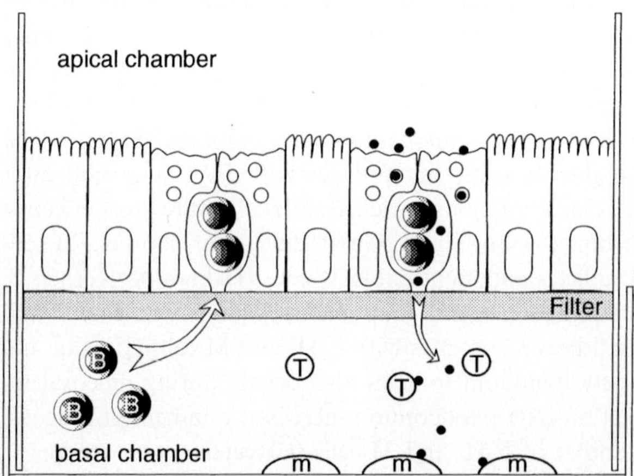

Fig. 2. Diagram of a co-culture system recently reported by Kernéis et al. (1997). Lymphocytes, added to the basolateral side of Caco-2 cell monolayers cultured on permeable filters, migrated into the monolayers. The lymphocytes induced conversion of some epithelial cells to an M cell-like phenotype that transported particles and bacteria. This system could be used in the future to test the interactions of transported pathogens with other cells types such as T cells or macrophages (*m*)

et al. 1997). Murine Peyer's-patch lymphocytes, when applied to the basolateral side of Caco-2 cell monolayers, entered the epithelium to reside in intra-epithelial pockets reminiscent of those under M cells in vivo (Fig. 2). The presence of lymphoid cells in the epithelium induced disorganization of the Caco-2 cell brush borders and the loss of cell surface sucrase-isomaltase, and this was paralleled by the appearance of active transcytosis of latex beads and bacteria (*Vibrio cholerae*) from the apical to the basolateral compartment. Although both mouse T cells and B cells entered the monolayer, information from previous studies using B or T cell-deficient mice (cited in Sect. 6.2 above) suggested a primary role for B cells in this phenomenon. Indeed, human B (Raji), but not T (Jurkat), lymphoid cells were able to induce M-cell features in Caco-2 cell monolayers (KERNÉIS et al. 1997). This system provides a new opportunity to dissect the molecular mechanisms that mediate the conversion of enterocytes in a pre-existing FAE to M cells, and the induction of M cells in follicle-associated crypts.

Acknowledgements. The contributions of current and former members of my laboratory are gratefully acknowledged. The author is especially indebted to Dr. Jean-Pierre Kraehenbuhl for his long-standing collaboration and friendship. The author's laboratory is supported by NIH Research Grants HD17557, AI34757, AI33384 and NIH Center Grant DK34854 to the Harvard Digestive Diseases Center.

References

Allan CH, Mendrick DL, Trier JS (1993) Rat intestinal M cells contain acidic endosomal-lysosomal compartments and express Class II major histocompatibility complex determinants. Gastroenterology 104:698–708

Amerongen HM, Weltzin, RA, Farnet CM, Michetti P, Haseltine WA, Neutra MR (1991) Transepithelial transport of HIV-1 by intestinal M cells: a mechanism for transmission of AIDS. J Acquir Immune Defic Syndr 4:760–765

Amerongen HM, Wilson GAR, Fields BN, Neutra MR (1994) Proteolytic processing of reovirus is required for adherence to intestinal M cells. J Virol 68:8428–8432

Apodaca G, Katz LA, Mostov KE (1994) Receptor-mediated transcytosis of IgA in MDCK cells is via apical recycling endosomes. J Cell Biol 12:67–86

Bass DM, Trier JS, Dambrauskas R, Wolf JL (1988) Reovirus type 1 infection of small intestinal epithelium in suckling mice and its effect on M cells. Lab Invest 55:226–235

Bliska JB, Galan JE, Falkow S (1993) Signal transduction in the mammalian cell during bacterial attachment and entry. Cell 73:903–920

Bockman DE, Cooper MD (1973) Pinocytosis by epithelium associated with lymphoid follicles in the bursa of Fabricius, appendix, and Peyer's patches. An electron microscopic study. Am J Anat 136:455–478

Borghesi C, Regoli M, Bertelli E, Nicoletti C (1996) Modifications of the follicle-associated epithelium by short-term exposure to a non-intestinal bacterium. J Pathol 180:326–332

Butcher EC (1991) Leukocyte-endothelial cell recognition: three (or more) steps to specificity or diversity. Cell 67:1033–1036

Bye WA, Allan CH, Trier JS (1984) Structure, distribution and origin of M cells in Peyer's patches of mouse ileum. Gastroenterology 86:789–801

Childers NK, Denys FR, McGhee JF, Michalek SM (1990) Ultrastructural study of liposome uptake by M cells of rat Peyer's patches: an oral vaccine system for delivery of purified antigen. Regional Immunol 3:8–16

Clark MA, Jepson MA, Simmons NL, Booth TA, Hirst BH (1993) Differential expression of lectin-binding sites defines mouse intestinal M-cells. J Histochem Cytochem 41:1679–1687

Corthesy B, Kaufmann M, Phalipon A, Peitsch M, Neutra MR, Kraehenbuhl JP (1996) A pathogen-specific epitope inserted into recombinant secretory immunoglobulin is immunogenic by the oral route. J Biol Chem 52:33670–33677

Cuvelier CA, Quatacker J, Mielants H, de Vos M, Veys E, Roels H (1993) M cells are damaged and increased in number in inflamed human ileal mucosa. Eur J Morphol 31:87–91

Druben DG, Nelson WJ (1996) Origins of cell polarity. Cell 84:335–344

Eckmann L, Kagnoff MF, Fierer J. (1995) Intestinal epithelial cells as watchdogs for the natural immune system. Trends Microbiol 3:118–120

Ermak TH, Owen RL (1987) Phenotype and distribution of T lymphocytes in Peyer's patches of athymic mice. Histochemistry 87:321–325

Ermak TH, Owen RL (1994) Differential distribution of lymphocytes and accessory cells in mouse Peyer's patches. Am J Trop Med Hyg 50:14–28

Ermak TH, Steger HJ, Pappo J (1990) Phenotypically distinct subpopulations of T cells in domes and M-cell pockets of rabbit gut-associated lymphoid tissues. Immunology 71:530–537

Ermak TH, Dougherty EP, Bhagat HR, Kabok Z, Pappo J (1995) Uptake and transport of copolymer biodegradable microspheres by rabbit Peyer's patch M cells. Cell Tissue Res 279:433–436

Falk P, Roth KA, Gordon JI (1994) Lectins are sensitive tools for defining the differentiation programs of epithelial cell lineages in the developing and adult mouse gastrointestinal tract. Am J Physiol 266:G987-G1003

Fantini J, Cook DG, Nathanson N, Spitalnik SL, Gonzalez-Scarano F (1993) Infection of colonic epithelial cell lines by type 1 human immunodeficiency virus (HIV-1) is associated with cell surface expression of galactosyl ceramide, a potential alternative gp120 receptor. Proc Natl Acad Sci USA 90:2700–2704

Farstad IN, Halstensen TS, Fausa O, Brandtzaeg P (1994) Heterogeneity of M cell-associated B and T cells in human Peyer's patches. Immunology 83:457–464

Finlay BB, Falkow S (1997) Common themes in microbial pathogenesis revisited. Micrbiol Mol Biol Rev 61:136–139

Finzi G, Cornaggia M, Capella C, Fiocca R, Bosi F, Solcia E (1993) Cathepsin E in follicle associated epithelium of intestine and tonsils: localization to M cells and possible role in antigen processing. Histochemistry 99:201–211

Frey A, Lencer WI, Weltzin R, Giannasca KT, Giannasca PJ, Neutra MR (1996) Role of the glycocalyx in regulating access of microparticles to apical plasma membranes of intestinal epithelial cells: implications for microbial attachment and oral vaccine targeting. J Exp Med 184:1045–1060

Fujimura Y, Kihara T, Mine H (1992) Membranous cells as a portal of Yersinia pseudotuberculosis entry into rabbit ileum. J Clin Electron Microsc 25:35–45

Gebert A, Hach G (1993) Differential binding of lectins to M cells and enterocytes in the rabbit cecum. Gastroenterology 105:1350–1361

Gebert A, Hach G, Bartels H (1992) Co-localization of vimentin and cytokeratins in M-cells of rabbit gut-associated lymphoid tissue (GALT). Cell Tissue Res 269:331–340

Giannasca PJ, Giannasca KT, Falk P, Gordon JI, Neutra MR (1994) Regional differences in glycoconjugates of intestinal M cells in mice: potential targets for mucosal vaccines. Am J Physiol 267:G1108-G1121

Gordon JI, Hermiston ML (1994) Differentiation and self-renewal in the mouse gastrointestinal epithelium. Curr Opin Cell Biol 6:795–803

Griffiths SL, Finkelstein RA, Critchley DR (1986) Characterization of the receptor for cholera toxin and Escherichia coli heat-labile toxin in rabbit intestinal brush borders. Biochem J 238:313–322

Grutzkau A, Hanski C, Hahn H, Riecken EO (1990) Involvement of M cells in the bacterial invasion of Peyer's patches: a common mechanism shared by Yersinia enterocolitica and other enteroinvasive bacteria. Gut 3:1011–1015

Hultgren SJ, Abraham S, Caparon M, Falk P, St. Geme III JW, Normark S (1993) Pilus and nonpilus adhesins: assembly and function in cell recognition. Cell 73:887–901

Inman LR, Cantey JR (1983) Specific adherence of Escherichia Coli (strain RDEC-1) to membranous (M) cells of the Peyer's patch in Escherichia Coli diarrhea in the rabbit. J Clin Invest 71:1–8

Ito S (1974) Form and function of the glycocalyx on free cell surfaces. Philos Trans R Soc Lond (Biol) 268:55–66

Jarry A, Robaszkiewicz M, Brousse N, Potet F (1989) Immune cells associated with M cells in the follicle-associated epithelium of Peyer's patches in the rat. Cell Tissue Res 225:293–298

Jepson MA, Clark MA, Simmons NL, Hirst BH (1993) Epithelial M cells in the rabbit caecal lymphoid patch display distinctive surface characteristics. Histochemistry 100:441–447

Jones BD, Falkow S (1996) Salmonellosis–host immune responses and bacterial virulence determinants. Annu Rev Immunol 14:533–561

Jones BD, Ghori N, Falkow S (1994) Salmonella typhimurium initiates murine infection by penetrating and destroying the specialized epithelial M cells of the Peyer's patches. J Exp Med 180:15–23

Jung HC, Eckmann L, Yang SK, Panja A, Fierer J, Morzycka-Wroblewska E, Kagnoff MF (1995) A distinct array of proinflammatory cytokines is expressed in human colon epithelial cells in response to bacterial invasion. J Clin Invest 95:55–65

Kernéis S, Bogdanova A, Colucci-Guyon E, Kraehenbuhl JP, Pringault E (1996) Cytosolic distribution of villin in M cells from mouse Peyer's patches correlates with the absence of a brush border. Gastroenterology 110:515–521

Kernéis S, Bogdanova A, Kraehenbuhl JP, Pringault E (1997) Conversion by Peyer's patch lymphocytes of human enterocytes into M cells that transport bacteria. Science 277:949–952

Kohbata S, Yokobata H, Yabuuchi E (1986) Cytopathogenic effect of Salmonella typhi GIFU 10007 on M cells of murine ileal Peyer's patches in ligated ileal loops: an ultrastructural study. Microbiol Immunol 30:1225–1237

Madara JL (1994) Migration of neutrophils through epithelial monolayers. Trends Cell Biol 4:4–7

Madara JL, Nash S, Moore R, Atisook K (1990) Structure and function of the intestinal epithelial barrier in health and disease. Monogr Pathol 31:306–324

Matter K, Mellman I (1994) Mechanisms of cell polarity: sorting and transport in epithelial cells. Curr Opin Cell Biol 6:545–554

Maury J, Nicoletti C, Guzzo-Chambraud L, Maroux S (1995) The filamentous brush border glycocalyx, a mucin-like marker of enterocyte hyper-polarization. Eur J Biochem 228:323–331

Mayer L, Panja A, Li Y, Siden E, Pizzimenti A, Gerardi F, Chandswang N (1992) Unique features of antigen presentation in the intestine. Ann N Y Acad Sci 664:39–46

McDaniel RV, Sharp K, Brooks D, McLaughlin AC, Winiski AP, Cafiso D, McLaughlin S (1986) Electrokinetic and electrostatic properties of bilayers containing gangliosides G_{M1}, G_{D1a} or G_{T1}. Comparison with a nonlinear theory. Biophys J 49:741–752

Mendelsohn CL, Wimmer E, Racaniello VR (1989) Cellular receptor for poliovirus: molecular cloning, nucleotide sequence, and expression of a new member of the immunoglobulin superfamily. Cell 56:855–865

Mooseker M (1985) Organization, chemistry and assembly of the cytoskeletal apparatus of the intestinal brush border. Ann Rev Cell Biol 1:209–241

Morin MJ, Warner A, Fields BN (1994) A pathway for entry of reoviruses into the host through M cells of the respiratory tract. J Exp Med 180:1523–1527

Mostov KE, Cardone MH (1995) Regulation of protein traffic in polarized epithelial cells. Bioessays 17:129–138

Neutra M (1979) Linear arrays of intramembrane particles in microvilli in primate large intestine. Anat Rec 193:367–381

Neutra MR, Kraehenbuhl JP (1996) Antigen uptake by M cells for effective mucosal vaccines. In: Kiyono H, Kagnoff M, McGhee G (eds) Mucosal Vaccines. Academic Press, New York, pp 41–55

Neutra MR, Phillips TL, Mayer EL, Fishkind DJ (1987) Transport of membrane-bound macromolecules by M cells in follicle-associated epithelium of rabbit Peyer's patch. Cell Tissue Res 247:537–546

Neutra MR, Wilson JM, Weltzin RA, Kraehenbuhl JP (1988) Membrane domains and macromolecular transport in intestinal epithelial cells. Am Rev Respir Dis 138:S10-S16

Neutra MR, Michetti P, Kraehenbuhl JP (1994) Secretory immunoglobulin A: induction, biogenesis and function. In: Johnson LR (ed) Physiology of the Gastrointestinal Tract. Raven Press, New York, pp 685–708

Neutra MR, Giannasca PJ, Giannasca KT, Kraehenbuhl J-P (1995) M cells and microbial pathogens. In: Blaser M, Smith PD, Ravdin JI, Greenberg HB, Guerrant RL (eds) Infections of the Gastrointestinal Tract. Raven Press, New York, pp 163–178

Neutra MR, Pringault E, Kraehenbuhl JP (1996a) Antigen sampling across epithelial barriers and induction of mucosal immune responses. Annu Rev Immunol 14:275–300

Neutra MR, Frey A, Kraehenbuhl JP (1996b) Epithelial M cells: gateways for mucosal infection and immunization. Cell 86:345–348

Nibert ML, Furlong DB, Fields BN (1991) Mechanisms of viral pathogenesis. J Clin Invest 88:727–734

Orlandi PA, Critchley DR, Fishman PH (1994) The heat-labile enterotoxin of Escherichia coli binds to polytactosaminoglycan-containing receptors in Caco-2 human intestinal epithelial cells. Biochemistry 33:12886–12895

Owen RL (1977) Sequential uptake of horseradish peroxidase by lymphoid follicle epithelium of Peyer's patches in the normal unobstructed mouse intestine: an ultrastructural study. Gastroenterology 72:440–451

Owen RL, Bhalla DK (1983) Cytochemical analysis of alkaline phosphatase and esterase activities and of lectin-binding and anionic sites in rat and mouse Peyer's patch M cells. Am J Anat 168:199–212

Owen RL, Pierce NF, Apple RT, Cray WCJ (1986) M cell transport of Vibrio cholerae from the intestinal lumen into Peyer's patches: a mechanism for antigen sampling and for microbial transepithelial migration. J Infect Dis 153:1108–1118

Owen RL, Bass DM, Piazza AJ (1990) Colonic lymphoid patches. A portal of entry in mice for type I reovirus administered anally. Gastroenterology 98:A468

Pappo J, Ermak TH (1989) Uptake and translocation of fluorescent latex particles by rabbit Peyer's patch follicle epithelium: a quantitative model for M cell uptake. Clin Exp Immunol 76:144–148

Pappo J, Owen RL (1988) Absence of secretory component expression by epithelial cells overlying rabbit gut-associated lymphoid tissue. Gastroenterology 95:1173–1177

Pimplikar SW, Simons K (1993) Role of heterotrimeric G proteins in polarized membrane transport. J Cell Sci Suppl 17:27–32

Rosenshine I, Donnenberg MS, Kaper JB, Finlay BB (1992) Signal transduction between enteropathogenic Escherichia coli (EPEC) and epithelial cells: EPEC induces tyrosine phosphorylation of host cell proteins to initiate cytoskeletal rearrangement and bacterial uptake. EMBO J 11:3551–3560

Roy MJ, Varvayanis M (1987) Development of dome epithelium in gut-associated lymphoid tissues: association of IgA with M cells. Cell Tissue Res 248:645–651

Savidge TC, Smith MW (1995) Evidence that membranous (M) cell genesis is immunoregulated. In: Mestecky J (ed) Advances in Mucosal Immunology. Plenum, New York, pp 239–241

Savidge TC, Smith MW, James PS, Aldred P (1991) Salmonella-induced M-cell formation in germ-free mouse Peyer's patch tissue. Am J Pathol 139:177–184

Schmidt GH, Wilkinson MM, Ponder BAJ (1985) Cell migration pathway in the intestinal epithelium: an in situ marker system using mouse aggregation chimeras. Cell 40:425–429

Semenza G (1986) Anchoring and biosynthesis of stalked brush border membrane glycoproteins. Annu Rev Cell Biol 2:255–314

Sicinski P, Rowinski J, Warchol JB, Jarzcabek Z, Gut W, Szczygiel B, Bielecki K, Koch G (1990) Poliovirus type 1 enters the human host through intestinal M cells. Gastroenterology 98:56–58

Siebers A, Finlay BB (1996) M cells and the pathogenesis of mucosal and systemic infections. Trends Microbiol 4:22–29

Sixma TK, Pronk SE, Kalk KH, Wartna ES, van Zanten BAM, Witholt B, Hol WGJ (1991) Crystal structure of a cholera toxin-related heat-labile enterotoxin from E. coli. Nature 351:371–377

Smith MW, James PS, Tivey DR (1987) M cell numbers increase after transfer of SPF mice to a normal animal house environment. Am J Pathol 128:385–389

Uchida J (1987) An ultrastructural study on active uptake and transport of bacteria by microfold cells (M cells) to the lymphoid follicles in the rabbit appendix. J Clin Electron Microsc 20:379–394

Wagner N, Lohler J, Kunkel EJ, Ley K, Leung E, Krissansen G, Rajewsky K, Muller W (1996) Critical role for beta-7 integrins in formation of the gut-associated lymphoid tissue. Nature 382:366–370

Walker RI, Schauder-Chock EA, Parker JL (1988) Selective association and transport of Campylobacter jejuni through M cells of rabbit Peyer's patches. Can J Microbiol 34:1142–1147

Wassef JS, Keren DF, Mailloux JL (1989) Role of M cells in initial antigen uptake and in ulcer formation in the rabbit intestinal loop model of Shigellosis. Infect Immun 57:858–863

Weltzin RA, Lucia Jandris P, Michetti P, Fields BN, Kraehenbuhl JP, Neutra MR (1989) Binding and transepithelial transport of immunoglobulins by intestinal M cells: demonstration using monoclonal IgA antibodies against enteric viral proteins. J Cell Biol 108:1673–1685

Winner LS III, Mack J, Weltzin RA, Mekalanos JJ, Kraehenbuhl JP, Neutra MR (1991) New model for analysis of mucosal immunity: intestinal secretion of specific monoclonal immunoglobulin A from hybridoma tumors protects against Vibrio cholerae infection. Infect Immun 59:977–982

Wolf JL, Rubin DH, Finberg RS, Kauffman RS, Sharpe AH, Trier JS, Fields BN (1981) Intestinal M cells: a pathway for entry of reovirus into the host. Science 212:471–472

Zhou F, Kraehenbuhl JP, Neutra MR (1995) Mucosal IgA response to rectally administered antigen formulated in IgA-coated liposomes. Vaccine 13: 637–644

Dendritic Cells and Langerhans Cells in the Uptake of Mucosal Antigens

G.G. MacPherson[1] and L.M. Liu[1,2]

[1]Sir William Dunn School of Pathology, Univeristy of Oxford, England

[2]*Present Address*: Center for Neurologic Diseases, Brigham and Women's Hospital, Harvard Medical School, 221 Longwood Ave., Boston, MA 02115, USA

1 What Is a Dendritic Cell? – An Overview of Dendritic Cell Biology

1.1 Introduction

Dendritic cells (DC) represent heterogenous populations of bone marrow-derived cells that have central roles in the initiation and regulation of immune responses (reviewed in Steinman 1991). The cells under consideration are quite distinct from the follicular DCs found in B-cell areas of lymphoid tissues. It is, however, now clear that there is not a single DC lineage, but that cells with DC properties may arise from at least three lineages and can exist in different states of maturation and activation. The relationships between these lineages and their functional significance in immune responses are far from clear.

The major defined function of DCs is to capture antigens (Ags) in peripheral tissues and transport them to secondary lymphoid tissues for presentation to T lymphocytes. Presentation to T cells involves processing protein Ags to produce major histocompatability complex (MHC)-binding peptides, but we have evidence that, in addition, DCs can retain and transport native Ags for presentation to B cells (Wykes et al. J. Immunol. In press). Moreover, there is increasing evidence that some DCs may have a regulatory role in immune responses, for example, CD8$\alpha\alpha$+ splenic DCs can kill the T cells they are stimulating (Suss and Shortman 1996), while Ags directly targeted to DCs in vivo can be tolerogenic (Finkelman et al. 1996).

It is also clear that DC function can be modulated in vivo and in vitro by a variety of stimuli, particularly those relating to inflammation or 'danger' (Matzinger 1994), and that this plasticity is a potential hazard in the interpretation of experiments involving in vitro manipulation of DCs.

The unravelling of DC biology is still in its early stages. DCs are rare cells in peripheral tissues, and the difficulty of extracting them from tissues has hampered their study. DCs can be grown in vitro and some cell lines with DC properties have been described (Elbe et al. 1994; Lutz et al. 1994; Xu et al. 1995; Volkmann et al. 1996), but it is often difficult to relate the properties of these cells to DCs in vivo. Thus, great caution is necessary in the interpretation of studies of DCs, but a full understanding of these fascinating cells is essential for vaccine development, the understanding of mucosal immune responses and mucosal tolerance, and the regulation of immune responses in general.

In this chapter we will review the properties of DCs in general and then discuss studies that have started to define the properties of mucosal DCs.

1.2 Anatomical Distribution of DCs

1.2.1 DCs in Peripheral Tissues

In situ, DCs can be tentatively identified by their irregular morphology in conjunction with immunostaining for MHC class II and other surface markers that are more or less specific for DCs [it should be noted that there seem to be no monoclonal antibodies (mAbs) that recognize all DCs or only DCs]. Expression of MHC class II and an irregular morphology is not, however, sufficient for DC identification; we have evidence that a proportion of such cells in the lamina propria (LP) of rats are in fact macrophages and, in inflamed tissues, many macrophages may express MHC class II.

Given these caveats, cells with the characteristics of DCs have been identified in all tissues, with the exception of much of the central nervous system (CNS), but have been isolated and examined functionally from relatively few. In some tissues, they appear to be randomly distributed while in others, e.g., epidermis or respiratory epithelia, they appear to have an organized distribution that may facilitate their interaction with invading pathogens. The density of DCs also varies in different tissues, and in different regions of the same tissue. Again, this may relate to the probability of infection occurring at a particular site. Finally, inflammatory stimuli may alter the density of DCs, particularly in mucosal tissues (McWilliam et al. 1996).

1.2.2 Central Lymphoid Tissues

Within organized secondary lymphoid tissues, DCs are concentrated particularly in T-cell areas. Thus, they are found in relatively large numbers as interdigitating cells (IDCs) in the paracortex of lymph nodes and the periarteriolar lymphoid sheath of the spleen, but are also present in the sub-capsular sinus and interfollicular areas of nodes (which may represent DCs in the process of migration), and in the marginal zone of the spleen. Marginal zone DCs, however, express surface markers that differ from those of IDCs. Thus, in the mouse, marginal zone DCs are 33D1+, NLDC145-, while IDCs express a reciprocal phenotype (Agger et al. 1990). The distribution of DCs in mucosal-associated lymphoid tissues will be discussed in Sect. 2.2.

1.3 DC Lineages

The first DC to be identified were isolated from mouse spleen and were thought to correspond to IDCs, but it now seems probable that these DCs are derived from the marginal zone and that IDCs are much more difficult to extract. DCs with similar properties can be grown from murine peripheral blood mononuclear cells. This blood precursor does not have the properties of a monocyte. DCs which may be analogous to this population can be grown from human CD34+ bone marrow or blood cells (reviewed in Young and Steinman 1996). In contrast, culture of human peripheral blood mononuclear cells (PBMNC) with interleukin-4 (IL-4) and tumor necrosis factor alpha (TNF-α) generates a population of DCs that develops from monocytes without cell division (Sallusto and Lanzavecchia 1994).

A third lineage of DCs has been identified in murine thymus. These DCs derive from a thymocyte population in which differentiation is restricted to DCs and T cells (Ardavin et al. 1993; Saunders et al. 1996; Wu et al. 1996). Recently it has been shown that human CD4+ plasmacytoid 'lymphocytes' can develop into cells with DC properties if stimulated with CD40-ligand (Grouard et al. 1997). How these DC lineages relate to each other within and across species is largely unknown, but this heterogeneity suggests that the functions of DCs in immune responses may be more complex than was previously thought.

1.4 DC Maturation

Within at least some lineages, it is clear that DCs can exist in different states of maturation. Much information has derived from the study of epidermal Langerhans cells (LC), the DCs of the epidermis. Thus, fresh LC are actively endocytic and phagocytic for a variety of particles (Reis-e-Sousa et al. 1993), can acquire and process protein Ags, are actively synthesizing MHC class II, but express low levels of B7 and are weak stimulators of resting T cells (Schuler and Steinman 1985). After in vitro culture, LC lose the ability to process protein Ags, shut down MHC class II biosynthesis, but increase expression of B7 and become powerful stimulators of resting T cells (Romani et al. 1989). Other peripheral DCs (heart, kidney, intestine) also show immature features that, similarly, change in culture (Austyn et al. 1994), but in the case of rat LP DCs, freshly isolated cells are able to stimulate resting T cells to an intermediate degree (Liu and MacPherson 1995b).

While it is generally thought that the changes seen in cultured DCs represent physiological maturation, it is not clear that they do in fact occur under normal steady-state conditions. It is becoming apparent that removal of DCs from their natural microenvironment may be sufficient to deliver 'activation' signals that initiate a pattern of differentiation that, in vivo, only occurs in inflamed or damaged tissues. Thus, we have shown that lymph DCs (L-DCs) that have just exited the small intestine and would normally migrate to the mesenteric nodes, if cultured, rapidly express markers that are not seen on DCs freshly extracted from the nodes (MacPherson 1989; MacPherson et al. 1989). These observations suggest that any in vitro study of DCs needs to be interpreted with caution; even the processes involved in isolating DCs from tissues may deliver activation signals that do not exist in steady-state conditions in vivo.

1.5 Antigen (Ag) Capture by DCs

1.5.1 In Vitro Studies

DCs isolated from peripheral or lymphoid tissues, or grown in vitro have been tested for their ability to endocytose soluble Ags and particles. The picture that has emerged is that peripheral or immature DCs are actively endocytic, but that DCs from lymphoid tissues are much less active. LC freshly isolated from murine epidermis can phagocytose a variety of particles, including zymosan, yeast, bacteria and latex, but after culture, phagocytosis is markedly reduced (REIS-E-SOUSA et al. 1993). Splenic DCs were, likewise, more phagocytic when freshly isolated than after culture. It was shown that phagocytosis by fresh LC was at least in part mediated via a mannose receptor.

Much of the current understanding of DC endocytosis has come from Lanzavecchia's group (SALLUSTO and LANZAVECCHIA 1994; SALLUSTO et al. 1995; and reviewed in LANZAVECCHIA 1996; CELLA et al. 1997). They have concentrated on one type of DC: the human blood monocyte-derived DCs. These cells exist in immature and mature states, the major stimuli to maturation being TNF-α, lipopolysaccharide (LPS) or ligation of CD40. In the immature state they are highly endocytic, utilizing the mannose receptor and also macropinocytosis, a process whereby cells engulf very large volumes of fluid into conspicuous cytoplasmic vacuoles. This enables such DCs to accumulate Ags at very low concentrations and makes them highly efficient antigen-presenting cells (APCs). In these immature DCs, most MHC class II is present in cytoplasmic vacuoles and is turning over rapidly. After maturation, macropinocytosis ceases and most MHC class II is expressed on the plasma membrane and is very stable, with a half-life of days.

It is, however, important to realize that the phenomena described for monocyte-derived DCs have only been seen in these DCs and in murine bone-marrow derived DCs (NORBURY et al. 1997). Normally, DCs would exist in connective tissue or in epithelia that have very little free water in them. It is difficult to see how macropinocytosis could occur under these circumstances. If, however, tissues become inflamed, increased vascular permeability leads to massive fluid exudation from venules and, under these circumstances, macropinocytosis could operate efficiently.

1.5.1.1 Interaction of Pathogens with DCs

The specific immune response has, of course, evolved as a defense against infectious disease, but relatively little is known about the interactions of DCs with living pathogenic organisms. It is clear that DCs can be infected with a variety of viruses, but interestingly there is evidence that, in the case of influenza and vaccinia, the infection may not be productive and that in vaccinia only early-stage proteins may be synthesized (BHARDWAJ et al. 1994; BRONTE et al. 1997; MAJOR, SMITH and MACPHERSON, unpublished observations).

The interaction of DCs with human immunodeficiency virus (HIV) is controversial. Some human DCs are CD4+, permitting infection by HIV, but DCs might also be infected following endocytosis of HIV. Early studies claimed that HIV could replicate in DCs, but more recent work suggests that HIV can be retained by DCs, without undergoing replication, and that when such a DC interacts with a T cell, HIV is transferred very efficiently to the latter (Pope et al. 1994; Pope et al. 1995; and reviewed in Cameron et al. 1994; Cameron et al. 1996). The role of mucosal DCs in HIV infection is discussed in Sect. 2.3.4.

The ability of immature DCs to phagocytose particles suggests that they may be able to interact directly with bacteria and parasites. Immature DCs cultured from murine bone marrow can endocytose baccille Calmette Guerin (BCG) and subsequently present BCG peptides to sensitized T cells (Inaba et al. 1993). More recently, it has been shown that a number of other bacteria can infect DCs (Guzman et al. 1994; Guzman et al. 1995; Conlan 1996; Filgueira et al. 1996).

1.5.1.2 Ag Capture for MHC Class-I Expression

It is becoming increasingly clear that, in some cell types at least, exogenous Ags can be captured, processed and expressed on MHC class I. This pathway has been well described in murine macrophages (reviewed in York and Rock 1996) and recently in murine bone marrow-derived DCs (Norbury et al. 1997). In these DCs, such exogenous ovalbumin processing was dependent on proteasomes and the transporter for antigen processing (TAP), showing that the classical MHC class I pathway was used. It is not clear how Ags escape from the endocytic pathway into the cytoplasm, nor whether this pathway can be utilized by all DCs, but it has great potential significance for the initiation of immune responses to viruses.

1.5.2 Ag Capture by DCs in Vivo

Early experiments showed that if a soluble Ag was injected i.v., DCs extracted from the spleen could stimulate sensitized T cells and prime naive T cells in vivo (Crowley et al. 1990). DCs were the only cells in the spleen with this ability. Other experiments showed that if contact sensitizers were painted onto murine skin, DCs extracted from draining nodes could present the Ag to sensitized T cells and could transfer sensitization to naive animals (Macatonia et al. 1987). These data were interpreted as showing that DCs captured Ags in the epidermis and transported them to the nodes, but recent experiments (Liddington and Austyn, personal communication) suggest that labelled DCs extracted from nodes had acquired Ags in the nodes, the Ag having reached the nodes in lymph.

Several experiments have investigated the interaction of DCs and pathogenic microorganisms in vivo. Conlan (1996) has shown that *Listeria monocytogenes* infect cells in the marginal zone of the spleen and that these cells then translocate to the T-cell areas, a migration very similar to that induced by LPS (De Smedt et al. 1996). Moll, in important studies, has shown that *Leishmania major* infects Langerhan's cells (LC) in the epidermis, that these LCs then transport the parasite to draining nodes; LCs in the node providing a long-term reservoir for the parasite, perhaps permitting continual stimulation of specific T cells (Blank et al. 1993; Moll et al. 1993; Moll et al. 1995).

1.6 DC Migration

The initiation of immune responses occurs in organized secondary lymphoid tissues, but pathogens primarily invade peripheral tissues. An essential part of DC function is the transport of Ags from peripheral tissues to lymph nodes. DCs migrate in afferent lymph from all tissues that have lymphatic drainage (reviewed in Yoffey and Courtice 1970), but patterns of migration are different in normal steady-state conditions and following inflammatory stimulation.

1.6.1 Steady-State Migration

DCs are continually migrating from peripheral tissues in afferent lymph in the absence of antigenic or inflammatory stimulation. All afferent lymphatics, e.g., renal, hepatic and testicular, that have been cannulated in sheep contain migratory veiled DCs (reviewed in Yoffey and Courtice 1970). In steady-state conditions, DCs spend a variable time in peripheral tissues before exiting into lymph. Thus, we have shown that the interval between the final division of a DC precursor in bone marrow and its appearance in lymph can vary between 48 h and more than 5 days (Pugh et al. 1983). In contrast, LCs spend weeks or months in the epidermis. The parameters that determine the period of residence are unknown, but there is evidence that e-cadherin is involved in the retention of LCs in the epidermis (Tang et al. 1993). The DCs that exit peripheral tissues show a mature phenotype (Pugh et al. 1983; MacPherson 1989; MacPherson et al. 1989; Liu and MacPherson 1993; Liu and MacPherson 1995a)(see Sect. 2.3.3).

1.6.2 Stimulated Migration

Several different stimuli have been shown to induce increased release of DCs from peripheral tissues. The most powerful stimulus is bacterial LPS. Systemic administration of LPS stimulates release of DCs from heart and kidney (Roake et al. 1995), and intestine (MacPherson et al. 1995) (see Sect. 1.6.2). This release is probably not due to a direct effect of LPS on DCs, rather it depends on LPS stimulating the release of other mediators, primarily TNF-α and IL-1. Thus, local injection of TNF-α induces release of LCs from the epidermis. LPS can also modulate DC migration within lymphoid tissues, notably the spleen, where systemic LPS induces DCs to migrate from the marginal zone into T-cell areas (De Smedt et al. 1996).

1.6.3 Molecular Basis of DC Migration

There is much current interest in the role of chemokines in directing the migration of leukocytes. Human DCs derived from CD34+ cord blood cells respond to MCP-1, MCP-2, MIP-1α, MIP-1β and RANTES (Xu et al. 1996). A novel chemokine, MDC, expressed in macrophages and blood-derived DCs is chemotactic for monocyte-derived DCs and activated natural killer (NK) cells (Godiska et al. 1997). Clearly, there is still much to learn in these areas.

1.7 Interaction of DCs with T Cells

DCs present MHC-associated peptides to CD4+ and CD8+ T cells in secondary lymphoid tissues. To do this, they must interact with recirculating T cells. A recent study has identified a chemokine DC-CK1 that is a specific chemoattractant for CD45RA+, naive T cells (Adema et al. 1997). One of the central properties of DC is that they are able to activate naive and resting T cells. Activation depends on adhesion interactions and DCs can form Ag-independent clusters with resting T cells (Inaba et al. 1989). Although lymphocyte function-associated antigen (LFA-1) is involved in this interaction, other as yet unidentified molecules are critical for adhesion. Following adhesion, activation of the T cell depends on the expression by DCs of co-stimulatory molecules, primarily B7.1 and B7.2. DCs are the only cells in unstimulated secondary lymphoid tissues that express B7 constitutively, and this explains their possible unique ability to initiate Ag-specific T-cell responses.

Following the initial activation of a resting T cell, other molecular interactions come into play. Thus, it has been shown that the interaction of CD40 on DCs with CD40 ligand, expressed by activated T cells, alters DC function, upregulating MHC class-II expression and, importantly, stimulating IL-12 production by the DCs (Caux et al. 1994; Sallusto et al. 1995; Cella et al. 1996; Kelsall and Stuber 1996; McLellan et al. 1996; Pinchuk et al. 1996). It is, thus, possible that DCs may have a role not just in the initial activation of T cells, but also in the modulation of immune responses, affecting Th1/Th2 differentiation. Other evidence that DCs may modulate the quality of immune responses is discussed in Sect. 2.4.

1.8 Activation and Secretory Properties of DCs

Recently, it has been suggested that the initiation of an immune response may not depend primarily on the recognition of foreign Ags, but rather on the modulation of APC (particularly DC) behavior by inflammatory or danger signals (Janeway and Bottomly 1994; Matzinger 1994). Indeed, there is considerable evidence that in steady-state conditions, Ags delivered to DCs may be ignored or may even tolerize T cells. Thus, i.v. Ags are delivered to DCs (Crowley et al. 1990), but are often tolerogenic. However, if splenic DCs that have acquired Ags in vivo are injected into naive mice, they can activate naive T cells. Similarly, we have shown that intestinal lymph DCs acquire enteric Ags and that such cells can activate naive T cells in vivo (Liu and MacPherson 1993), but oral Ags are often also tolerogenic. Recently, it has been shown that a soluble rat anti-mouse DC mAb, injected i.v., tolerizes mice to rat IgG, whereas an aggregated mAb is immunogenic (Finkelman et al. 1996).

Recent evidence suggests that a variety of pro-inflammatory stimuli can alter DC behavior and properties. Among the most important are TNF-α, IL-1 and ligation of CD40. These stimuli induce changes such as upregulation of MHC class II, expression or upregulation of B7 expression and secretion of Il-12.

It is well recognized that macrophages can be induced to secrete nitric oxide (NO). Recently it has been shown that bone-marrow-derived DCs can be induced to secrete NO (Lu et al. 1996; Powell and MacPherson, unpublished observations), and we have shown that immature (FcR+) DCs can inhibit the ability of mature DCs to stimulate an allogeneic mixed lymphocyte reaction (MLR) (Powell and MacPherson, unpublished observations) via a mechanism that is partially NO-dependent, suggesting that DCs themselves may have a role in the regulation of immune responses.

2 Mucosal DCs

2.1 Introduction

Most infectious agents either infect mucosal surfaces directly or gain entry to the body via mucosae or mucosal lymphoid tissue. In view of this, it is perhaps surprising that so little is known of the properties and functions of mucosal DCs in comparison with other DCs. Much of this relates to the difficulty of obtaining sufficient numbers of DCs from mucosal tissues to enable experiments to be carried out; we routinely use six rat small intestines to prepare LP DCs! However, given the specialization of mucosal immune responses, it is critical that the roles of DCs in their initiation and regulation are understood. What is known about mucosal DCs suggests that while they share many of the properties described for DCs in other sites, they may have specializations that suggest important regulatory roles.

DCs are found throughout mucosal tissues, in both non-lymphoid epithelia and connective tissues, and in mucosa-associated lymphoid tissue (MALT), including the associated draining nodes. This section will review the distribution and properties of DCs in lymphoid and non-lymphoid mucosal tissues, separately.

2.2 Distribution of DCs in Mucosal Tissues

2.2.1 Non-lymphoid Tissues

DCs, identified by morphological and immunocytochemical criteria, have been described in all mucosal tissues that have been examined. There are, however, as discussed in Sect. 1.2.1, difficulties with the identification of DCs by these criteria alone. This is exemplified by our studies of LP DCs in the rat. The LP contains a large population of cells that are irregular and MHC class II+. Only a proportion of these are also MRC OX62+ [OX62 recognizes an Ag, probably an integrin, that is present on all DCs in lymph draining the small intestine (Brenan and Puklavec 1992)]. When fluorescent Ags or latex particles are injected intra-intestinally, some MHC class II+ cells in the LP accumulate the marker, but double labelling shows that such cells are MRC OX62-. These cells might represent immature DCs in "Ag capture mode", but we have never seen labelled DCs in intestinal pseudo-afferent lymph from such animals, suggesting strongly that the cells that accumulate the marker do not leave the LP and are, thus, very unlikely to be DCs (Liu and MacPherson, unpublished observations). There may be alternative explanations for these observations, but the point is that identification of DCs in situ is difficult and may be unreliable.

Given these provisos, DCs appear to be relatively abundant in mucosae. In tissues expressing a stratified squamous epithelium, DCs with the characteristic features of LCs are present in the suprabasal layers. In non-squamous stratified epithelia, notably the respiratory tract, a population of intra-epithelial DCs has also been identified. Using a technique of tangential sectioning to examine the rat, murine and human respiratory tract, Holt's group has shown that these DCs form an intricate branching network, very similar to that of LCs in the epidermis (Holt et al. 1989; Holt et al. 1990). They are all MHC class II+ and in the rat are CD4+. The density of intra-epithelial DCs is maximal in the trachea and decreases steadily down the tract.

In the intestine, lined by columnar epithelium, there is evidence for a population of DCs that resides above the basement membrane, and there is also much evidence that DCs below the basement membrane can extend processes up between epithelial cells (Maric et al. 1996).

In addition to these DC populations that are associated directly with epithelium, other DCs lie in connective tissue without apparent direct epithelial contact. It is not clear from these in situ studies whether phenotypic differences exist between DCs in different sites, but microdissection studies (see Sect. 2.3.3) suggest that there are, and that the DCs at different sites may be functionally different.

2.3 Functional Properties of Mucosal DCs

2.3.1 Isolation of Mucosal DCs

DCs are a relatively rare cell types in all tissues and their isolation is fraught with difficulties. Most techniques involve mincing the tissue, enzymatic digestion, differential adherence, and negative or positive selection. The yields are low and the proportion of DCs that is recovered is unknown. The procedures used to isolate DCs are associated with two major problems – they may be selective for a subpopulation(s) of DCs and they may induce changes in the isolated DCs that do not represent normal in vivo events.

An alternative approach has been in use in our laboratory for some years. DCs are normally removed from the lymph in the first node they reach; mesenteric lymphadenectomy in the rat, sheep and mouse (Mayrhofer et al. 1986; Pugh et al. 1983; Rhodes 1985; Bujdoso et al. 1989) has been used to remove this blockage to migration, with the result that DCs can be collected by cannulation of central lymphatics. In the rat the thoracic duct is cannulated. Such DCs have recently left the intestine physiologically, can be collected in the cold and can be concentrated by simple density-gradient centrifugation. They are as close to a physiological population of DCs as can be acquired at present. These L-DCs are difficult to collect in large numbers and represent DCs at just one stage in their life history. In addition, we do not know whether these DCs derive from the LP, Peyer's patches (PP) or both. The study of them has, however, given a number of important insights into DC properties and functions.

2.3.2 Kinetic Studies

2.3.2.1 Steady State

The ability of DCs to act as APCs depends on their migration to the draining lymph node. We have studied the migration and turnover of DCs in the normal rat small intestine (Pugh et al. 1983). Dividing DC precursors in the bone marrow can be labelled by means of an i.v. injection of a DNA precursor, usually tritiated thymidine (^{3}H-TdR) or bromo-deoxyuridine (BrdU). As there is no evidence for significant DC division in the intestine, the minimal time taken for labelled DCs to appear in pseudo-afferent intestinal lymph represents the time from the last division in the marrow and includes the time taken to traverse the intestine. Labelled DCs appear in lymph within 48 h, with peak numbers arriving at 3–4 days. Thus, DCs spend a minimum of 48 h and a modal time of 3–4 days in the intestine, before migrating to the nodes. We cannot accurately estimate the maximum time in the intestine because the rate of decline of labelled DC appearance in lymph is affected both by input into the gut of cells that were labelled early in their differentiation in marrow and those that have spent longer periods in the gut. These data show that intestinal DCs turn over much more rapidly than LCs, but with similar kinetics to murine splenic DCs (Steinman et al. 1974).

In the respiratory tract of rats, a different approach, measuring the kinetics of DC reconstitution after depletion by corticosteroids or irradiation, suggests that the average half-life of DCs is about 2 days (Holt et al. 1994). Thus, it appears that in the absence of any known stimulation, DCs spend only short times in mucosae before migrating to draining nodes.

2.3.2.2 Stimulated Migration

A variety of non-specific stimuli can affect DC migration dramatically, possibly acting via a final common pathway involving TNF-α. In the intestine, i.v. endotoxin induces a rapid increase in the numbers of DCs migrating in lymph. This effect occurs within 6 h, peaks at 12–24 h and is over within 48 h, resulting in an approximate 10-fold increase in the numbers of DCs that can be collected over that period. The source of the migrating DCs appears to be the LP as, at 24 h, the numbers of OX62+ cells in the LP is much reduced. It seems unlikely that the LPS was acting directly on DCs, but we suspect that TNF-α is involved because an anti-TNF Ab markedly inhibited the effects of LPS. The DCs released following LPS administration did not differ from steady-state DCs in MLR stimulation, but other properties were not investigated. In other models, e.g., murine heart and kidney (Roake et al. 1995), IL-1 is also involved in stimulating DC migration.

In addition to inducing migration of DCs from mucosae, inflammatory stimuli also cause a rapid influx of DCs and/or DC precursors. Thus, Holt's group has shown that large numbers of DC precursors enter respiratory epithelium within hours of challenge with bacterial, viral or protein Ags. These cells spend up to 48 h in the epithelium and then migrate to the draining nodes (McWilliam et al. 1996). Interestingly, similar stimuli did not result in DC accumulation in the peritoneal cavity or epidermis.

Now that it is becoming clear that different DC lineages exist, and that DCs can be "activated" in different ways, it is increasingly important to characterize different DC populations functionally. We are not aware of any experiments in which this has been carried out for DCs activated in vivo, and it is not known how the DCs that accumulate in inflamed tissues relate to those that traffic through mucosae under steady-state conditions.

2.3.3 Properties of DCs Isolated from Mucosae

2.3.3.1 Gastrointestinal Tract

Relatively few studies have examined the phenotypic and functional properties of DCs isolated from the gut. Pavli et al. (1990) isolated DCs from murine PP and LP. They found that these cells resembled splenic DCs in phenotype and function. We used similar approaches to isolate DCs from rat PP and LP (Liu and MacPherson 1995b). Yields were small and DCs could not be enriched to more than 30–40% purity. DCs freshly isolated expressed high levels of MHC class II and, in contrast to freshly isolated LCs (Schuler and Steinman 1985) and heart or kidney DCs (Austyn et al. 1994), gave intermediate levels of stimulation in a MLR compared with lymph DCs. After overnight culture with granulocyte macrophage–cerebrospinal fluid (GM-CSF), both PP and LP DCs became as potent MLR stimulators as lymph DCs (Liu and MacPherson 1995b). It is interesting that PP DCs were poor MLR stimulators, as we would have predicted that DCs from the

T-cell areas would give good stimulation. It is possible that, as with spleen, DCs from T areas are more difficult to isolate, and we may have been isolating the sub-epithelial dome DCs described in the mouse (KELSALL and STROBER 1996; RUEDL et al. 1996) (see Sect. 2.3.3.3).

DCs isolated from the thoracic duct lymph of mesenteric-lymphadenectomized rats have recently left the small intestine and represent DCs actively involved in Ag transport (PUGH et al. 1983; LIU and MACPHERSON 1991; LIU and MACPHERSON 1993). These cells appear to differ functionally from DCs isolated either from peripheral tissues or secondary lymphoid tissues. Thus, they are fully mature in terms of MLR stimulation; their potency does not change for at least 72 h in culture, but they retain the ability to process native Ag for the same period (LIU and MACPHERSON 1995a). A partial explanation of these observations comes from recent experiments (LIU et al. J. Immunol. In press). We have shown that in rat intestinal lymph, DC sub-populations can be distinguished by their expression of CD4. CD4+ DCs are more effective APCs for naive and sensitized T cells, survive better in culture, but do not lose the ability to process Ags in culture. CD4- DCs are weak APCs, contain phagolysosomes, but do not express detectable Fc receptors (FcR), survive poorly in culture, and completely lose the ability to process native Ags in culture, while becoming as strong stimulators of a MLR as the CD4+ cells. Thus, functionally, CD4- DCs resemble LCs, whereas the CD4+ DCs do not. We do not know how these two populations relate to each other, but in vivo kinetic studies suggest that the CD4+ DC are not precursors of CD4- cells.

APCs have been isolated from human colon (MAHIDA et al. 1988). These cells gave good stimulation in a MLR, but the authors considered that they had both macrophage and DC characteristics. In contrast, another study (PAVLI et al. 1993) showed that when macrophages and DC were separately isolated from human colon, most MLR stimulation could be attributed to DCs.

2.3.3.2 Respiratory Tract

Many reports have described the properties of DCs isolated from the respiratory tract. Early studies showed that DCs could be isolated from rat (VAN DER BRUGGE-GAMELKOORN et al. 1985) human and murine tissues (SERTL et al. 1986), and that the murine DCs could present ovalbumin (OVA) to a specific hybridoma. More recently, studies have focused on DC heterogeneity in the respiratory tract and on the regulation of DC activity.

KRADIN et al. (1993) found that DCs could be distinguished by FcR expression. Both sets could present Ags to sensitized T cells, but the FcR+ cells were less effective stimulators of naive T cells. The FcR+ DCs could phagocytose latex particles and contained phagolysosomes. The relationships of the populations were unclear. GONG et al.(1992) used microdissection to isolate airway epithelium and showed that the DCs present in the epithelium (EDCs) differed from those isolated from lung parenchyma (PDCs). EDCs expressed less MHC class II, did not express ICAM-1 and more than 50% were FcR+. Most PDCs were ICAM-1+ and less than 5% were FcR+. Functionally, EDCs were more efficient at presentation of soluble and particulate Ags to T cells, but less efficient stimulators of a MLR. Again, the relationships between the populations were unclear. Although the EDCs express a more "immature" phenotype, this does not mean they are the precursors of the PDCs.

Holt's group has studied respiratory tract DCs extensively. Functionally, they have shown that DCs isolated from lung slices or airway epithelium are efficient APCs for T cells, and that the interaction of DCs and T cells could be downregulated by pulmonary macrophages (Holt et al. 1988). They also showed that following inhalation of an OVA aerosol, epithelial DCs could activate OVA-sensitized T cells in vitro. Other workers showed that immune responses to inhaled Ags were stimulated if alveolar macrophages were eliminated (Thepen et al. 1989; Thepen et al. 1991; Thepen et al. 1992; Strickland et al. 1993). More recently, it has been shown that the downregulation of DC APC activity by macrophages is mediated by NO (Holt et al. 1993), and that if rats were given an inhibitor of NO synthase, administration of Ags to sensitized animals by inhalation led to a dramatically increased acute inflammatory response in the lung (Thepen et al. 1994).

2.3.3.3 Peyer's Patches

PP and their equivalent in the respiratory tract are the major portals by which Ags in the intestinal lumen are made available to the cells of the immune system. DCs have been described at two sites in PP: the T-cell areas, in which DCs are thought to correspond to IDCs in lymph nodes, and in the sub-epithelial area underlying the dome. This is the region into which the Ag is delivered by M cells, and it is an attractive hypothesis that DCs in this area capture Ags and transport them to T-cell areas for presentation to recirculating T cells. In the mouse, DCs in the two sites differ in their expression of surface markers. Sub-dome DCs are negative for NLDC145 and CD11c, whereas T area DCs express both markers (Kelsall and Strober 1996). In a recent study (Ruedl et al. 1996), it was shown that CD11c- PP DCs were functionally immature in terms of T cell activation, but were actively endocytic and could phagocytose latex particles. After culture with GM-CSF, TNF-α, or anti-CD40 mAbs, the CD11c-, NLDC145- DCs expressed both markers, lost the ability to process native Ags, downregulated MHC class II and invariant chain synthesis, upregulated B7 and became potent stimulators of resting T cells, thus acquiring the properties of mature DCs. It is suggested that the sub-dome DCs migrate to T-cell areas, but there is no direct proof of this. This differentiation is similar to that described for LCs in culture, but it is not known whether it represents normal maturation in the absence of inflammatory stimuli.

2.3.4 Ag Acquisition by Mucosal DCs in Vivo

Although mucosal DCs are strategically placed to acquire Ags via mucosal surfaces, there is very little concrete evidence that they can do this, and even less that they do so with pathogenic organisms. We have shown that soluble Ag, administered by gavage or intra-intestinal injection, is acquired by DCs in the intestinal wall, and that Ag-bearing DCs appear in intestinal lymph within 6 h (Liu and MacPherson 1991; Liu and MacPherson 1993). We have not been able to demonstrate the presence of Ag directly, but DCs from Ag-fed rats can stimulate sensitized T cells and, more importantly, can sensitize naive T cells following subcutaneous injection. It was important in these experiments to show that the injected DCs were presenting Ags directly, and this was done by injecting parental strain Ag-bearing DCs into an F1 recipient, showing that T cells were only sensitized to the MHC of the injected DCs.

We could not determine from these experiments the origin of the Ag-bearing DCs; they could have arisen from PP, LP or both. Recently, Kelsall and Strober (1996) have shown that CD11c+ DCs, isolated from murine PP after feeding OVA, are able to activate OVA-specific naive T cells in vitro. Thus, as expected, PP is clearly one route for delivery of Ags to mucosal DCs, but a route via LP cannot be excluded.

In contrast to studies with soluble Ags, essentially nothing is known about the interaction of mucosal DCs with particulate Ags or pathogens in situ. Mayrhofer et al. (1986) showed that, in rats infected with *S. typhimurium*, cells with DC characteristics in pseudo-afferent intestinal lymph contained Salmonella Ags. A preliminary report has shown that, following infection of mice with *S. typhimurium* expressing green fluorescent protein, bacteria co-localize in PP with cells expressing DC characteristics (N418 expression, lack of macrophage markers) (SA Hopkins and JP Kraehenbuhl, personal communication). In studies of influenza infection in mice, it has been shown that soon after intra-nasal infection, the virus appears in the draining nodes and is associated with DCs, macrophages and B cells (Hamilton Easton and Eichelberger 1995), but it is not known if DCs were responsible for the capture of the virus in the epithelium and its transport to the nodes or whether the free virus infected APCs already present in the nodes.

Clearly, there remains much uncertainty concerning the roles of DCs in mucosal infections and there is currently much interest in the role of mucosal DCs in HIV infection.

2.3.5 Mucosal DC and Human Immunodeficiency Virus

HIV is most often acquired through infection of mucosal surfaces, and DCs are strategically placed to encounter the virus, transport it to lymphoid tissues and infect T cells. It is, of course, impossible to investigate the role of DCs as HIV transporters in humans, but some important evidence has come from studies of simian immunodeficiency virus (SIV) in rhesus macaques (Spira et al. 1996). After intra-vaginal inoculation of SIV, viral DNA was detected by in situ reverse-transcriptase polymerase-chain reaction (RT PCR). Viral DNA was first detected in LP cells with the characteristics of DCs in both the vagina and the cervix. Within 2 days, infected cells were detected in the sub-capsular sinus and paracortex (T-cell area) of the draining nodes. Interestingly, no infected cells were seen in vaginal or cervical epithelium, even though many LCs are present in vaginal epithelium. It is possible that LCs bind or endocytose virus without it replicating; this would prevent detection by PCR. This study provides very suggestive evidence that DCs are the primary target of SIV (and HIV) in mucosal infection.

DCs can clearly be infected in the mucosal lymphoid tissues of infected individuals. Thus, FRANKEL et al. (1996) have shown that infected syncytia at the surface of the tonsil express DC markers, and that such syncytia can be present in patients without symptoms of acquired immunodeficiency syndrome (AIDS).

There is also evidence that HIV infection can decrease numbers of DCs in mucosal tissues. Thus, LIM et al. (1993) showed a decrease in cells with DC phenotype in the duodenum, and SPINILLO et al. (1993) in the cervical epithelium of HIV-infected patients. This might lead to increased susceptibility to intercurrent infections.

2.4 Role of DCs in Regulation of Mucosal Immune Responses

Immune responses initiated at mucosal surfaces differ from systemic responses in several ways and the roles of APCs in determining the outcome of mucosal immunization are poorly understood. IgA synthesis is initiated in PP and it appears that the switch of B cell sIg from IgM to IgA is under T-cell control, but there is evidence that DCs may also have a role in isotype switching. Early studies (SPALDING et al. 1984; SPALDING and GRIFFIN 1986) showed that polyclonal activation of sIgM B cells, in the presence of mixtures of DCs and T cells from PP, led to increased IgA secretion, and there were indications that it was the DCs and not the T cells that determined the outcome. Recent work from Bancherau's group supports the concept that DCs may have a direct role in isotype switching. They showed that polyclonal activation of naive, sIgM+ human tonsillar B cells, in the presence of CD40 ligand and blood-derived DCs, led to a skewing of the response towards IgA (FAYETTE et al. 1997). The mechanisms by which DCs influence B-cell isotype switching are unknown, but we have evidence that it may occur as a direct result of DC/B cell contact, prior to T-cell involvement (WYKES, M. et al. J. Immunol. in press).

EVERSON et al. (1996) produced evidence to suggest that cytokine secretion by mucosal DCs may influence T-cell activation. He showed that activation of T cells by mucosal DCs induced a Th2 pattern of cytokine secretion, whereas splenic DCs induced Th1 cytokines. Again, the mechanisms by which DCs influence cytokine secretion by T cells are unknown. Little information is available on cytokine and chemokine secretion by mucosal DCs, but it is an attractive hypothesis that they regulate B and T cell activation by a combination of cell surface and secreted molecules. The effects of TGF-β and other locally available factors on DC differentiation and function in mucosae are important areas for future research.

3 Conclusions and Future Directions

Understanding the biology of mucosal DCs is still in its infancy. In broad terms, we know where they are and what they do, but in most cases, not where different sub-populations are or what they do. We do not understand the role of DCs in steady-state conditions; do they serve to maintain tolerance to self Ags present in peripheral tissues? We do not know the roles of different sub-populations in Ag uptake and presentation and we know very little about their interaction with pathogenic microorganisms. We do not know what signals direct DCs to mucosal tissues, what molecular interactions retain them in those tissues or, in most cases, what induces them to leave. We do not know what directs them to T-cell areas in lymphoid tissues or which molecules enable them to interact with T cells. We do not understand much of the molecular basis of DC activation. Despite these obvious gaps, the recognized role of DCs as APCs, and their potential role in immunotherapy has stimulated much research of their properties and functions. We can confidently expect that progress will henceforth be rapid, but we must retain a healthily skeptical viewpoint when attempts are made to generalize about what is clearly a highly complex cell type.

References

Adema GJ, Hartgers F, Verstraten R, et al. (1997) A dendritic-cell-derived C-C chemokine that preferentially attracts naive T cells. Nature 387:713–717

Agger R, Crowley MT, Witmer-Pack MD (1990) The surface of dendritic cells in the mouse as studied with monoclonal antibodies. Int Rev Immunol 6:89–101

Ardavin C, Wu L, Li CL, et al. (1993) Thymic dendritic cells and T cells develop simultaneously in the thymus from a common precursor population. Nature 362:761–763

Austyn JM, Hankins DF, Larsen CP, et al. (1994) Isolation and characterization of dendritic cells from mouse heart and kidney. J Immunol 152:2401–2410

Bhardwaj N, Bender A, Gonzalez N, et al. (1994) Influenza virus-infected dendritic cells stimulate strong proliferative and cytolytic responses from human CD8+ T cells. J Clin Invest 94:797–807

Blank C, Fuchs H, Rappersberger K, et al. (1993) Parasitism of epidermal Langerhans cells in experimental cutaneous leishmaniasis with Leishmania major. J Infect Dis 167:418–425

Brenan M, Puklavec M (1992) The MRC OX-62 antigen: a useful marker in the purification of rat veiled cells with the biochemical properties of an integrin. J Exp Med 175:1457–1465

Bronte V, Carroll MW, Goletz TJ, et al. (1997) Antigen expression by dendritic cells correlates with the therapeutic effectiveness of a model recombinant poxvirus tumor vaccine. Proc Natl Acad Sci USA 94:3183–3188

Bujdoso R, Hopkins J, Dutia BM, et al. (1989) Characterisation of sheep afferent lymph dendritic cells and their role in antigen carriage. J Exp Med 170:1285–1302

Cameron P, Pope M, Granelli Piperno A, Steinman RM (1996) Dendritic cells and the replication of HIV-1. J Leukoc Biol 59:158–171

Cameron PU, Lowe MG, Crowe SM, et al. (1994) Susceptibility of dendritic cells to HIV-1 infection in vitro. J Leukoc Biol 56:257–265

Caux C, Massacrier C, Vanbervliet B, et al. (1994) Activation of human dendritic cells through CD40 cross-linking. J Exp Med 180:1263–1272

Cella M, Scheidegger D, Palmer K, et al. (1996) Ligation of CD40 on dendritic cells triggers production of high levels of interleukin-12 and enhances T cell stimulatory capacity: T-T help via APC activation. J Exp Med 184:747–752

Cella M, Sallusto F, Lanzavecchia (1997) Origin, maturation and antigen presenting function of dendritic cells. Curr Opin Immunol 9:10–16

Conlan JW (1996) Early pathogenesis of Listeria monocytogenes infection in the mouse spleen. J Med Microbiol 44:295–302

Crowley M, Inaba K, Steinman R (1990) Dendritic cells are the principal cells in mouse spleen bearing immunogenic fragments of foreign proteins. J Exp Med 1990 172:383–386

De Smedt T, Pajak B, Muraille E, et al. (1996) Regulation of dendritic cell numbers and maturation by lipopolysaccharide in vivo. J Exp Med 184:1413–1424

Elbe A, Schleischitz S, Strunk D, Stingl G (1994) Fetal skin-derived MHC class I+, MHC class II- dendritic cells stimulate MHC class I-restricted responses of unprimed CD8+ T cells. J Immunol 153:2878–2889

Everson MP, McDuffie DS, Lemak DG, Koopman WJ, McGhee JR, Beagley KW, et al. (1996) Dendritic cells from different tissues induce production of different T cell cytokine profiles. J Leukoc Biol 1996 Apr, 59(4):494–498

Fayette J, Dubois B, Vandenabeele S, Bridon JM, Vanbervliet B, Durand I, Banchereau J, Caux C, Briere F, et al. (1997) Human dendritic cells skew isotype switching of CD40-activated naive B cells towards IgA1 and IgA2. J Exp Med. 1997 Jun 2, 185(11):1909–1918

Filgueira L, Nestle FO, Rittig M, et al. (1996) Human dendritic cells phagocytose and process Borrelia burgdorferi. J Immunol 157:2998–3005

Finkelman FD, Lees A, Birnbaum R, et al. (1996) Dendritic cells can present antigen in vivo in a tolerogenic or immunogenic fashion. J Immunol 157:1406–1414

Frankel SS, Wenig BM, Burke AP, et al. (1996) Replication of HIV-1 in dendritic cell-derived syncytia at the mucosal surface of the adenoid. Science 272:115–117

Godiska R, Chantry D, Raport CJ, et al. (1997) Human macrophage-derived chemokine (MDC), a novel chemoattractant for monocytes, monocyte-derived dendritic cells, and natural killer cells. J Exp Med 185:1595–1604

Gong JL, McCarthy KM, Telford J, et al. (1992) Intraepithelial airway dendritic cells: a distinct subset of pulmonary dendritic cells obtained by microdissection. J Exp Med 175:797–807

Grouard G, Rissoan MC, Filgueira L, et al. (1997) The enigmatic plasmacytoid T cells develop into dendritic cells with interleukin (IL)-3 and CD40-ligand. J Exp Med 185:1101–1111

Guzman CA, Rohde M, Bock M, Timmis KN (1994) Invasion and intracellular survival of Bordetella bronchiseptica in mouse dendritic cells. Infect Immun 62:5528–5537

Guzman CA, Rohde M, Chakraborty T, et al. (1995) Interaction of Listeria monocytogenes with mouse dendritic cells. Infect Immun 63:3665–3673

Hamilton Easton A, Eichelberger M (1995) Virus-specific antigen presentation by different subsets of cells from lung and mediastinal lymph node tissues of influenza virus-infected mice. J Virol 69:6359–6366

Holt PG, Schon-Hegrad MA, Oliver J (1988) MHC class II antigen-bearing dendritic cells in pulmonary tissues of the rat. Regulation of antigen presentation activity by endogenous macrophage populations. J Exp Med 167:262

Holt PG, Schon-Hegrad MA, Philips MJ, McMenamin PG (1989) Ia-positive dendritic cells form a tightly meshed network within the human airway epithelium. Clin Exp Allergy 19:597–601

Holt P, Schon HM, Oliver J, et al. (1990) A contiguous network of dendritic antigen presenting cells within the respiratory epithelium. Int Arch Allergy Appl Immunol 91:155–159

Holt PG, Oliver J, Bilyk N, et al. (1993) Downregulation of the antigen presenting cell function(s) of pulmonary dendritic cells in vivo by resident alveolar macrophages. J Exp Med 177:397–407

Holt PG, Haining S, Nelson DJ, Sedgwick JD (1994) Origin and steady-state turnover of class II MHC-bearing dendritic cells in the epithelium of the conducting airways. J Immunol 153:256–261

Inaba K, Romani N, Steinman RM (1989) An antigen-independent contact mechanism as an early step in T cell-proliferative responses to dendritic cells. J Exp Med 170:527–542

Inaba K, Inaba M, Naito M, Steinman RM (1993) Dendritic cell progenitors phagocytose particulates, including bacillus Calmette-Guerin organisms, and sensitize mice to mycobacterial antigens in vivo. J Exp Med 178:479–488

Janeway CA, Jr, Bottomly K (1994) Signals and signs for lymphocyte responses. Cell 76:275–285

Kelsall BL, Strober W (1996) Distinct populations of dendritic cells are present in the subepithelial dome and T cell regions of the murine Peyer's patch. J Exp Med 183:237–247

Kelsall BL, Stuber E, Neurath M, Strober W (1996) Interleukin-12 production by dendritic cells. The role of CD40-CD40L interactions in Th1 T-cell responses. Ann N Y Acad Sci 795:116–126
Kradin RL, Xia W, McCarthy K, Schneeberger EE (1993) FcR+/−subsets of Ia+ pulmonary dendritic cells in the rat display differences in their abilities to provide accessory co-stimulation for naive (OX-22+) and sensitized (OX-22-) T cells. Am J Pathol 142:811–819
Lanzavecchia A (1996) Mechanisms of antigen uptake for presentation. Curr Opin Immunol 8:348–354
Lim SG, Condez A, Poulter LW, et al. (1993) Mucosal macrophage subsets of the gut in HIV: decrease in antigen-presenting cell phenotype. Clin Exp Immunol. 1993 Jun, 92(3):442–447
Liu LM, MacPherson GG(1991) Lymph-borne (veiled) dendritic cells can acquire and present intestinally administered antigens. Immunology 73:281–286
Liu LM, MacPherson GG (1993) Antigen acquisition by dendritic cells: intestinal dendritic cells acquire antigen administered orally and can prime naive T cells in vivo. J Exp Med 177:1299–1307
Liu LM, MacPherson GG (1995a) Antigen processing: cultured lymph-borne dendritic cells can process and present native protein antigens. Immunology 84:241–246
Liu LM, MacPherson GG (1995b) Rat intestinal dendritic cells: immunostimulatory potency and phenotypic characterisation Immunology 85:88–93
Lu L, Bonham CA, Chambers FG, et al. (1996) Induction of nitric oxide synthase in mouse dendritic cells by IFN-gamma, endotoxin, and interaction with allogeneic T cells: nitric oxide production is associated with dendritic cell apoptosis. J Immunol 157:3577–3586
Lutz MB, Granucci F, Winzler C, et al. (1994) Retroviral immortalization of phagocytic and dendritic cell clones as a tool to investigate functional heterogeneity. J Immunol Methods 174:269–279
Macatonia SE, Knight SC, Edwards AJ, et al. (1987) Localization of antigen on lymph node dendritic cells after exposure to the contact sensitizer fluorescein isothiocyanate. Functional and morphological studies. J Exp Med 166:1654–1667
MacPherson GG (1989) Properties of lymph-borne (veiled) dendritic cells in culture I Modulation of phenotype, survival and function: partial dependence on GM-CSF. Immunology 68:102–107
MacPherson GG, Fossum S, Harrison B (1989) Properties of lymph-borne (veiled) dendritic cells in culture II. Expression of the IL-2 receptor: role of GM-CSF. Immunology 68:108–113
MacPherson GG, Jenkins CD, Stein MJ, Edwards C (1995) Endotoxin-mediated dendritic cell release from the intestine. Characterization of released dendritic cells and TNF dependence. J Immunol 154:1317–1322
Mahida YR, Wu KC, Jewell DP (1988) Characterization of antigen-presenting activity of intestinal mononuclear cells isolated from normal and inflammatory bowel disease colon and ileum. Immunology 65:543–549
Maric I, Holt PG, Perdue MH, Bienenstock J (1996) Class II MHC antigen (Ia)-bearing dendritic cells in the epithelium of the rat intestine. J Immunol 156:1408–1414
Matzinger P (1994) Tolerance, danger, and the extended family. Annu Rev Immunol 12:991–1045
Mayrhofer G, Holt PG, Papdimitriou JM (1986) Functional characteristics of the veiled cells in afferent lymph from the rat intestine. Immunology 58:379–387
McLellan AD, Sorg RV, Williams LA, Hart DN (1996) Human dendritic cells activate T lymphocytes via a CD40: CD40 ligand-dependent pathway. Eur J Immunol 26:1204–1210
McWilliam AS, Napoli S, Marsh AM, et al. (1996) Dendritic cells are recruited into the airway epithelium during the inflammatory response to a broad spectrum of stimuli. J Exp Med 184:2429–2432
Moll H, Flohe S, Rollinghoff M (1995) Dendritic cells in Leishmania major-immune mice harbor persistent parasites and mediate an antigen-specific T cell immune response. Eur J Immunol 25:693–699
Moll H, Fuchs H, Blank C, et al. (1993) Langerhans cells transport Leishmania major from the infected skin to the draining lymph node for presentation to antigen-specific T cells. Eur J Immunol 23:1595–1601
Norbury CC, Chambers BJ, Prescott AR, et al. (1997) Constitutive macropinocytosis allows TAP-dependent major histocompatibility complex class I presentation of exogenous soluble antigen by bone marrow-derived dendritic cells. Eur J Immunol 27:280–288
Pavli P, Woodhams C, Doe W, Hume D (1990) Isolation and characterization of antigen presenting dendritic cells from the mouse intestinal lamina propria. Immunology 70:40–47
Pavli P, Hume DA, van de Pol E, Doe WF (1993) Dendritic cells, the major antigen-presenting cells of the human colonic lamina propria. Immunology 78:132–141
Pinchuk LM, Klaus SJ, Magaletti DM, et al. (1996) Functional CD40 ligand expressed by human blood dendritic cells is up-regulated by CD40 ligation. J Immunol 157:4363–4370
Pope M, Betjes MG, Romani N, et al. (1994) Conjugates of dendritic cells and memory T lymphocytes from skin facilitate productive infection with HIV-1. Cell 78:389–398

Pope M, Gezelter S, Gallo N, et al. (1995) Low levels of HIV-1 infection in cutaneous dendritic cells promote extensive viral replication upon binding to memory CD4+ T cells. J Exp Med 182:2045–2056

Pugh CW, MacPherson GG, Steer HW (1983) Characterization of nonlymphoid cells derived from rat peripheral lymph. J Exp Med 157:1758–1779

Reis-e-Sousa C, Stahl PD, Austyn JM (1993) Phagocytosis of antigens by Langerhans cells in vitro. J Exp Med 178:509–519

Rhodes JM (1985) Isolation of large mononuclear Ia-positive veiled cells from the mouse thoracic duct. J Immunol Methods 85:383–392

Roake JA, Rao AS, Morris PJ, et al. (1995) Dendritic cell loss from nonlymphoid tissues after systemic administration of lipopolysaccharide, tumor necrosis factor, and interleukin 1. J Exp Med 181:2237–2247

Romani N, Koide S, Crowley M, et al. (1989) Presentation of exogenous protein antigens by dendritic cells to T cell clones. Intact protein is presented best by immature, epidermal Langerhans cells. J Exp Med 169:1169–1178

Ruedl C, Rieser C, Bock G, et al. (1996) Phenotypic and functional characterization of CD11c+ dendritic cell population in mouse Peyer's patches. Eur J Immunol 26:1801–1806

Sallusto F, Lanzavecchia A (1994) Efficient presentation of soluble antigen by cultured human dendritic cells is maintained by granulocyte/macrophage colony-stimulating factor plus interleukin 4 and downregulated by tumor necrosis factor alpha. J Exp Med 179:1109–1118

Sallusto F, Cella M, Danieli C, Lanzavecchia A (1995) Dendritic cells use macropinocytosis and the mannose receptor to concentrate macromolecules in the major histocompatibility complex class II compartment: downregulation by cytokines and bacterial products (comments). J Exp Med 182:389–400

Saunders D, Lucas K, Ismaili J, et al. (1996) Dendritic cell development in culture from thymic precursor cells in the absence of granulocyte/macrophage colony-stimulating factor. J Exp Med 184:2185–2196

Schuler G, Steinman RM (1985) Murine epidermal langerhans cells mature into potent immunostimulatory dendritic cells in vitro. J Exp Med 161:526–546

Sertl K, Takemura T, Tschachler E, et al. (1986) Dendritic cells with antigen presenting capability reside in airway epithelium, lung parenchyma, and visceral pleura. J Exp Med 163:436

Spalding DM, Griffin JA (1986) Different pathways of differentiation of pre-B cell lines are induced by dendritic cells and T cells from different lymphoid tissues. Cell 44:507–515

Spalding DM, Williamson SI, Koopman WJ, McGhee JR, et al. (1984) Preferential induction of polyclonal IgA secretion by murine Peyer's patch dendritic cell-T cell mixtures. J Exp Med. 1984 Sep 1, 160(3):941–946

Spinillo A, Tenti P, Zappatore R, De Seta F, Silini E, Guaschino S, et al. (1993) Langerhans' cell counts and cervical intraepithelial neoplasia in women with human immunodeficiency virus infection. Gynecol Oncol. 1993 Feb, 48(2):210–213

Spira AI, Marx PA, Patterson BK, et al. (1996) Cellular targets of infection and route of viral dissemination after an intravaginal inoculation of simian immunodeficiency virus into rhesus macaques. J Exp Med 183:215–225

Steinman RM (1991) The dendritic cell system and its role in immunogenicity. Annu Rev Immunol 9:271–296

Steinman RM, Lustig DS, Cohn ZA (1974) Identification of a novel cell type in peripheral lymphoid organs of mice III. Functional properties in vivo. J Exp Med 139:1431–1435

Strickland DH, Thepen T, Kees UR, et al. (1993) Regulation of T-cell function in lung tissue by pulmonary alveolar macrophages. Immunology 80:266–272

Suss G, Shortman K (1996) A subclass of dendritic cells kills CD4 T cells via Fas/Fas-ligand-induced apoptosis. J Exp Med 183:1789–1796

Tang A, Amagai M, Granger LG, et al. (1993) Adhesion of epidermal Langerhans cells to keratinocytes mediated by E-cadherin. Nature 361:82–85

Thepen T, Van Rooijen N, Kraal G (1989) Alveolar macrophage elimination in vivo is associated with an increase in pulmonary immune response in mice. J Exp Med 170:499–509

Thepen T, McMenamin C, Oliver J, et al. (1991) Regulation of immune response to inhaled antigen by alveolar macrophages: differential effects of in vivo alveolar macrophage elimination on the induction of tolerance vs immunity. Eur J Immunol 21:2845–2850

Thepen T, Hoeben K, Breve J, Kraal G (1992) Alveolar macrophages down-regulate local pulmonary immune responses against intratracheally administered T-cell-dependent, but not T-cell-independent antigens. Immunology 76:60–64

Thepen T, Kraal G, Holt PG (1994) The role of alveolar macrophages in regulation of lung inflammation. Ann N Y Acad Sci 725:200–206

van der Brugge-Gamelkoorn GJ, van de Ende MB, Sminia T (1985) Nonlymphoid cells of bronchus associated lymphoid tissue in the rat in situ and in suspension. With special reference to interdigitating and follicular dendritic cells. Cell Tissue Res 239:177

Volkmann A, Neefjes J, Stockinger B (1996) A conditionally immortalized dendritic cell line which differentiates in contact with T cells or T cell-derived cytokines. Eur J Immunol 26:2565–2572

Wu L, Li CL, Shortman K (1996) Thymic dendritic cell precursors: relationship to the T lymphocyte lineage and phenotype of the dendritic cell progeny. J Exp Med 184:903–911

Xu S, Ariizumi K, Caceres Dittmar G, et al. (1995) Successive generation of antigen-presenting, dendritic cell lines from murine epidermis. J Immunol 154:2697–2705

Xu LL, Warren MK, Rose WL, et al. (1996) Human recombinant monocyte chemotactic protein and other C-C chemokines bind and induce directional migration of dendritic cells in vitro. J Leukoc Biol 60:365–371

Yoffey JM, Courtice C (1970) Lymphatics, lymph and the lymphomyeloid complex. Academic Press, New York, pp 548

York IA, Rock KL (1996) Antigen processing and presentation by the class I major histocompatibility complex. Annu Rev Immunol 14:369–396

Young JW, Steinman RM (1996) The hematopoietic development of dendritic cells: a distinct pathway for myeloid differentiation. Stem Cells 14:376–387

Epithelial Cells in Antigen. Sampling and Presentation in Mucosal Tissues

D. Kaiserlian

1 Introduction

Mucosal surfaces and the skin are tissues that are exposed directly to environmental antigens. They are both characterized by the presence of an epithelium exerting a barrier function and by the presence of dendritic/Langerhans' cells which capture and transport antigens from the tissue for presentation to naive T cells in draining lymph nodes. Mucocutaneous tissues belong to integrated immune systems, including "MALT" (mucosal associated lymphoid tissues) and "SALT" (the skin associated lymphoid tissues). In contrast to the skin epithelium, which is a stratified

Inserm U404, "Immunité et Vaccination", Batiment Pasteur, Avenue Tony Garnier, 69365 Lyon CX 07, France

squamous keratinizing malpighian epithelium, the epithelia overlying mucosal surfaces at different sites in the body differ dramatically in their organization and their antigen-sampling strategies.

Stratified (i.e., esophagus, rectum, vagina) and pseudostratified (nasal cavity, trachea) epithelia contain a network of dendritic cells resembling epidermal Langerhans' cells, which protrude their dendrites to the luminal side of the epithelium and sample antigen for subsequent presentation in draining or distant organized secondary lymphoid tissues. Simple epithelia (stomach, intestine, bronchiole) are composed of a follicle associated epithelium containing microfold epithelial (M) cells and a villus-associated epithelium covering the lamina propria. M cells represent a phenotype of epithelial cells covering the lymphoid follicles specialized in the delivery of particulate material from the lumen to underlying antigen-presenting cells (dendritic cells/macrophages) for presentation to T cells within the mucosa-associated lymphoid follicle. Alternatively, the villus-associated, simple columnar epithelium overlying the diffuse lymphoid tissues of the mucosal lamina propria has an absorptive function and is also involved in the sampling of soluble antigens, including dietary proteins, peptides and haptens penetrating via the oral route.

The intestinal villus epithelium is endowed with some unique phenotypic and functional characteristics, including (a) an absorptive function associated with polarized transcytosis of luminal proteins through a degradative or a direct pathway; (b) a constitutive expression of major histocompatibility complex (MHC) class-II molecules dependent on the age, the microbial flora and the diet; and (c) basolateral contacts with T cells, including intraepithelial lymphocytes (IELs) (mostly CD8+T cells) and T lymphocytes located in the lamina propria (mostly CD4+ T cells).

The recent observations that classical and non-classical MHC molecules are expressed by the intestinal epithelium and that mucosal T cells, including IEL and lamina propria lymphocytes (LPL), utilize activation pathways distinct from those of peripheral lymph-node T cells suggest that epithelial cells of the villi may have a role in antigen sampling and presentation to mucosal T cells in the intestine. Over the past 10 years, evidence has accumulated showing that intestinal epithelial cells express molecules which are found on antigen-presenting cells; these include MHC antigens and adhesion molecules involved in binding to specific ligands expressed on mucosal lymphocytes.

At variance with antigen-presenting cells of hematopoietic origin (i.e., dendritic cells, macrophages and B cells), intestinal epithelial cells (IECs) transcytose antigen in both a native and peptide form, lack co-stimulatory molecules and express non-polymorphic restriction elements, such as CD1d molecules. Taken together, these features suggest that IECs may exert some unique function in mucosal immunity and raise three major questions: (1) Are IECs capable of presenting exogenous or self antigens to T cells?; (2) Do they represent a non-classical antigen-presenting cell type, endowed with unique processing pathways and ligands specifically involved in mucosal T-cell activation?; and (3) What is the role of IECs in infectious immunity and oral tolerance?

2 Protein Transcytosis Through the Villus Epithelium

The epithelium of the intestine is a single cell layer that separates the highest concentration of exogenous antigens from the largest population of lymphocytes in the body. Columnar, absorptive epithelial cells (i.e., enterocytes) covering the villi of the small intestine represent a selective barrier allowing absorption of nutrients after luminal degradation, but preventing, in general, penetration of macromolecules which have escaped enzymatic hydrolysis. These include food proteins, but also whole viruses and bacteria., This barrier, however, is not impenetrable in as much as small amounts of macromolecules resistant to digestive enzymes (such as bovine milk β-lactoglobulin and lactoferrin) may cross the epithelium by transcytosis through epithelial cells.

Although absorption of intact proteins by transcytosis has no nutritional value, it may represent a means to activate immunocompetent cells in the epithelium or in the underlying lamina propria. Immunohistochemical studies of in vitro uptake of β-lactoglobulin by enterocytes in organ culture of normal human duodenal mucosa explants showed that β-lactoglobulin is present on apical microvillus membranes of epithelial cells, after 5 min of culture, and is transported to basolateral membranes toward the base of the villi within 20 min, demonstrating that intact or partially degraded protein is rapidly transported through enterocytes (Wheeler et al. 1993). Electron-microscopy analysis of the uptake of electron-dense peroxidase revealed that the protein transcytosis occurs through endocytic vesicles present beneath the brush-border microvilli and in the apical cytoplasm (Hugon and Borgers 1967; Varshaw et al. 1971; Owen 1977; Wheeler et al. 1993).

Studies using isolated fragments of intestine or monolayers of intestinal cell lines, mounted on Ussing chambers, showed that soluble-phase proteins can be absorbed through the epithelium via two distinct transcellular pathways: a major lysosomal pathway through which 90% of the protein is degraded into amino acids (50%) and small peptides (40%), and a minor direct-transcytosis pathway allowing 10% of the protein to reach the basal side as native intact protein (Heyman and Desjeux 1996). In the normal gut, intercellular transport of large proteins is limited by tight junctions which are present even at sites of cell desquamation at the villus tip, although the size of the pores of tight junctions may allow paracellular transport of peptides of less than 10 amino acids (Mayrhofer 1994). Alternatively, the barrier function of the epithelium may be altered during intestinal inflammation, resulting in an increase in permeability to intact protein (Heyman and Desjeux 1996). Freeze-fracture electron-microscopy analysis of HT29 cell monolayers showed that tight junctional complexes are destroyed upon 48-h incubation with tumor necrosis factor alpha (TNF-α) and interferon gamma (IFN-γ), suggesting that paracellular leakage of macromolecules through tight junctions may occur at sites of pro-inflammatory cytokines released into the mucosa. It should be noted that IFN-γ stimulates the flux of native antigen through epithelial cells without affecting the relative proportion of its breakdown products (Terpend et al. 1997).

The endocytic vesicular pathways of protein processing in enterocytes are not well characterized. Ex vivo studies using bovine serum albumin (BSA) injected into a ligated loop of rat jejunum showed that, in enterocytes, BSA co-localized with MHC class II in small endocytic vesicules present at the apical cytoplasm underlying the terminal web (perhaps early endosomes), along membranes of multi-vesicular bodies in the perinuclear cytoplasm and in vesicles of the basal cytoplasm. A patchy distribution of co-localized antigen was also found along basolateral membranes of most enterocytes (Gonella et al. 1993). Whether absorbed antigen reaches the basolateral membrane by direct transport from endosomes or multi-vesicular bodies awaits further studies to determine the route and extent of antigen processing in enterocytes. However, recent studies of untreated celiac disease demonstrated that active translocation of gliadin to HLA-DR-containing late endosomes/lysosomes occurs only in the jejunal epithelium of untreated celiac patients, but is not observed in patients with a gluten-free diet (Zimmer et al.1995). This suggested that transport of gliadin to a class-II compartment is a prerequisite for activation of specific T cells associated with the disease. Whether antigens handled by enterocytes in normal or pathological situations are presented by the epithelial cells themselves or delivered to other antigen presenting cells (APCs) of the underlying lamina propria for presentation to T cells is not clear at present.

3 Phenotype of Intestinal Epithelial Cells (IECs)

IECs constitutively express a number of cell surface and cytoplasmic molecules involved in antigen processing and presentation (Table 1).

3.1 MHC Class II Molecules

MHC class II is constitutively expressed on normal IECs in the mouse, rat and human (for review see Bland 1988), but is absent on bronchial epithelium (Suda et al. 1995). MHC class II on gut epithelium has some unique characteristics: it is expressed at low density on the basolateral side of mature enterocytes covering the upper two-thirds of the villi in the small intestine, but is absent in crypt epithelial cells from which they differentiate (Parr and McKenzie 1979). The epithelium of the large intestine is class-II negative under normal conditions, but may become positive in inflammatory states, including inflammatory bowel disease (IBD) and celiac disease (Salomon et al. 1991). Although class-II expression is expressed on adult enterocytes, it appears only after birth and increases progressively to adult levels, parallel to the number of IEL (Cerf-Bensussan et al. 1984), and coincides with the introduction of a complex diet and normal gut flora (Bland 1996). In this respect, IFN-γ production by concanavalin A (Con-A)-activated IEL can induce

Table 1. Molecules expressed by intestinal epithelial cells

MHC and associated antigens	Class Ia (polymorphic)	+
	Class Ib (nonpolymorphic)	+
	CD1d	+
	TL	+
	Qa	+
	Class II	
	Mls	+
Antigen processing and presentation molecules	Invariant chain, Ii	±
	DM	?
	Cathepsin B, L, H, D	+
	Cathepsin E (FAE)	+
	CD26	+
Adhesion molecules	ICAM-I (cell lines)	+
	LFA-3	+
	E-cadherin	+
	CD23	+
Costimulatory molecules	B7.1-3	–
	HSA (CD24)	–
	CD40	–

de novo expression of class-II molecules on the rat intestinal epithelial cell (IEC) line IEC 17, which is constitutively class-II negative (Cerf-Bensussan et al. 1984).

More recent studies reported that interleukin (IL)-18, an IFN-γ-inducing cytokine, is constitutively expressed in the cytoplasm of IECs from fetal to adult age (Takeuchi et al. 1997), thus indicating that epithelial cells may promote the production of IFN-γ by IELs. Therefore, exogenous stimuli provided by the microbial flora, dietary proteins and pro-inflammatory cytokines released by local T cells seem to be responsible for triggering the development of epithelial class II in the mouse. Ultrastructural analysis of MHC class II in subcellular compartments of rat enterocytes showed that class II are present in the endocytic pathway, including multi-vesicular bodies just deep to the terminal web, and in the apical cytoplasm, lysosomes and the supranuclear cytoplasm (Mayrhofer and Spargo 1990). The situation may, however, be quite different in humans, since epithelial expression of class-II molecules begins at 17 weeks of gestation and is present both on the apical brush border and at basolateral membranes of enterocytes (Mayrhofer and Spargo 1989). Given the fact that HLA-D is strongly associated with celiac disease and that humans are highly sensitive to staphylococcal enterotoxins, it is possible that apical class-II molecules can directly bind gluten or enterotoxins, and may play a pivotal role in antigen absorption and contribute to the pathogenesis of celiac disease and food enteropathy.

3.2 MHC Class I-Like Molecules

IECs constitutively express classical MHC class-I molecules (class Ia) and class Ib molecules. In contrast to class-Ia molecules, which are polymorphic and have a ubiquitous distribution, class-Ib molecules represent a family of non-polymorphic ligands expressed in selected tissue niches, including the thymus, the skin and the intestine (Calabi and Bradbury 1991). Class-Ib molecules expressed by the intestinal epithelium include the mouse TL cell leukemia antigens (Hershberg et al. 1990; Wu et al. 1991), a low affinity RFcIgG receptor on neonatal enterocytes of rat and mouse (Simister and Rees 1985) and the CD1d molecule (Blumberg et al. 1996). These non-classical class-I molecules may be involved in the presentation of distinct classes of antigens to distinct types of T cells. In this respect, TL of enterocytes has been proposed to serve as a ligand for IEL expressing a restricted $\gamma\delta$ TcR repertoire (Hershberg et al. 1990; Wu et al. 1991).

CD1 molecules in humans comprise a family of five members, four of which are known to be expressed *in vivo*: CD1a and CD1c have been described in normal human intraepithelial lymphocytes, but not on epithelial cells of the intestine. Alternatively, CD1d and its mouse homologues CD1.1 and CD1.2 have been identified in the small-intestinal epithelium of human (Blumberg et al. 1991 ; Canchis et al. 1993; Balk et al. 1994) and Balb/C mice (Bleicher et al. 1990), respectively, using the 3C11- and 1H1 CD1d-specific IgM monoclonal antibodies (mAbs) (which cross react with human CD1d). Separate studies using other mCD1 specific mAbs (1B1 and 9C7), which also recognize both CD1.1 and CD1.2 molecules, failed to detect CD1 on epithelial cells of the intestine in C57BL/6 mice (Brossay et al. 1997). Whether these discrepancies relate to differences in the isotype and/or epitope specificity of the different CD1 specific mAbs remains unclear. It is possible that mouse IEC express only the CD1.2 isoform, which is lacking in C57BL/6 mice due to a point mutation in the CD1.2 gene. Besides constitutive expression, human CD1d can be specifically upregulated by IFN-γ in vitro (Colgan et al. 1996) or during intestinal inflammation (Canchis et al. 1993).

At variance with other CD1d-expressing cells in the skin and thymus, normal human and mouse enterocytes express a non-β2m-associated 37-kDa membrane isoform of CD1d that lacks *N*-linked carbohydrates, indicating tissue-specific processing of CD1d (Balk et al. 1994). The lack of association with β2m may be dictated by the conformation of non-glycosylated CD1d and indicates that conventional sites for binding peptidic antigen may not be present. In this respect, CD1d can be recognized by CD8+ T cells (Balk et al. 1991; Panja et al. 1993). Specific recognition of CD1d-transfected target cells by CD8+ intraepithelial T-cell clones suggested that CD1d is a ligand for CD8+ T cells (Balk et al. 1991). Anti-CD1d (3C11) mAbs could block T-cell proliferation induced by human IECs (shown to result in expansion of CD8+ T cells), but did not affect that induced by airway epithelial cells (resulting in CD4+ T-cell expansion), suggesting CD1d expressed on human IECs in vivo is the ligand that induces proliferation of CD8+ T cells (Panja et al. 1993).

The intestinal mucosa is a specialized immunological compartment composed largely of cytolytic CD8+ T cells with oligoclonal TcR repertoire (Balk et al. 1991) which may be specific for a limited array of antigens presented on CD1d. Alternatively, CD1d is a ligand for IL-4 secreting CD4+ NK1.1+ TcR αβ+ T cells (Bendelac et al. 1995) and may also serve as a ligand for activation of helper T lymphocyte (Th)2-type regulatory T cells in the mucosa. Studies of soluble mouse

CD1 molecules have identified a hydrophobic peptide binding motif which, at variance with class-Ia molecules can bind long peptides (CASTANO et al. 1995). Thus, CD1d molecules may be specialized in presentation of hydrophobic peptides and non-peptidic lipid or glycolipid bacterial antigens, as described for CD1b molecules (SIELING et al. 1995).

3.3 Minor Lymphocyte Stimulating Antigens

Minor lymphocyte stimulating (Mls) antigens comprise a family of self superantigens, encoded by endogenous mouse retrovirus, which are involved in negative selection of T cells expressing certain Vβ families of the T-cell receptor. We previously reported that normal mouse enterocytes constitutively express the Mls antigen encoded by the *Orf* region of the endogenous Mtv-7 retrovirus, since they were able to activate Mls-specific T-cell hybridomas (KAISERLIAN et al. 1993b). This finding is consistent with expression of RAG-1 and RAG-2 genes in the human intestinal epithelium (LYNCH et al. 1995) and suggests that enterocytes may participate in the shaping of the repertoire of mucosal T cells in the mucosa. It should be emphasized that the intestinal epithelium may also be involved in extrathymic T-cell maturation from bone-marrow progenitors in as much as it expresses the hematopoietic factor c-kit (KLIMPEL et al. 1995) and is able to induce the differentiation of bone-marrow stem cells into double-positive (CD4+CD8+) and single-positive (CD4+ or CD8+) TcRαβ$^+$ T cells (MARIC et al. 1996).

In addition, normal human IECs constitutively express the CD23 molecule (i.e., the low affinity FcεRII, also a receptor for the CD21 molecule) (KAISERLIAN et al. 1993a; GROSJEAN et al. 1994), an adhesion molecule whose soluble form (sCD23) is endowed with thymocyte maturation function. Taken together, these studies are consistent with reports showing that extrathymic differentiation of T cells can take place in the intestine (POUSSIER and JULIUS 1994) and that intestinal epithelium may be involved in this process (KLEIN 1996). Thus, the cell-surface molecules shared by IEC and antigen-presenting cells (i.e., MHC class-I and class-II, CD1 and Mls antigens) may serve as restriction elements for presentation of both self and non-self antigens in the gut and as ligand for distinct sets of mucosal T cells.

3.4 Dendritic Cell-Specific Antigens

We have reported that the murine intestinal epithelium is heterogeneous and comprises at least three subsets defined by expression of molecules selectively expressed on antigen-presenting dendritic cells. On the basis of MHC class II, DEC-205 and S100 expression, three subsets of gut epithelial cells can be identified: mature villus enterocytes are class II+, DEC-205+ and S100+; immature crypt epithelial cells are negative for all three markers; and epithelial cells of the Peyer's patch dome region are class II-, DEC-205- and S100+ (VIDAL et al. 1989). The dendritic cell-specific mAb NLDC-145, which recognizes a mannose receptor,

DEC-205, exclusively described on dendritic cells and thymic epithelial cells (Kraal et al. 1986), stains both epithelial cells of intestinal villi and dendritic cells in the lamina propria (Vidal et al. 1989). The molecule recognized by this antibody has been recently identified as a mannose receptor (DEC-205) involved in antigen processing (Jiang et al. 1995).

In contrast to the exclusive cell-surface expression of DEC-205 on splenic dendritic cells and epidermal Langerhans cells, expression in enterocytes is prominent in the cytoplasm of enterocytes. DEC-205 can capture and endocytose an antigen and deliver it to an endocytic compartment that is abundant in dendritic cells and resembles class II-containing vesicles. Thus DEC-205 could represent a mannose receptor involved in antigen uptake and transcytosis and/or processing by IECs. Another dendritic cell molecule expressed on villus but not crypt epithelium is the S100 molecule (Vidal et al. 1989), a calcium-binding protein with a tissue distribution limited to Langerhans' cells of the skin, melanocytes and glial cells. The functional significance of S100 and DEC-205 in the antigen presentation function of gut epithelial cells is not known.

3.5 Adhesion Molecules

Mucosal epithelial cells can express ICAM-1, LFA-1, E-cadherin and CD23 molecules. Enterocytes are closely associated in situ with IELs and may also make contact with lamina-propria T cells. Both T-cell types express a phenotype of activated memory cells characterized by expression of the CD45RO molecule. We have previously showed that ICAM-1 is constitutively expressed on human gut epithelial cell lines (HT29 and Caco-2), is upregulated by IFN-γ and IL-1β and mediates adhesion to activated T lymphocytes through ICAM-1/LFA-1 interactions (Kaiserlian et al. 1991).

Although ICAM-1 is not found on normal mature epithelial cells, it is induced on crypt epithelial cells and on the rectal epithelium during graft-versus-host (GVH) concurrent with the presence of surrounding LFA-1-positive intraepithelial cells (Norton et al. 1992). This argues that immune activation and/or inflammation can cause de novo expression of ICAM-1 which may retain leukocytes at inflammatory sites. That such a mechanism may contribute to lympho–epithelial interactions in other mucosae as well is supported by the observations that ICAM-1 is induced on human nasal epithelial cells during respiratory syncytial (RSV) infection and causes adhesion of phytohemagglutinin A (PHA)-activated tonsilar T cells to RSV-infected epithelial cells and an increase in IL-4 and IL-5 production (Matsuzaki et al. 1996). Likewise parainfluenza virus infection of human bronchial epithelial cells in vitro increases ICAM-1 expression (Tosi et al. 1992). Finally, ICAM-1 was found on the airway epithelium in a primate model of chronic inflammation, in which injection of blocking anti-ICAM-1 antibodies diminished the inflammatory response to allergen exposure (Wegner et al. 1990).

In contrast to ICAM-1, E-cadherin is an adhesion molecule constitutively expressed by the epithelium of the normal intestine. E-cadherin mediates the interaction between epithelial cells and IELs through binding to the $\alpha_E\beta7$ (CD103) ligand (Cepek et al. 1994), which allows IEL cytotoxicity toward epithelial cells (Roberts and Kilshaw 1993).

Normal human enterocytes constitutively express at their basal and apical pole the CD23 molecule, i.e., the low affinity receptor for the Fc portion of IgE, also a receptor for the adhesion molecule CD21. CD23 expression is enhanced in both the small and large intestine in IBD and autoimmune enteropathy (Kaiserlian et al. 1993a). Furthermore, heterotypic CD23–CD21 interaction between APCs and T cells is required for efficient antigen presentation of allergens to specific class-II-restricted CD4+ T-cell clones (Grosjean et al. 1994). It is not known at present whether CD23 contributes to epithelial–T cell interactions during inflammation or enhances allergen-specific T-cell responses or whether it is merely a consequence of local release of pro-inflammatory cytokines.

3.6 Co-stimulatory Molecules

Co-stimulatory molecules, including B7, CD40 and HSA, are expressed neither by normal IECs, nor by the human epithelial cell lines HT29 and T84, even after stimulation with IFN-γ alone or in combination with GM-CSF or TNF-α (Hershberg et al. 1997), suggesting that these cells may induce anergy rather than activate naive T cells.

4 Antigen Processing by IECs

Efficient processing and presentation of exogenous antigens by MHC class-II molecules to CD4+ T cells generally requires expression of the invariant chain (Ii), the HLA-DM$\alpha\beta$ heterodimer and a series of proteases, required for both proteolysis of endocytosed antigens and normal processing of class II/Ii complex to an intermediate that is a substrate for HLA-DM (review in Wolf and Ploegh 1995). The intestinal epithelium represents an interesting exception compared with B cells and macrophages, in as much as MHC class-II molecules are transported at the cell surface, despite low to undetectable levels of Ii, depending on the mouse strain. Immunohistochemical staining of mucosal tissues using the rat mAb In-1, specific for the cytoplasmic tail of Ii, showed that Ii expression is low in the jejunal epithelium and absent in epithelial cells of the colon, liver and bronchi (Momburg et al. 1996). Although Ii mRNA is present in normal mouse enterocytes of the duodeno-jejunum, depending on the mouse strain, expression of cytoplasmic Ii is either missing (C3H enterocytes) (Vidal et al. 1993b) or detectable at low levels (C57BL/6 mice enterocytes) (Kaiserlian, unpublished data). The human epithelial cell lines, HT29 and T84, which have a phenotype of immature crypt epithelial cells, also lacked constitutive expression of Ii and HLA-DM, although both can be induced by IFN-γ. (Hershberg et al. 1997).

Proteolytic enzymes are present in enterocytes. Ectopeptidases, such as DPPIV (CD26) are present on brush border microvilli of human and mouse enterocytes (Fleischer 1994). However, these enzymes participate exclusively in luminal degradation of proteins, by generating dipeptides that can be readily absorbed by the cell, and are unlikely to play a significant role in vesicular antigen processing. The lysosomal proteases, cathepsin B, D, H and L are the major ubiquitous enzymes responsible for antigen processing in conventional antigen-presenting cells (Buus and Werdelin 1986), while the non-lysosomal aspartic protease cathepsin E is found exclusively in Langerhans'/dendritic cells, but not macrophages (Solcia et al. 1993), and is essential for antigen processing for class-II presentation (Bennett et al. 1992). The follicular-associated epithelium of the intestine, the appendix and the tonsils express cathepsin E, but not D, which co-localizes with HLA-DR in endomomes (Finzi et al. 1993). The enterocytes of the rat duodenum contain cathepsin B, L and H (Fukuhashi et al. 1991), and the immature colon epithelial cell lines HT29 and T84 constitutively co-express aspartic proteinases as well as the cysteine proteases cathepsin B, H and L (Hershberg et al. 1997). Thus, gut epithelial cells are equipped with the lysosomal and non-lysosomal proteases, but the precise vesicular pathway which operates to process proteins into immunogenic peptides remains to be defined.

Initial studies by Bland (1989) reported that antigen processing of ovalbumin by purified rat enterocytes could be blocked by lysosomotropic agents, such as chloroquine, monensin and ammonium chloride, but not by the cysteine protease inhibitor leupeptine, suggesting that processing by enterocytes required alkaline proteases but not acidic proteases. More recently, studies by Hershberg et al. (1997) supported and extended this observation by demonstrating that HLA-DR-transfected colon adenocarcinomas (HT29 and T84) can utilize two distinct pathways of processing by class-II molecules, depending on the presence or the absence of stimulation with IFN-γ.

A classical pathway, involving protein degradation by cysteine proteases in an acidic compartment inhibited by leupeptine, bafilomycin and chloroquine, could be demonstrated only when the cells were stimulated with IFN-γ, was associated with de novo expression of invariant chain and HLA-DM$\alpha\beta$ and resulted in efficient presentation of tetanus toxoid to specific T cells. The other pathway, was independent of Ii and HLA-DM$\alpha\beta$, since it operated in the absence of IFN-γ, was leupeptin-insensitive and was associated with a low efficacy to present tetanus toxoid. Such a pathway of Ii-independent processing is not specific to IECs and is likely to be peptide-epitope specific (Pinet et al. 1994; Ceman and Sant 1995).

These studies suggest that the efficacy of antigen presentation to CD4+ T cells by enterocytes depends on the context of mucosal inflammation which allows processing and loading onto class II of peptide epitopes that differ from those generated at homeostasis. This latter hypothesis is reminiscent of the observation that ovalbumin uptake and degradation in slower in rat enterocytes than splenic macrophages (Bland and Whiting 1989) and that antigens processed by enterocytes could not be presented by paraformaldhyde-fixed splenic adherent cells, indicating that the set of peptides generated in enterocytes could not bind to the class-II groove of conventional antigen-presenting cells.

5 Antigen Presentation by IECs

5.1 Activation of CD8+ T Cells by IECs

Recently, the ability of purified MHC class-II-expressing mature IECs from the rat, human and mouse intestine to present soluble exogenous protein antigens to specific (memory) T cells has been documented (Bland and Warren 1986; Mayer and Shlien 1987; Kaiserlian et al. 1989). The studies in rat and human showed that freshly isolated gut epithelial cells can stimulate proliferation of ovalbumin (OVA)-primed (Bland and Warren 1986) or tetanus toxoid-primed (Mayer and Shlien 1987) T cells, respectively. In both systems, T-cell proliferation could be blocked by anti-class-II (but not anti-class-I) mAbs and was inhibited by chloroquine, indicating that enterocytes could process and present exogenous proteins as peptides associated with cell surface class-II molecules. Furthermore, it was reported that a human colon adenocarcinoma, DLD1, which expresses class-II molecules only when pretreated with IFN-γ, could stimulate proliferation of allogeneic T cells, only if class II was induced by prior IFN-γ treatment.

Surprisingly, both studies described that the in vivo-primed T cells, which expanded in response to in vitro stimulation with either syngeneic gut epithelial cells and exogenous antigens or with autologous or allogeneic gut epithelial cells, were CD8+ T cells. Furthermore, these CD8+ T cells could mediate antigen non-specific suppression of both T- and B-cell mediated immune responses, and the authors proposed that gut epithelial cells may play a role in oral tolerance. In contrast, parallel studies from the same group demonstrated that Crohn's disease epithelium predominantly stimulated CD4+ T cells in allogeneic mixed-lymphocyte reaction (MLR) (Mayer and Eisenhardt 1990). This difference led to the speculation that the intestinal epithelium participates in oral tolerance via stimulation of CD8+ T cells with suppressive function, and that in the setting of IBD, epithelial presentation to CD4+ T cells may be responsible for breakdown of oral tolerance relevant to the pathogenesis of this condition.

However, the mechanism for the class II-dependent stimulation of CD8+ T cells remains obscure, since CD8 usually binds to class-I, but not class-II molecules. It is possible that CD8+ T cell proliferation is dependent on class II-restricted CD4+ T (perhaps helper) cells, although there is no evidence for either antigen-specific activation of CD4+ T cells or anti-CD4 mAb inhibition of CD8+ T cell expansion induced by allogeneic epithelial cells. It was subsequently proposed that anti-class II-mAb inhibition of epithelial-cell presentation of nominal antigen and allogeneic class II, was due to a blocking of class-II molecules present on human activated T cells. However, this explanation does not fit with either the data obtained from the DLD1 cell line, the data obtained in the rat species in which activated T cells do not express class-II, or with a mechanism by which T-cell proliferation can be blocked by masking class II on the T cells. The selective expansion of CD8+ T cells could be due to a bystander help from CD4+ T cells through cytokine production. Alternatively, antigen-pulsed epithelial cells may trigger two independent phenomena: (i) antigen-specific activation of CD4+ T cells through class-II molecules and (ii) antigen non-specific activation of CD8+ T cells by, as yet, undefined ligands.

In this respect, recent data suggested that CD1d molecules of gut epithelial cells were responsible for CD8+ T cell expansion in as much as anti-CD1d (3C11) and anti-CD8 mAbs blocked proliferation of peripheral T cells induced by allogeneic IECs (Panja et al. 1993). Since CD1d is a non-polymorphic molecule, it would be interesting to determine whether anti-CD1d mAbs could also inhibit autologous T-cell proliferation induced by IECs and, more importantly, whether CD8+ T-cell proliferation could be blocked by anti-TcR antibodies. Although triggering of CD8+ regulatory T cells by epithelial cells appeared as a unique property of intestinal (but not airway) epithelial cells, the in vivo relevance of these findings remains to be established, because it is observed only with peripheral blood but not lamina propria T cells. Alternatively, activation of T cells from the lamina propria by allogeneic intestinal epithelia results in proliferation of both CD4 and CD8+ T cells and is not blocked by either anti-CD1d, anti-class II or anti-class I mAbs, suggesting that alternate ligands are recognized by mucosal T cells (Panja et al. 1994).

In contrast with gut epithelial cells, normal airway epithelial cells (including lobar bronchus and nasal turbinates obtained from patients undergoing thoracotomy resection for a tumor or surgical resection for a mechanical obstruction), express MHC class-II molecules which are able to activate allogeneic CD4+ T-cell proliferation (Kalb et al. 1991). Interestingly, anti-CD1d mAb has no effect on T-cell proliferation induced by airway epithelial cells (Panja et al. 1993).

5.2 Antigen Presentation by IECs to Class II-Restricted CD4$^+$ T Cells

The issue concerning the ability of IECs to present antigens on MHC class-II molecules to CD4+ T cells is still highly controversial. This is due to the divergent results obtained from at least five different laboratories. The basis of these discrepancies most likely relates to the technical difficulty of isolating pure populations of mature and viable epithelial cells that are not contaminated with traces of class II+ antigen-presenting cells (i.e., dendritic cells) and also to the various experimental protocols and read-out systems used for the assessment of CD4+ T-cell activation. In addition, utilization of intestinal tissues from different species (human vs rodents), different segments of the gut (human colon vs rodents' small intestine), different strains and, in the case of mice, strains with different H-2 haplotypes and, finally, housing conditions in animal facilities, all render the data difficult to compare.

In order to examine the ability of IECs to process and present antigens on MHC class-II molecules to specific class II-restricted CD4+ T cells, we have analyzed the capacity of highly purified class II+ IECs isolated from the mouse small intestine (KAISERLIAN et al. 1989) and of the duodenal epithelial cell line, MODE-K (VIDAL et al. 1993a) to present exogenous or self superantigens, nominal proteins and peptides to specific class II-restricted T-cell hybridomas (Table 2).

T-cell hybridomas offer the advantage over primed polyclonal T cells from immunized mice or antigen-specific T-cell clones to be less dependent on co-stimulation, therefore, allowing direct analysis of protein processing and peptide loading onto class-II molecules. Titration of the levels of IL-2 produced in the 24 h supernatant of co-cultures containing epithelial cells (or syngeneic spleen cells), the antigen and a specific class II-restricted CD4+ T-cell hybridoma, was used as a read-out system to test for T-cell activation. Depending on the nature of the antigen, two mechanisms of T-cell activation occur. In the case of protein antigens, T-cell activation is class II-restricted and is induced if the T-cell receptor recognizes a conformational complex of the processed peptide bound in the polymorphic groove of syngeneic class-II molecules. Alternatively, superantigens can cross link monomorphic regions of the class-II molecule (outside the peptide binding groove) to the variable region of the β chain of the T-cell receptor; thus, T-cell activation by superantigens is class-II dependent (but not class-II restricted) and does not require processing into peptides.

Table 2. Antigen presentation by mature intestinal epithelial cells and by IFN-γ-treated MODE-K cells

T-cell hybridomas	Specificity	Class-II restriction	IL-2 EC[a]	MODE-K[b]
	Superantigens			
KS47.1	SEB	All haplotypes	+	nt
FJ8.1	Mls-1[a]	All haplotypes	+	+
	Proteins			
CAK1.22.1	KLH	I-A^{d}	+	nt
3A9	HEL	I-A^{k}	–	+
TS12	RNAse	I-A^{k}	–	+
3DO.54.8	OVA	I-A^{d}	–	nt
B8P3	Insulin	I-A^{d}	–	nt
T44.23.5	Apamin	I-A^{d}	–	nt
2QO23–24	I-A^{k}		–	nt
3QO23–24.4	I-E^{k,d}		–	nt
	Peptides			
3A9	HEL peptides	I-A^{k}	+	+
T14.120	GAT	I-A^{d}	–	nt

[a]Antigen-presentation assays using mature villus enterocytes (*EC*) were performed using purified enterocytes from the duodeno-jejunum of either Balb/c (H-2^{d}) or C3H (H-2^{k}) mice, except for Mls-presentation assays for which enterocytes from Mls1^{a+} (CBA and DBA/2, H-2^{k}) strains were used
[b]The MODE-K cell line was established from immature (midvillus or crypt) enterocytes from C3H mice (H-2^{k}) and were treated with 100 U/ml IFN-γ for 5–6 days before use as antigen-presenting cells. Untreated MODE-K cells do not constitutively express MHC class-II molecules and did not present antigen (not shown)

5.2.1 Superantigens

Normal gut epithelial cells from Balb/C (H-2^d) and C3H (H-2^k) mice efficiently present the exogenous superantigen *Staphylococcus enterotoxin* B (SEB) to the specific T-cell hybridoma KS47.1. Epithelial cells from CBA (Mls-1^{a+}) mice can present the self superantigen Mls, encoded by the endogenous retrovirus Mtv-7. Addition of either the anti-class-II mAb, antibody CD311 or a mAb specific for a peptide of the *Orf* of Mtv-7 strongly inhibited the ability of IEC to induce proliferation of T cells from Mls-1a strains (Kaiserlian et al. 1993b). These data clearly demonstrate that gut epithelial cells can present both endogenous and exogenous superantigens in association with cell-surface MHC class-II molecules to CD4+ T cells.

5.2.2 Native Proteins

We initially reported that freshly isolated IECs could present keyhole limpet hemocyanin (KLH) to a specific I-A^k-restricted CD4+ T-cell hybridoma (CAK 1.2), and that the anti-class-II mAb CD311 could block IL-2 production by the T cells (Kaiserlian et al. 1989). In subsequent studies, we tried to confirm and extend this observation using a panel of different protein antigens and specific class II-restricted CD4+ hybridomas. However, we found that gut epithelial cells were poorly efficient at presenting either HEL, OVA, RNAse, insulin, apamin, or even peptides (Table 2) to specific IL-2-secreting T-cell hybridomas.

Several hypotheses may explain these discrepant results. First, normal mature gut epithelial cells may either be unable to process native proteins into peptides or could use a unique pathway of processing, distinct from that of conventional APCs (B cells, macrophages, dendritic cells), generating peptides which cannot be recognized by the TcR of specific T cells which have been generated using conventional APCs. That enterocytes could present KLH could be due to either production of the appropriate peptide or the fact that the epitope recognized by CAK1.2 is a linear epitope that does not require processing, as it has been previously reported. Second, CAK 1.2. may be less dependent on co-stimulation than the other hybridomas used. Third, different housing conditions may trigger focal release of endogenous cytokines (such as IFN-γ), and has been recently shown to cause site-specific qualitative and quantitative differences in epithelial class-II expression in discrete segments of the small intestine (Sidhu et al. 1992) and to convert enterocytes into efficient presenting cells by triggering an Ii and DM-dependent processing pathway (Hershberg et al. 1997).

Hoyne et al. (1995) reported that mouse epithelial cells were unable to present native proteins, but could present peptides to primed T cells. However, although they were unable to demonstrate IL-2 production induced by epithelial cells, they found IL-3 production, reminiscent of the characteristics of antigen-driven T-cell anergy. Furthermore, only one of three peptides of similar length (24 amino acids), presented by spleen cells, was efficiently presented by gut epithelial cells. Thus, their results tend to show that (a) gut epithelial cells cannot generate the same set of peptides as those generated by processing in conventional APCs and (b) that only some peptides that normally bind to class II can also bind to enterocytes of class II, but even then, T cell activation does not result in IL-2 production. In keeping with these data, we observed that normal gut epithelial cells were also unable to activate IFN-γ production or proliferation of a CD4+ T-cell clone when stimulated with the relevant hemoglobin peptide (Hb_{64-76}). Thus, IEC from the small intestine appeared strikingly different from conventional antigen-presenting cells in that they are able to present superantigens in association with class II, but are poorly efficient at presenting native antigens or peptides on class-II molecules which can be recognized by class II-restricted CD4+ T cells.

5.3 Mature IECs Express a Distinct Conformation of MHC Class-II Molecules

The observation that normal gut epithelial cells, although expressing cell-surface class-II molecules, are functionally distinct from APC of hematopoietic origin prompted us to analyze more thoroughly the class-II molecules constitutively expressed by these cells. Immunohistochemical, biochemical and molecular analysis revealed that mature enterocytes express an abnormal conformation of MHC class-II molecules (Vidal et al. 1993b).

Both fluorescence-activated cell sorter (FACS) analysis and immunohistochemical staining of cryostat sections of normal mouse duodenum revealed that antibodies directed against a conserved (monomorphic) epitope of class-II molecules strongly stained enterocytes, whereas mAbs specific for polymorphic conformational epitopes located in the class-II peptide-binding groove of I-A or I-E molecules were virtually unable to react with gut epithelial cells (Fig. 1.). This abnormal antigenicity of class-II molecules was not related to structural differences within the class-II molecule itself in as much as: (i) IECs have mRNA for α and β chains of I-A and I-E, (ii) the cDNA sequences encoding for the extracellular (polymorphic) domains of the Eβ chain are identical to those found in B cells, (iii) I-A and I-E heterodimers of the expected molecular weight can be immunoprecipitated from the cell surface and (iv) the anti-class-II antibodies, which stain IECs in normal mice, fail to react with IECs of class-II knock-out mice (D. Kaiserlian, unpublished data). Thus, IECs encode for I-A and I-E molecules, without evidence of alternative spliced variants, but express a cell surface conformation of class II serologically distinct from that of conventional APC.

Two hypotheses have been tested to explain the molecular basis for serological and functional abnormalities of IEC class-II molecules. The first hypothesis stated that invariant chain deficiency may be responsible for an altered conformation of class II, which impairs anti-class-II reactivity with IEC, as reported with Ii-deficient class-II-transfected L cells (Peterson and Miller 1990). Although enterocytes indeed exhibited either low levels or a total lack of the class-II-associated invariant chain, depending on the mouse strain, such a hypothesis is unlikely since (i) Ii-deficient APCs are still efficient at presenting peptides (Viville et al. 1993) and (ii) the intensity of CD311 staining of enterocytes from Ii knock-out mice is dra-

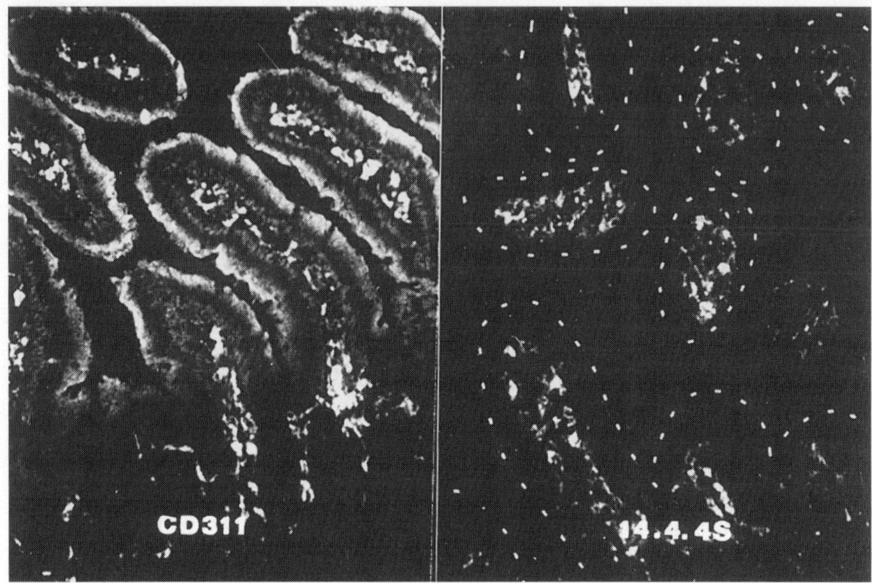

Fig. 1. Immunohistochemical analysis of class II-molecules expressed by the small intestinal epithelium of normal mice. *Left Panel:* epithelial cells of upper two thirds of intestinal villi are stained by the rat mAb CD311, specific for a monomorphic epitope of class II molecules, which also stains macrophages and B cells in the lamina propria. *Right Panel:* lack of staining of the intestinal epithelium with the mouse mAb 14.4.4S specific for a polymorphic epitope of I-Eα, and with a panel of anti-class-II murine mAbs specific for conformational epitopes of class II (see VIDAL et al. 1993b). Note that the lack of enterocyte-staining contrasts with the positive staining of hematopoietic cells in the lamina propria

matically decreased arguing that abnormal class-II conformation is not due to a lack of Ii (D. KAISERLIAN, unpublished data). The second hypothesis stated that class-II loading of both exogenous and self peptides, processed in endosomal/lysosomal compartments is impaired. Support for this hypothesis is provided by failure of enterocytes to stimulate alloreactive T-cell hybridomas, which recognize a conformational complex formed by self peptide bound in the class-II groove (VIDAL et al. 1993b). Furthermore, we recently observed that the mAb Y-ae, which recognizes a major self peptide of the I-E^k molecule present in the I-A^d groove of B cells in B10BR (I-E^k I-A^d) congenic mice (RUDENSKY et al. 1991), does not react with enterocytes of these congenics.

Finally, a T-cell hybridoma specific for a self peptide of hemoglobin, constitutively bound in the groove of I-A ^{k} (EVAVOLD et al. 1992) can be activated by spleen cells but not by enterocytes from CBA mice (KAISERLIAN, unpublished data). Thus, enterocyte class-II molecules are not loaded with self peptides which are normally present in the class-II groove of conventional antigen-presenting cells. One possible explanation is that the class-II groove of IEC is occupied by class II-associated invariant chain peptide (CLIP), a proteolytic fragment of the invariant chain, which is cleaved by DM when class II reaches peptide-containing endocytic vesicles. DM is a heterodimeric molecule with an α and a β chain encoded by the

DMA and DMB genes, respectively, which are conserved in mice and humans and play a critical role in facilitating antigen presentation by class-II molecules (reviewed in ROCHE 1995).

Interestingly, antigen-presenting cells from DM-knock-out mice have an abnormal conformation of mouse class II, due to the presence of CLIP, associated with a failure to stimulate proliferation of allogeneic T cells (FUNG-LEUNG et al. 1996). Moreover, recent studies reported that the colon adenocarcinomas HT29 and T84, transfected with HLA-DR4, which lack constitutive expression of both DM molecules and Ii, have an abnormal DR conformation and are poorly efficient at processing and presenting tetanus toxoid to a specific DR4-restricted T-cell hybridoma. Thus, a constitutive lack of DM may explain both the abnormal class-II conformation and inefficient presentation of both native protein and peptides to CD4+ T cells.

5.4 The Intestinal Epithelial Cell Line MODE-K: A Model for Antigen Presentation by Epithelial Cells at Sites of Inflammation

The MODE-K cell line, obtained by immortalization of normal mouse small-IECs by SV40 large T-gene transfer (VIDAL et al. 1993a), expresses cytokeratins, villin, poly-Ig and VIP receptor and class-I molecules and has a phenotype of immature IECs. Similarly to crypt epithelial cells, MODE-K cells have only cytoplasmic class II, but lack Ii, and can be induced for expression of cell surface class-II molecules and cytoplasmic Ii by IFN-γ treatment in vitro. In contrast to normal mature enterocytes, IFN-γ-treated MODE-K cells express a normal conformation of class II at the cell surface and are very efficient at presenting both native exogenous proteins and peptides (Table 2). These findings are consistent with the recent studies of HERSHBERG et al. (1997) and suggest that epithelial cells of the midvillus or the crypts may be capable of presenting foreign antigens under pathological situations associated with the local release of (class-II-inducing) pro-inflammatory cytokines (Fig. 2).

6 Role of Epithelial Cells in Tolerance

Considering that dietary proteins may cross the epithelium at the level of both Peyer's patches and the mucosal villi, a possible function of mature IECs in local immune response to dietary antigens would be to prevent IL-2 production and expansion of memory (CD45RO+) T-cells present in the lamina propria and the epithelium, that may have been activated by antigen in the Peyer's patches (i.e., before their emigration to the mucosa). In this scenario, enterocytes may exert a local downregulation of proliferation of either primed T cells, activated in the

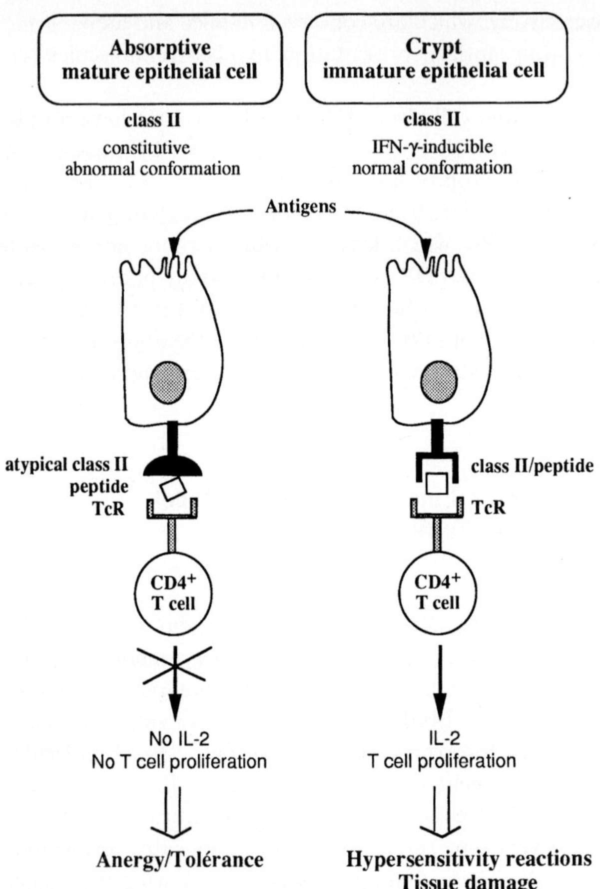

Fig. 2. Hypothetical scheme of the outcome of antigen presentation by MHC class-II molecules on intestinal epithelial cells in a normal and a pathological situation. *Left Panel:* normal mature enterocytes of the small intestine exhibit an abnormal conformation of class II (VIDAL et al. 1993b), also identified in HLA-DR transfected human colon adenocarcinoma cell lines (HT29, T84) as a result of DM deficiency (HERSHBERG et al. 1997); in both situations the ability to present exogenous antigen to CD4+ T cells is weak or absent and could result in local T-cell anergy and/or tolerance. *Right Panel*: IFN-γ treatment of immature "crypt-like" intestinal epithelial cells, including the mouse cell line MODE-K cells (VIDAL et al. 1993a) and HLA-DR transfected HT29 or T84 cells (HERSHBERG et al. 1997), induces a normal conformation of stable class-II dimer at the cell surface and a high efficiency to present native proteins and peptides to specific CD4+ T cells. This suggests that class-II negative enterocytes of the lower villi or the crypts, which become class-II positive in pathological situations associated with a local release of pro-inflammatory cytokines (inflammatory bowel disease, celiac disease), may also become able to present antigen to specific CD4+ T cells in the mucosa, resulting in a breakdown of oral tolerance

Peyer's patches and homing to the mucosa, or naive T cells present in the mucosa, in response to large quantities of various luminal soluble antigens that may cross the epithelium.

Experiments have recently been carried out to test this hypothesis in a mouse model of oral tolerance to haptens, in which intragastric feeding with the hapten dinitro-chloro-benzene (DNCB) prevents contact sensitivity to this hapten induced by subsequent epicutaneous immunization. (GALLIAERDE et al. 1995). We have observed that ingested hapten is rapidly captured by both villus epithelial cells and by cells with macrophage-dendritic morphology in the Peyer's patches and in the lamina propria. Using hapten-specific T cells from peripheral lymph nodes of mice epicutaneously primed with DNCB, we observed that Peyer's patch APC cells, but not enterocytes, harvested at various time points (i.e., 15 min to 2 h) after DNCB feeding, could promote proliferation of hapten-specific T cells. This indicated that the hapten captured by Peyer's patch APCs could activate specific CD4+ T cells, while enterocytes could not. Instead, enterocytes from hapten-fed mice could block in vitro proliferation of hapten-specific T cells induced by hapten-pulsed splenic APC. This inhibitory effect of IEC was mediated by IL-10 and transforming growth factor beta (TGF-β), secreted by enterocytes, and resulted in a T-cell anergy which could be reversed by IL-2 and IL-4. As expected, this inhibitory effect of enterocytes was antigen non-specific in as much as enterocytes from DNCB-fed mice could inhibit T-cell proliferation in response to an irrelevant hapten.

Thus, villus enterocytes exhibit three main features which allow us to postulate that they may contribute to local mechanisms of oral tolerance: (1) they lack co-stimulatory molecules required for priming of naive T cells by potentially immunogenic dietary proteins; (2) they express an abnormal conformation of class II which is poorly efficient at presenting processed antigens to antigen-specific memory T cells (less dependent on co-stimulation compared with virgin T cells); and (3) they can inhibit T-cell mediated immune responses through secretion of immunosuppressive cytokines, including IL-10 and TGF-β (both of which are produced by gut epithelial cells in vitro) (KOYAMA and PODOLSKY 1989; PANJA et al. 1995; D. KAISERLIAN, personal communication). In the process of homeostasis, enterocytes may limit the development of hypersensitivity reactions and untoward immune responses directed against dietary or environmental antigens and maintain tissue integrity (Fig. 3).

It is most likely that, in addition to local (possibly antigen non-specific) mechanisms, peripheral (antigen-specific) mechanisms are involved in systemic tolerance to orally encountered proteins. Indeed, enterocytes are unlikely to play a direct role in the systemic inhibition of antigen-specific responses, since they do not migrate and their cytokine-mediated bystander inhibitory effect is antigen non-specific. Other mechanisms, including antigen handling in the Peyer's patches and its transport to peripheral tissues, may be responsible for peripheral tolerance induced by oral antigen administration. In this respect, we recently reported that class-II-restricted T cells are involved in oral tolerance (DESVIGNES et al. 1996), since MHC class-II knock-out mice and normal mice treated with a depleting anti-CD4 mAb, cannot develop oral tolerance for hapten-specific contact sensitivity. Moreover, a single intragastric feeding with the hapten, in the absence of cutaneous sensitization, is able to prime for a hapten-specific contact-sensitivity response. This latter finding provides an experimental basis for the hypothesis that oral tolerance accounts for the lack of systemic immunity in response to oral immunization with

Intestinal epithelial cells in immunity and tolerance

mature EC

1. **Lack of costimulatory molecules (B7, HSA, CD40)**
2. **Abnormal conformation of class II**
   Can present superantigens
   but not most soluble proteins to $CD4^+$ T cells
3. **ProduceT cell inhibitory cytokines**
   Induce Th1 anergy

-> role in local mechanism of oral tolerance ?

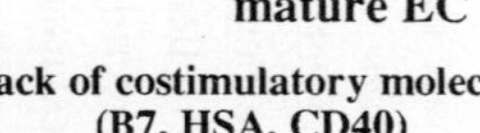

immature EC

1. **Do not express MHC class II molecules constitutively**
2. **Express a normal conformation of class II induced by IFN-γ**
3. **Can present superantigens, exogenous proteins and peptides to $CD4^+$ T cells**

-> role in breakdown of oral tolerance ?

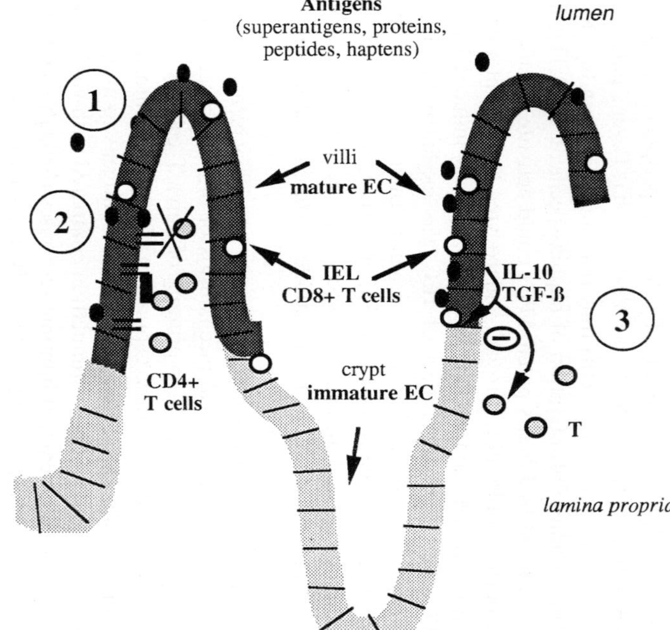

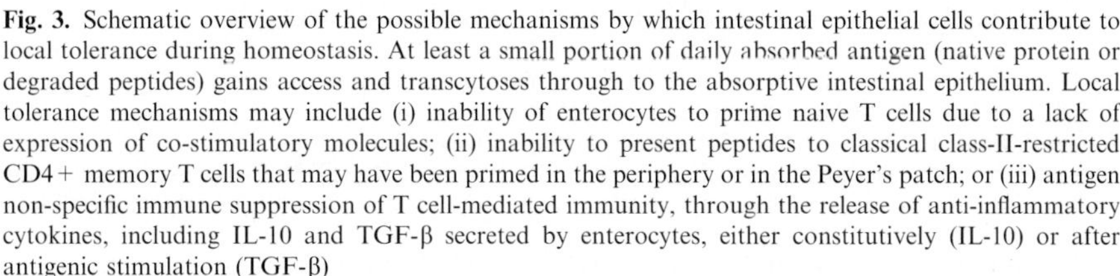

Fig. 3. Schematic overview of the possible mechanisms by which intestinal epithelial cells contribute to local tolerance during homeostasis. At least a small portion of daily absorbed antigen (native protein or degraded peptides) gains access and transcytoses through to the absorptive intestinal epithelium. Local tolerance mechanisms may include (i) inability of enterocytes to prime naive T cells due to a lack of expression of co-stimulatory molecules; (ii) inability to present peptides to classical class-II-restricted CD4+ memory T cells that may have been primed in the periphery or in the Peyer's patch; or (iii) antigen non-specific immune suppression of T cell-mediated immunity, through the release of anti-inflammatory cytokines, including IL-10 and TGF-β secreted by enterocytes, either constitutively (IL-10) or after antigenic stimulation (TGF-β)

soluble antigens. It also shows that downregulatory CD4+ T cells may be the key players in the mechanism of systemic tolerance induced by an oral tolerogen.

The fact that cytokines, such as IL-10 and TGF-β produced by enterocytes, are able to suppress Th1 cell-mediated hypersensitivity reactions, while promoting IgA switch and expansion of surface IgA+ B cells, is reminiscent of the functional duality of the intestinal immune system, which appears to be geared toward T-cell tolerance and B-cell immunity. An exception to this general tone may be dictated by pathological situations in which mucosal inflammation and/or infection triggers novel pathways of antigen presentation by epithelial ligands which have the ability to activate mucosal T cells.

References

Balk SP, Ebert EC, Blumenthal RL, McDermott FV, Wucherpfennig KW, Landau SB, Blumberg RS (1991) Oligoclonal expansion and CD1 recognition by human intestinal intraepithelial lymphocytes. Science 253:1411–1415

Balk SP, Burke S, Polischuk JE, Frantz ME, Yang L, Porcelli S, Colgan SP, Blumberg RS (1994) β2-microglobulin-independent MHC class 1b molecule expressed by human intestinal epithelium. Science 265:259–262

Bendelac A, Lantz O, Quimby ME, Yewdell JW, Bennink JR, Brutkiewicz RR (1995) CD1 recognition by mouse NK1+ T lymphocytes. Science 268:863–865

Bennett K, Levine T, Ellis JS, Peanasky RJ, Samloff IM, Kay J, Chain BM (1992) Antigen processing for presentation by class II major histocompatibility complex requires cleavage by cathepsin E. Eur J Immunol 22:1519–1524

Bland P (1988) MHC class II expression by the gut epithelium. Immunol Today 9:174–178

Bland PW (1996) Control of MHC class II expression by enterocytes. In: Kaiserlian D (ed) Antigen presentation by intestinal epithelial cells. RG Landes Company, Texas, pp 53–75

Bland PW, Warren LG (1986) Antigenic presentation by epithelial cells of the rat small intestine. 1. Kinetics, antigen specificity and blocking by anti-Ia antisera. Immunology 58:1–7

Bland PW, Whiting CV (1989) Antigen processing by isolated rat intestinal villus enterocytes. Immunology 68:497–502

Bleicher PA, Balk SP, Hagen SJ, Blumberg RS, Flotte TJ, Terhorst C (1990) Expression of murine CD1 on gastrointestinal epithelium. Science 250:679–682

Blumberg RS, Terhorst C, Bleicher P, McDermott FV, Allan CH, Landau SB, Trier JS, Balk SP (1991) Expression of a nonpolymorphic MHC class I-like molecule, CD1d, by human intestinal epithelial cells. J Immunol 147:2518–2524

Blumberg RS, Colgan SP, Morales V, Kim H, Balk S (1996) In: Kaiserlian D (ed) Antigen presentation by intestinal epithelial cells. RG Landes Company, Texas, pp 77–93

Brossay L, Jullien D, Cardell S, Sydora BC, Burdin N, Modlin RL, Kronenberg M (1997) Mouse CD1 is mainly expressed on hemopoietic-derived cells. J Immunol 159:1216–1224

Buus S, Werdelin O (1986) A group-specific inhibitor of lysosomal cysteine proteinases selectively inhibits both proteolytic degradation and presentation of the antigen dinitrophenyl-poly-L-lysine by guinea pig accessory cells to T cells. J immunol 136:452–458

Calabi F, Bradbury A (1991) The CD1 system. Tissue Antigens 37:1

Canchis PW, Bhan AK, Landau SB, Yang L, Balk SP, Blumberg RS (1993) Tissue distribution of the non-polymorphic major histocompatibility complex class I-like molecule, CD1d. Immunology 80:561–565

Castano AR, Tangri S, Miller JEW, Holcombe HR, Jackson MR, Huse WD, Kronenberg M, Peterson PA (1995) Peptide binding and presentation by mouse CD1. Science 269:223–226

Ceman S, Sant AJ (1995) The function of invariant chain in class II-restricted antigen presentation. Semin Immunol 7:373–387

Cepek KL, Shaw SK, Parker CM, Russell GJ, Morrow JS, Rimm DL, Brenner MB (1994) Adhesion between epithelial cells and T lymphocytes mediated by E-cadherin and the aEb7 integrin. Nature 372:190–193

Cerf-Bensussan N, Quaroni A, Kurnick JT, Bhan AK (1984) Intraepithelial lymphocytes modulate Ia expression by intestinal epithelial cells. J Immunol 132:2244–2252

Colgan SP, Morales VM, Madara JL, Polischuk JE, Balk SP, Blumberg RS (1996) IFN-g modulates CD1d surface expression on intestinal epithelia. Am J Physiol 271:C276-C283

Desvignes C, Bour H, Nicolas JF, Kaiserlian D (1996) Lack of oral tolerance but oral priming for contact sensitivity to dinitrofluorobenzene in major histocompatibility complex class II-deficient mice and in CD4+ T cell-depleted mice. Eur J Immunol 26:1756–1761

Evavold BD, Williams SG, Hsu BL, Buus S, Allen PA (1992) Complete dissection of the Hb(64–76) determinant using T helper 1, T helper 2 clones, and T cell hybridomas. J Immunol 149:347–353

Finzi G, Cornaggia M, Capella C, Fiocca R, Bosi F, Solcia E, Samloff IM (1993) Cathepsin E in follicle associated epithelium of intestine and tonsils: localization to M cells and possible role in antigen processing. Histochemistry 99:201–211

Fleischer B (1994) CD26: a surface protease involved in T-cell activation. Immunol Today 15:180–184

Fukuhashi A, Nakahara A, Fukutomi H, Kominami E, Grube D, Uchiyama Y (1991) Immunocytochemical localization of cathepsins B, H and L in the rat gastro-duodenal mucosa. Histochemistry 95:231–239

Fung-Leung W-P, Surh CD, Liljedahl M, Pang J, Lefurcq D, Peterson PA, Webb SR, Karlsson L (1996) Antigen presentation and T cell development in H2-M-deficient mice. Science 271:1278–1281

Galliaerde V, Desvignes C, Peyron E, Kaiserlian D (1995) Oral tolerance to haptens: intestinal epithelial cells from DNCB-fed mice inhibit hapten-specific T cell activation in vitro. Eur J Immunol 25:1385–1390

Gonnella PA, Wilmore DW (1993) Co-localization of class II antigen and exogenous antigen in the rat enterocyte. J Cell Science 106:937–940

Grosjean I, Lachaux A, Bella C, Aubry J-P, Bonnefoy J-Y, Kaiserlian D (1994) CD23/CD21 interaction is required for presentation of soluble protein antigen by lymphoblastoid B cell lines to specific CD4+ T cell clones. Eur J Immunol 24:2982–2986

Hershberg R, Eghtesady P, Sydora B, Brorson K, Cheroutre H, Modlin R, Kronenberg M (1990) Expression of the thymus leukemia antigen in mouse intestinal epithelium. Proc Natl Acad Sci USA 87:9727–9731

Hershberg RM, Framson PE, Cho DH, Lee LY, Kovats S, Beitz J, Blum JS, Nepom GT (1997) Intestinal epithelial cells use two distinct pathways for HLA class II antigen processing. J Clin Invest 100:204–215

Heyman M, Desjeux JF (1996). Antigen handling by epithelial cells. In: Kaiserlian D (ed) Antigen presentation by intestinal epithelial cells. RG Landes Company,Texas, pp 1–19

Hoyne GF, Callow MG, Kuo MC, Thomas WR (1995) Presentation of peptides and proteins by intestinal epithelial cells. Immunology 80:204–208

Hugon JS, Borgers M (1967) Abstorption of horseradish peroxidase by the mucosal cells of the duodenum of mouse. I. The fasting animal. J Histochem Cytochem 14:229–237

Jiang W, Swiggard WJ, Heufler C, Peng M, Mirza A, Steinman RM, Nussenzweig MC (1995) The receptor DEC-205 expressed by dendritic cells and thymic epithelial cells is involved in antigen processing. Nature 375:151–155

Kaiserlian D, Vidal K, Revillard JP (1989) Murine enterocytes can present soluble antigen to specific class II-restricted CD4+ T cells. Eur J Immunol 19:1513–1516

Kaiserlian D, Rigal D, Abello J, Revillard JP (1991) Expression, function and regulation of the intercellular adhesion molecule-1 (ICAM-1) on human intestinal epithelial cell lines. Eur J Immunol 21:2415–2421

Kaiserlian D, Lachaux A, Grosjean I, Graber P, Bonnefoy JY (1993a) Intestinal epithelial cells express the CD23/FceRII molecule: enhanced expression in enteropathies. Immunology 80:90–95

Kaiserlian D, Vidal K, McDonald HR, Grosjean I (1993b) Mouse intestinal epithelial cells express the self superantigen MlS1a. Eur J Immunol 23:2717–2720

Kalb TH, Chuang MT, Marom Z, Mayer L (1991) Evidence for accessory cell function by class II MHC antigen-expressing airway epithelial cells. Am J Respir Cell Mol Biol 4:320–329

Klein J (1996) Extrathymic maturation of IEL in the gut mucosa. In: Kaiserlian D (ed) Antigen presentation by intestinal epithelial cells. RG Landes Company, Texas, pp 107–119

Klimpel G, Chopra A, Langley K, Wypych J, et al. (1995) A role for stem cell factor and C-kit in the murine intestinal tract secretory response to cholera toxin. J Exp Med 182:1931–1942

Koyama S, Podolsky DK (1989) Differential expression of transforming growth factors α and β in rat intestinal epithelial cells. J Clin Invest 83:1768–1773

Kraal G, Breel M, Janse M, Bruin G (1986) Langerhans cells, veiled cells, and interdigitating cells in the mouse recognized by a monoclonal antibody. J Exp Med 163:981–997

Lynch S, Kelleher D, McManus R, O'Farrelly C (1995) RAG1 and RAG2 expression in human intestinal epithelium: evidence of extrathymic T cell differentiation. Eur J Immunol 25:1143–1147

Maric D, Kaiserlian D, Croituru K (1996) Intestinal epithelial cell line induction of T cell differentiation from bone marrow precursors. Cell Immunol 172:172–179

Matsuzaki Z, Okamoto Y, Sarashina N, Ito E, Togawa K, Saito I (1996) Induction of intercellular adhesion molecule-1 in human nasal epithelial cells during respiratory syncytial virus infection. Immunology 88:565–568

Mayer L, Shlien R (1987) Evidence for function of Ia molecules on gut epithelial cells in man. J Exp Med 166:1471–1483

Mayer L, Eisenhardt D (1990) Lack of induction of suppressor T cells by intestinal epithelial cells from patients with inflammatory bowel disease. J Clin Invest 86:1255–1260

Mayrhofer G (1994) Epithelial deposition of antigen. In: Goldie R (ed) Immunopharmarcology of epithelial barriers. Academy Press, London, pp 19–70

Mayrhofer G, Spargo LDJ (1989) Subcellular distribution of class II major histocompatibility antigens in enterocytes of the human and rat small intestine. Immunol Cell Biol 67:251–260

Mayrhofer G, Spargo LDG (1990) Distribution of class II major histocompatibility antigens in enterocytes of the rat jejunum and their association with organelles of the endocytic pathway. Immunology 70:11–19

Momburg F, Koch N, Möller P, Moldenhauer G, Butcher GW, Hämmerling GJ (1996) Differential expression of Ia and Ia-associated invariant chain in mouse tissues after in vivo treatment with IFN-γ. J Immunol 136:940–948

Norton J, Sloane JP, Al-Saffar N, Haskard DO (1992) Expression of adhesion molecules in human intestinal graft-versus-host disease. Clin Exp Immunol 87:231–236

Owen RL (1977) Sequential uptake of horseradish peroxidase by lymphoid follicle epithelium of Peyer's patches in the normal unobstructed mouse intestine: an ultrastructural study. Gastroenterology 72:440–452

Panja A, Blumberg RS, Balk SP, Mayer L (1993) CD1d is involved in T cell-intestinal epithelial cell interactions. J Exp Med 178:1115–1119

Panja A, Barone A, Mayer L (1994) Stimulation of lamina propria lymphocytes by intestinal epithelial cells: evidence for recognition of nonclassical restriction elements. J Exp Med 179:943–950

Panja A, Zhou Z, Mullin G, Mayer L (1995) Secretion and regulation of IL-10 by intestinal epithelial cells. Gastroenterology 108:A890

Parr EL, McKenzie IFC (1979) Demonstration of Ia antigens on mouse intestinal epithelial cells by immunoferritin labeling. Immunogenetics 8:499–508

Peterson M, Miller J (1990) Invariant chain influences the immunological recognition of MHC class II molecules. Nature 345:172

Pinet V, Malnati MS, Long EO (1994) Two processing pathways for the MHC class II-restricted presentation of exogenous influenza virus antigen. J Immunol 52:4852–4860

Poussier P, Julius M (1994) Thymus independent T cell development and selection in the intestinal epithelium. Annu Rev Immunol 12:521–553

Roberts K, Kilshaw PJ (1993) The mucosal T cell integrin aM290b7 recognizes a ligand on mucosal epithelial cell lines. Eur J Immunol 23:1630–1635

Roche PA (1995) HLA-DM: an in vivo facilitator of MHC class II peptide loading. Immunity 3:259–262

Rudensky AY, Rath S, Preston-Hurlburt P, Murphy DB, Janeway Jr CA (1991) On the complexity of self. Nature 353:660–662

Salomon P, Pizzimenti A, Pania A, Reisman A, Mayer L (1991) The expression and regulation of class II antigens in normal and inflammatory bowel disease peripheral blood monocytes and intestinal epithelium. Autoimmunity 9:141–149

Sidhu NK, Wright GM, Markham RJF, Ireland WP, Singh A (1992) Quantitative regional variation in the expression of major histocompatibility class II antigens in enterocytes of the mouse small intestine. Tissue Cell 24:221–228

Sieling PA, Chatterjee D, Porcelli SA, Prigozy TI, Mazzaccaro RJ, Soriano T, Bloom BR, Brenner MB, Kronenberg M, Brennan PJ, Modlin RL (1995) CD1-restricted T cell recognition of microbial lipoglycan antigens. Science 269:227–230

Simister NE, Rees AR (1985) Isolation and characterization of an Fc receptor from neonatal rat small intestine. Eur J Immunol 15:733

Solcia E, Paulli M, Silini E, Fiocca R, Finzi G, Kindl S, Boveri E, Bosi F, Cornaggia M, Capella C, Samloff IM (1993) Cathepsin E in antigen-presenting Langerhans and interdigitating reticulum cells. Its possible role in antigen processing. Eur J Histochem 37:19–26

Suda T, Sato A, Sugiura W, Chida K (1995) Induction of MHC class II antigens on rat bronchial epithelial cells by interferon-gamma and its effect on antigen presentation. Lung173:127–137

Takeuchi M, Nishizaki Y, Sano O, Ohta T, Ikeda M, Korimoto M (1997) Immunohistochemical and immuno-electron-microscopic detection of interferon-g-inducing factor ("inkerleukin-18") in mouse intestinal epithelial cells. Cell Tissue Res 289:499–503

Terpend K, Boisgerault F, Blatton MA, Desjeux JF, Heyman PM (1997) Protein transport and processing by human HT29–19A intestinal cells: effect on IFN-g. Gut (in press)

Tosi MF, Stark JM, Hamedani A, Smith CW, Gruenert DC, Huang YT (1992) Intercellular adhesion molecule-1 (ICAM-1)-dependent and ICAM-1-independent adhesive interactions between polymorphonuclear leukocytes and human airway epithelial cells infected with parainfluenza virus type 2. J Immunol 149:3345–3349

Varshaw AL, Walker WA, Cornell R, Isselbacher KJ (1971) Small intestine permeability to macromolecules. Transmission of horseradish peroxidase into mesenteric lymph and portal blood. Lab Invest 25:675–680

Vidal K, Kaiserlian D, Revillard JP (1989) Heterogeneity of murine gut epithelium: three subsets defined by dendritic cell markers. Reg Immunol 2:360–365

Vidal K, Grosjean I, Revillard J-P, Gespach C, Kaiserlian D (1993a) Immortalization of mouse intestinal epithelial cells by the SV40-large T gene. J Immunol Methods 166:63–73

Vidal K, Samarut C, Magaud J-P, Revillard JP, Kaiserlian D (1993b) Unexpected lack of reactivity of allogeneic anti-Ia monoclonal antibodies with MHC class II molecules expressed by mouse intestinal epithelial cells. J Immunol 151:4642–4650

Vivile S, Neefjes J, Lotteau V, Dierich A, Lemeur M, Ploegh H, Benoist C, Mathis D (1993) Mice lacking the MHC class-II invariant chain. Cell 72:635–648

Wegner CD, Gundel RH, Reilly P, Haynes N, Letts LG, Rothlein R (1990) Intercellular adhesion molecule-1 (ICAM-1) in the pathogenesis of asthma. Science 247:456–459

Wheeler EE, Challacombe DN, Kerry PJ, Pearson EC (1993) A morphological study of β-lactoglobulin absorption by cultured explants of the human duodenal mucosa using immunocytochemical and cytochemical techniques. J Pediatr Gastroentereol Nutr 16:157–164

Wolf PP, Ploegh HI (1995) How MHC class II molecules acquire peptide cargo: biosynthesis and trafficking through the endocytic pathway. Annu Rev Cell Dev Biol 267:306

Wu M, van Kaer L, Itohara S, Tonegawa S (1991) Highly restricted expression of the thymus leukemia antigens on intestinal epithelial cells. J Exp Med 174:213–218

Zimmer KP, Poremba C, Weber P, Ciclitira PJ, Harms E (1995) Translocation of gliadin into HLA-DR antigen containing lysosomes in coeliac disease enterocytes. Gut 36:703–709

Oral Tolerance and Anti-Pathological Vaccines

C. Czerkinsky[1], J.-B. Sun[2], and J. Holmgren[2]

1 Introduction

One of the primary goals in developing effective therapies against diseases caused by unwanted or tissue-damaging inflammatory immune responses is to specifically suppress or decrease, to an acceptable level, the intensity of untoward immune reactions without affecting the remainder of the immune system.

Induction of tolerance in mature pathogenic T cells represents an ideal form of specific immunotherapy in the treatment of autoimmune diseases, graft rejection, allergies and chronic inflammatory diseases associated with persistence of infectious microorganisms. Peripheral immunological tolerance can be induced consistently in experimental animals, but unfortunately our understanding of the underlying mechanisms is too rudimentary, and protocols to implement such an approach, clinically, are far from perfect. Three main modes of peripheral tolerance induction by antigens have been considered: (1) parenteral administration of antigens,

[1] INSERM Unit 364, Cellular and Molecular Immunology, Faculté de Médecine-Pasteur, Avenue de Valombrose, 06107 Nice Cedex 02, France
[2] Department of Medical Microbiology and Immunology, University of Göteborg, Sweden

(2) parenteral administration of antigen analogs which act as T cell receptor (TCR) antagonists, and (3) mucosal administration of antigens or so-called "oral tolerance". Mucosally induced immunological tolerance is clearly distinct, mechanistically, from tolerance induced by systemic administration of soluble antigen or TCR antagonists.

Mucosal administration of antigen is, in fact, a long-recognized method of inducing peripheral tolerance (WELLS 1911). The phenomenon, often referred to as "oral tolerance" (because initially documented by the effect of oral administration of antigen), is characterized by the fact that animals fed or having inhaled an antigen become refractory or have a diminished capability of developing an immune response when re-exposed to that very same antigen introduced by the systemic route, e.g., by injection. This effect is especially pronounced for Th1 cell-mediated immune responses and is regarded as an important natural physiological mechanism, whereby we avoid developing delayed-type (DTH) inflammatory immune reactions to environmental antigens, such as dietary and airborne antigens and products from commensal microorganisms. Mucosally-induced immunological tolerance can affect all types of adaptive immune responses, depending on the animal species, the age, the form and dose of antigen, and the route of mucosal administration (enteric, buccal, nasal, rectal).

2 Mechanisms of Induction and Expression of Peripheral Tolerance after Mucosal Delivery of Antigens

Mucosal uptake of antigens may result in the development of immunity or tolerance, or even both; the decision is being taken in the epithelium or underlying lymphoid tissue and is determined mainly by the nature and physicochemical form of the antigen (Fig. 1).

Depending on the dose of antigen administered, deletion (CHEN et al. 1995) or anergy of antigen-specific T cells (WHITACRE et al. 1991) and/or expansion of cells producing immunosuppressive cytokines (IL-4, IL-10 and TGF-β) (KHOURY et al. 1992; CHEN et al. 1994) may result in decreased T-cell responsiveness. It is interesting to note that the latter scenario involves cytokines that are also known to promote IgA isotype switching and IgA production (MURRAY et al. 1987; COFFMAN et al. 1987; VAN VLASSELAER et al. 1992; DEFRANCE et al. 1992), and is thus compatible with the observation that secretory humoral immune responses and systemic T-cell tolerance may develop concomitantly (CHALLACOMBE and TOMASI 1980). Because tolerance can be transferred by both serum and cells from tolerized animals, it is possible that humoral antibodies (perhaps IgA), circulating undegraded antigens or tolerogenic fragments and cytokines may act synergistically to confer T-cell unresponsiveness.

Without excluding the above possibilities, another mechanism that may be considered could involve antigen-driven attraction of inflammatory T cells from the

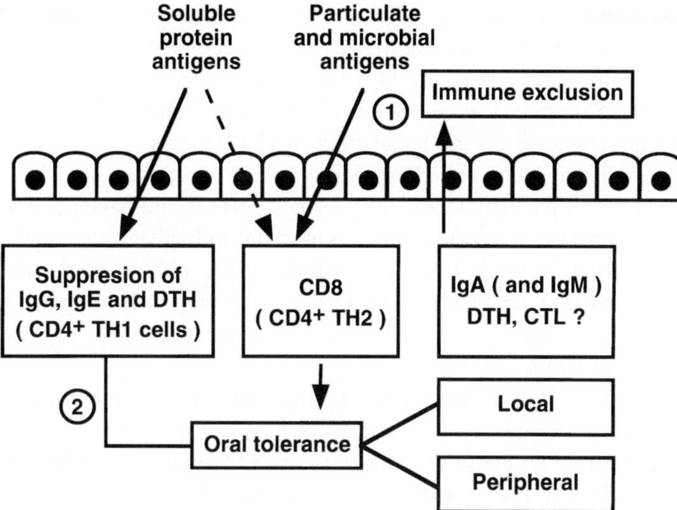

Fig. 1. Diagrammatic representation of the major adaptive defense mechanisms induced in the enteric mucosa

periphery into the mucosal microenvironment, where they could be rendered anergic, functionally skewed, deleted or even ignored. This form of antigen-driven "anatomic deviation" would imply that inflammatory T cells activated in peripheral tissues exhibit a promiscuous migratory behavior, allowing their entry not only into systemic organs, but also into mucosal tissues at the site of antigen uptake. The latter scenario has not yet been addressed experimentally, but is compatible with studies showing that auto-aggressive arthritogenic and diabetogenic T cells do express the cell-surface mucosal integrin $\alpha4\beta7$, whose ligand MadCAM-1 (mucosal addressin cellular adhesion molecule), is also expressed in inflamed pancreas and synovial tissues. In this respect, it is also interesting to note that induction of oral tolerance with a prototype soluble protein antigen has recently been shown to differentially affect the production by peripheral T cells of the β(C-C) chemokines macrophage inflammatory protein (MIP-1α) and monocyte chemotactic peptide (MCP-1) (Karpus and Lukacs 1996), which are potent chemoattractants for T cells.

Irrespective of the effector mechanism(s) involved, a major question that arises is where and how tolerance is induced, be it suppression, anergy, deletion, ignorance and/or anatomical deviation. To date, very little is known regarding the mechanisms governing induction of mucosal tolerance, and especially the intracellular pathways of entry of tolerogens, the nature of antigen-presenting cell (APC) elements involved, their tissue localization, and the characteristics of signals transduced from such cells to responding T cells.

At variance with systemically administered antigens, antigens handled in mucosal tissues have already been subjected to a variety of innate factors such as proteases, acids, salts and mucins, that have altered their form prior to uptake. As a result of this "extratissular conditioning", different epitopes may be exposed and their uptake and/or processing may involve many different cell types.

The observation that mucosally induced systemic tolerance depends on an intact epithelial barrier (Strobel et al. 1983; Bruce et al. 1987) suggests a central role for the epithelium. Specialized epithelial microfold (M) cells overlying organized lymphoid aggregates, such as the Peyer's patches or the tonsils, have been shown to uptake a variety of particulate antigens, such as viruses and bacteria, and to allow direct entry of invasive microorganisms in mucosal-inductive sites. Such a pathway is thought to result in the induction of secretory IgA immune responses. Although the ability of M cells to serve as APCs appears to be poorly supported, these cells could still, theoretically, be involved in an abortive form of antigen presentation, leading to tolerance induction.

The role of absorptive epithelial cells, such as intestinal enterocytes, in tolerance induction has been underscored by several studies and, in particular, by elegant experiments reported by Rubin et al. (1981). Reovirus type III enters M cells and elicits a protective IgA response, whereas reovirus type I infects enterocytes and induces tolerance. Most of the currently available data indicate that antigens sampled from the lumen by intestinal enterocytes are preferentially presented to CD8+ suppressor cells (Mayer et al. 1996). Epithelial enterocytes express co-stimulatory molecules, such as non-classical MHC class I (CD1d) molecules, involved in antigen presentation, to subpopulations of T cells and abnormal forms or levels of MHC class-II molecules (Kaiserlian 1995), leading to selective triggering of suppressive CD8+ T cells and/or abortive presentation to CD4+ T cells. In addition, epithelial enterocytes have been shown to produce cytokines, such as IL-10 and TGF β, which are particularly efficient at suppressing the inductive phase of CD4+ T cell-mediated responses.

All known types of classical APCs, including dendritic cells, macrophages and B cells, have been shown to populate mucosal tissues, but because of their heterogeneity and the difficulty in isolating pure subpopulations of APCs from mucosal tissues, their respective role in inducing or maintaining tolerance has not yet been elucidated. Although activated B cells and tissue macrophages are powerful APCs for memory T-helper cells, evidence suggests that antigen presentation by resting B cells results in T-cell tolerance (Eynon and Parker 1992; Fuchs and Matzinger 1992). Resting B cells lack critical co-stimulatory molecules, but are efficient at internalizing specific antigens. It should, however, be noted that B cells activated in vitro with bacterial lipopolysaccharide (LPS), a prominent component of the normal mucosal microflora, are capable of inducing tolerance when injected into naive hosts (Fuchs and Matzinger 1992).

Although functional dendritic cells have been identified in mucosal tissues such as the Peyer's patches (Spalding et al. 1983; Kelsall and Strober 1996), the mesenteric lymph (Liu and MacPherson 1993; MacPherson et al. 1995), the intestinal lamina propria (Pavli et al. 1990) and the airway mucosa (McWilliam et al. 1994), their role in activating rather than suppressing naive T cells has received strongest support. Interestingly, LPS, which is known to cause the rapid exit of dendritic cells, has also been shown to enhance tolerance induction (Khoury et al. 1990). However, Holt (1994) has proposed that a subpopulation of dendritic cells in the airway mucosa may leak out immunogenic peptides from MHC class-II to MHC class-I molecules or to non-classical restriction elements, resulting in subsequent presentation to CD8+ γδ T cells. The latter could secrete immunosuppressive cytokines [interferon gamma (IFN-γ)] that would prevent the proliferation of CD4+ T cells, especially Th2 cells. Such a hypothesis has been proposed to explain the suppressive effect of airborne antigens on induction of respiratory type-I allergic responses.

The above information places a strong emphasis on the site of entry and the intracellular pathway of processing of antigens administered to a mucous membrane in the induction of tolerance and/or immunity, and calls for the need to develop vaccine formulations with intrinsic immunomodulating and cellular targeting properties.

3 Mucosal Immunotherapy: Potential and Limitations

Since mucosally induced immunological tolerance is exquisitely specific to the antigen initially ingested or inhaled and, thus, does not influence the development of systemic immune responses against other antigens, its manipulation has become an increasingly attractive strategy for preventing and possibly treating illnesses associated with or resulting from the development of adverse immunological reactions against self and non-self antigens. The approach had been proposed earlier as a strategy to prevent or to reduce the intensity of allergic reactions to chemical drugs (Chase 1946). Later, the same rationale was followed in attempts to prevent or treat allergic reactions to common allergens (Wortmann 1977; Rebien et al. 1982). More recently, nasal administration of a synthetic peptide (Der P1) entailing a dominant T-cell epitope of house-dust mite allergen could inhibit T-cell and reaginic antibody responses in mice (Hoyne et al. 1993).

The phenomenon of mucosally induced systemic tolerance has likewise been utilized to reduce or suppress immune responses, not only against foreign antigens, but also against self antigens, i.e., components derived from host tissues. It has, thus, been possible to delay the onset and/or to decrease the intensity of experimental autoimmune diseases in a variety of animal systems by mucosal deposition of auto-antigens onto the intestinal (by feeding) or the respiratory (by aerosol or intranasal instillation of antigens) mucosa (reviewed by Weiner 1997). For instance, oral administration of collagen type II has been shown to delay the onset of autoimmune arthritides (Thompson and Staines 1986; Nagler-Anderson et al. 1986; Zang et al. 1990). Similarly, it has been possible to suppress an experimental form of autoimmune uveoretinitis by oral administration of the soluble retinal antigen (S-antigen). Experimental autoimmune encephalomyelitis, a chronic relapsing demyelinating disorder that can be induced in susceptible strains of rodents by injection of purified myelin autoantigens or crude spinal-cord homogenate together with adjuvant, can be suppressed partially or completely in animals fed myelin antigens or synthetic peptides.

Although the above examples indicate that mucosal administration of foreign and self antigens offers good promise for inducing specific immunological tolerance, the applicability of this approach in human and in veterinary medicine remains limited by practical problems. Indeed, to be broadly clinically applicable, mucosally induced immunological tolerance must also be effective in patients in whom the disease process has already established itself and/or in whom potentially tissue-damaging immune cells already exist. This is especially important when considering strategies of tolerance induction in patients suffering from or prone to an autoimmune disease, an allergic condition or a chronic inflammatory reaction to a persistent microorganism. Current protocols of mucosally induced tolerance have had limited success in suppressing the expression of an already established state of systemic immunological sensitization (Hansson et al. 1979; Staines et al. 1996). This may partly explain the disappointing results of recent clinical trials of oral tolerance in patients with multiple sclerosis and rheumatoid arthritis (Weiner 1997).

In addition, and by analogy with mucosal vaccines aimed at inducing immune responses to infectious pathogens, induction of immunological tolerance by mucosal application of most antigens requires administration of massive amounts of antigens or prolonged administration of relatively smaller amounts of antigens which are then only effective in rather narrow dose ranges. A likely explanation is that most antigens are extensively degraded before entering a mucosal tissue and/or are absorbed in insufficient quantities.

4 Cholera Toxin-B Subunit as a Mucosal Carrier-Immunomodulating System for Anti-pathological Vaccination

It has been widely assumed that strong mucosal immunogens, such as protease-resistant molecules with mucosa-binding properties, induce local and systemic immune responses in a characteristic fashion, without inducing immunological tolerance, when administered orally. Based on this assumption, mucosal administration of antigens coupled to mucosa-binding molecules, such as cholera toxin (CT) or its mucosa-binding fragment CTB, has been proposed as a strategy to induce local and systemic immune responses rather than tolerance (McKenzie and Halsey 1984; Nedrud et al. 1987; de Aizpurua and Russell-Jones 1988; Czerkinsky et al. 1989). Indeed, some years ago, CT and CTB attracted interest not only as potent mucosal immunogens and carrier molecules for oral delivery of foreign protein antigens, but also as agents capable of abrogating systemic immunological tolerance when co-administered with various antigens/tolerogens (Elson and Ealding 1984).

In the course of recent studies, we have observed that physical coupling of an antigen to CTB led to alterations that resulted in hitherto unexpected effects: when given by various mucosal (oral, intranasal, vaginal, rectal) routes, CTB induced a strong mucosal IgA immune response to itself and, in some cases, also to the conjugated antigen. But, instead of abrogating systemic tolerance to itself and to the conjugated antigens CTB enhanced it profoundly (SUN et al. 1994).

Based on this unexpected finding and on the results of other experiments with a variety of soluble protein antigens and particulate antigens (CZERKINSKY et al. 1996), we had good reasons to believe that such a system may be advantageous for suppressing systemic T cell responses. First, it minimizes by several hundred-fold the amount of antigen/tolerogen and drastically reduces the number of doses that would otherwise be required to induce tolerance orally. Second, but probably most important, this strategy appears to be applicable in preventing expression of an already established state of systemic immune sensitization. In the following subsections, the results are summarized of recent studies using this type of approach as a means to prevent or treat pathological immune responses associated with experimental autoimmune diseases, type-I allergies and allograft rejection.

4.1 Treatment of Organ-Specific Autoimmune Diseases

We have demonstrated that mucosal administration of relevant autoantigens linked to CTB could inhibit the development of clinical disease in animal models of experimentally inducible autoimmune diseases, such as allergic encephalomyelitis (SUN et al. 1996) and collagen-induced arthritis (TARKOWSKI et al. 1998). In the latter system, nasal administration of a collagen type-II–CTB conjugate could inhibit disease progression, even when treatment was initiated after onset of clinically overt disease (TARKOWSKI et al. 1998) (Fig. 2). Furthermore, oral treatment of female nonobese diabetic (NOD) mice with a CTB–insulin conjugate could suppress type-I diabetes (BERGEROT et al. 1997) even when given as late as 15 weeks post-birth (that is, at a time when all mice have evidence of insulitis). Taken together, these observations indicate that CTB-driven mucosal tolerance can affect not only the afferent, but also the efferent phase of systemic T cell-mediated inflammatory responses.

Protection against clinical disease was consistently associated with decreased IL-2 responses in lymph nodes draining the site of disease induction (SUN et al. 1994; SUN et al. 1996). However, depending on the nature of the conjugated antigen, the route of administration of the conjugate (oral, nasal) and the animal species used, this type of treatment variably affected the capacity of lymph-node T cells to produce stereotype Th1 (IFN-γ) or Th2 (IL-4, IL-5) cytokines on in vitro re-exposure to the fed or inhaled autoantigen.

The most striking observation in all three models of autoimmune diseases tested was the finding that treatment with the CTB-antigen suppressed leukocyte infiltration into the target organ (SUN et al. 1996; BERGEROT et al. 1997; TARKOWSKI et al. 1998). This observation suggests that the mechanisms governing induction of

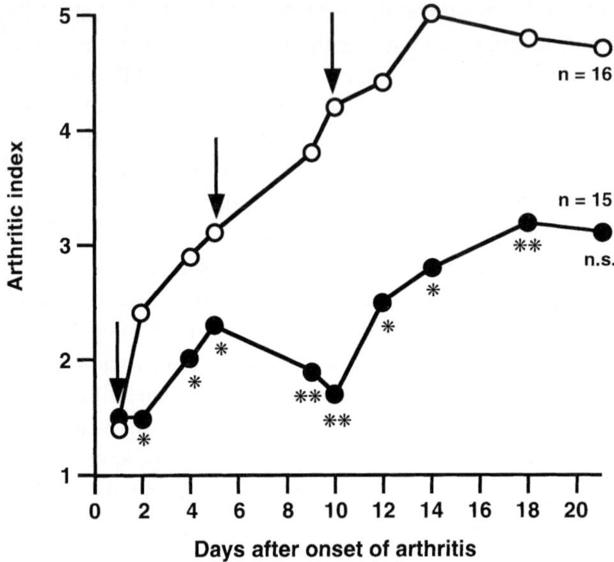

Fig. 2. Nasal treatment with CTB type-II collagen suppresses disease progression in mice with recent-onset arthritis. Mice were injected with type-II collagen in Freund's adjuvant on two consecutive occasions, and randomly assigned to treatment with CTB type-II collagen or CTB-OVA given at the indicated times (*arrows*) after onset of clinical arthritis (defined as arthritic index of 1 on two consecutive days). Asterisks denote significant differences between the two animal groups $*p<0.05$; $**\ p<0.01$

tolerance by feeding or inhaling CTB-linked antigens may involve modifications of the migratory behavior of inflammatory cells. Assuming that protective T cells induced by CTB-mediated mucosal delivery of autoantigens, unlike inflammatory T cells, do not migrate into the target tissue, they may still leave the mucosa and, via the blood, reach draining lymph nodes where they could interfere with the recruitment and migration of inflammatory leukocytes. Supporting this interpretation are the results of adoptive co-transfer experiments among congenic NOD mice. Thus, injection of irradiated male Thy 1.2 recipients with diabetogenic T cells from syngeneic female mice and T cells from congenic Thy 1.1 mice fed with CTB-insulin demonstrated a selective recruitment of Thy 1.1 donor cells in the peri-pancreatic lymph nodes concomitant with reduced islet-cell infiltration (Bergerot et al. 1997).

These results suggest that treatment with CTB-autoantigen induces the selective migration and retention of protective T cells into lymphoid tissues draining the site of organ injury. Conversely, and without arguing with the latter interpretation, treatment with CTB-antigen may also induce the migration of inflammatory T cells or their precursors from the periphery into mucosal effector tissues where they could be anergized, deleted or even ignored. Further studies of the expression of adhesion molecules on protective cells and of inflammatory cells could be valuable.

Recent studies have demonstrated that CTB itself, even when administered systemically, induces profound downregulation of systemic immune responses to co-administered antigens (Williams et al. 1997). Collectively, these observations indicate that CTB not only acts as a powerful carrier delivery system to facilitate mucosal uptake of co-administered antigens, but is also endowed with strong immunomodulating properties. Hence, this strategy of tolerance induction appears to involve mechanisms that are distinct from those governing induction of tolerance after mucosal administration of free antigens.

4.2 Prevention of Graft Rejection

By coupling thymocytes to cholera-B subunits and feeding this conjugate to mice, we have also been able to significantly prolong the survival of transplanted hearts in allogeneic mouse recipients. (Sun et al. unpublished observations). Recently, feeding CTB-conjugated donor keratinocytes has been shown to prevent corneal allograft rejection in mice (JR Niederkorn, personal communication).

4.3 Prevention of Type-I Allergies

Nasal administration of a soluble protein allergen (ovalbumin) linked to *Escherichia coli* heat-labile enterotoxin-B subunit, a GM1-binding analogue of CTB, has been found to suppress systemic delayed hypersensitivity responses and IgE antibody responses in mice (Tamura et al. 1997). This observation is consistent with the recent finding that interferon-γ, IL-4 and IL-6 responses are decreased together with IgG1 and IgG2a antibody responses in animals given a nasal CTB type-II collagen vaccine (Tarkowski et al. 1998), and indicates that, under certain conditions, this form of tolerance can affect both TH1- and TH2-driven responses.

5 Mucosal Vaccines for Simultaneous Induction of Anti-Infectious and Anti-Pathological Immunity: The Cholera-Toxin B Subunit Paradigm

Somewhat surprisingly, vaccinologists, in general, and mucosal immunologists, in particular, have usually believed a reciprocal relationship to exist between induction of immunity and tolerance. The observation that mucosal immunity, which is typified by secretory IgA antibodies, may develop concomitantly with systemic immunological tolerance (Challacombe and Tomasi 1980) has led to the belief that vaccines against mucosal pathogens should primarily stimulate immunity without inducing tolerance. However, from a theoretical standpoint, the possibility of manipulating the mucosal immune system toward both immunity and tolerance appears rather attractive when considering strategies aimed at protecting the host from colonization or invasion by mucosal pathogens, but also to interfere with the development of potentially harmful systemic immunological reactions against the same pathogens or their products.

The notion that immunological tolerance may provide the host with a protective mechanism against an infectious disease has been elegantly illustrated by recent studies of transgenic mice. Whereas mice from a susceptible (BALB/c) background develop an early Th2-driven IL-4 response and ultimately succumb to infection with *Leishmania major*, mice from the same background, but rendered centrally tolerant by transgenic expression in the thymus of the Leishmania homologue of receptors for activated C kinase (LACK, a protective surface antigen of Leishmania), fail to produce this early response and resolve their infection (Julia et al. 1996). Similarly, mice expressing an endogenous superantigen, which mediates the thymic deletion of LACK-specific T cells, do not develop progressive disease (Launois et al. 1997).

Very recently, we demonstrated that the tolerance of mature post-thymic, parasite-specific T cells can also be induced in the periphery after nasal administration of as little as 12μg of *L. major* LACK antigen conjugated to CTB (Mc Sorley et al. 1998). Such treatment markedly delayed the onset of lesion development in infected mice and was associated with decreased proliferative responses

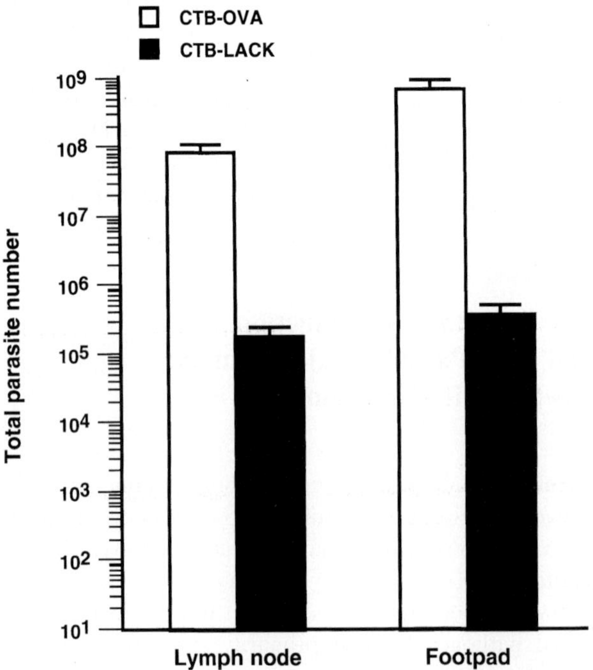

Fig. 3. Nasal treatment with a *Leishmania major* antigen (*LACK*) conjugated to CTB reduces parasite burden in the skin and draining nodes of infested mice. Animals were given CTB-LACK (12μg) or a control conjugate (CTB-OVA) 1 week prior to challenge with *L. major* promastigotes. Parasite numbers were determined 5 weeks later (Adapted from Mc Sorley et al. 1998)

to LACK. An examination of cytokine responsiveness to LACK after induction of mucosal tolerance with CTB-LACK revealed that while the Th1 response to LACK was suppressed, TH2 cytokine production (in terms of IL-4 and IL-5) was apparently unaffected. Most importantly, treatment with CTB-LACK reduced, by almost three logs, parasite burden in the skin and draining lymph nodes of infected mice (Fig. 3).

Similar findings have also been observed in mice infested with the parasite nematode, *Schistosoma mansoni* (Sun et al. unpublished observations). Thus, nasal treatment of mice with *S. mansoni* glutathione *S*-transferase (GST) conjugated to CTB suppressed granuloma formation, and decreased parasite burden and egg deposition in the liver of infested animals. Protection with this nasal CTB-GST vaccine was associated with decreased systemic T-cell proliferative responses to GST and reduced spontaneous proliferation of liver leukocytes. Hepatic production of IFN-γ, IL-5 and IL-3 were markedly reduced, whereas that of IL-4 remained unaffected. Most importantly, such treatment could significantly prolong the survival of animals, even when initiated as late as 6 weeks after initial infection, that is at a time when liver granulomatous reactions are most pronounced.

These results lend promise for the development of a novel class of therapeutic vaccines against diseases associated with inflammatory reactions caused by persistent microorganisms and their products.

Mucosally induced tolerance has the virtue of being a powerful natural and specific protective mechanism against adverse immune reactions that may result from mucosal intake of immunogenic matters. This property may be utilized to treat disorders associated with untoward immune responses to self and non-self antigens, such as certain autoimmune diseases, allergic reactions and graft rejection. Hence, mucosal immunomodulation via appropriate delivery of tolerogenic compounds may be where the future of anti-inflammatory vaccines lies.

References

Bergerot I, Ploix C, Petersen J, Moulin V, Rask C, Fabien N, Lindblad M, Mayer A, Czerkinsky C, Holmgren J, Thivolet C (1997) A cholera toxoid-insulin conjugate as oral vaccine against spontaneous autoimmune diabetes. Proc Natl Acad Sci U S A 94:4610–4614

Bruce MG, Strobel S, Hanson DG, Ferguson A (1987) Transferable tolerance for cell-mediated immunity after feeding is prevented by radiation damage and restored by immune reconstitution. Clin Exp Immunol 70:611–618

Challacombe SJ, Tomasi TB Jr. (1980) Systemic tolerance and secretory immunity after oral immunization. J Exp Med 152:1459–1472

Chase MW (1946) Inhibition of experimental drug allergy by prior feeding of the sensitizing agent. Proc Soc Exp Biol 61:257–259

Chen Y, Kuchroo VK, Inobe J-I, Haffler DA, Weiner HL (1994) Regulatory T cell clones induced by oral tolerance: suppression of autoimmune encephalomyelitis. Science. 265:1237–1240

Chen Y, Inobe J-I, Marks R, Gonnella P, Kuchroo VK, Weiner HL (1995) Peripheral deletion of antigen-reactive T cells in oral tolerance. Nature 376:177–180

Coffman RL, Shrader B, Carty J, Mossman TR, Bond MW (1987) A mouse T cell product that preferentially enhances IgA production. Biologic characterization. J Immunol 139:3685–3690

Czerkinsky C, Russell MW, Lycke N, Lindblad M, Holmgren J (1989) Oral administration of a streptococcal antigen coupled to cholera toxin B subunit evokes strong antibody responses in salivary glands and extramucosal tissues. Infect Immun 57:1072–1077

Czerkinsky C, Sun JB, Lebens M, Li BL, Rask C, Lindblad M, Holmgren J (1996) Cholera toxin B subunit as transmucosal carrier-delivery and immunomodulating system for induction of anti-infectious and anti-pathological immunity. Ann NY Acad Sci 778:185–193

de Aizpurua HJ, Russell-Jones GJ (1988) Oral vaccination. Identification of classes of proteins that provoke an immune response upon oral feeding. J Exp Med 167:440–451

Defrance T, Vanbervliet B, Briere F, Durand I, Rousset F, Banchereau J (1992) Interleukin 10 and transforming growth factor β cooperate to induce anti-CD40-activated naive human B cells to secrete immunoglobulin A. J Exp Med 175:671–682

Elson CO, Ealding W (1984) Cholera toxin did not induce oral tolerance in mice and abrogated oral tolerance to an unrelated antigen. J Immunol 133:2892–2898

Eynon EE, Parker DC (1992) Small B cells as antigen-presenting cells in the induction of tolerance to soluble protein antigens. J Exp Med 175:131–138

Fuchs EJ, Matzinger P (1992) B cells turn off virgin but not memory T cells. Science 258:1156–1159

Hansson DG, Vaz NM, Rawlings LA, Lynch JM (1979) Inhibition of specific immune responses by feeding protein antigens. II. Effects of prior passive and active immunization. J Immunol 122:2261–2266

Holt PG (1994) Immunoprophylaxis of atopy: light at the end of the tunnel? Immunol Today 15:484–489

Hoyne GF, O'Hehir RE, Wraith DC, Thomas WR, Lamb JR (1993) Inhibition of T cell and antibody responses to house dust mite allergen by inhalation of the dominant T cell epitope in naive and sensitized mice. J Exp Med 178:1783–1788

Julia V, Rassoulzadegan M, Glaichenhaus N (1996) Resistance to Leishmania major induced by tolerance to a single antigen. Science 274:421–423

Kaiserlian D (1995) The intestinal epithelial cell: a nonconventional type of antigen-presenting cell. In: Auricchio S, Ferguson A, Troncone R (eds) Mucosal Immunity and the gut epithelium: interactions in health and disease. Karger, Basel, pp 32–39 (Dynamic Nutrition Research, Vol 4.)

Karpus WJ, Lukacs NW (1996) The role of chemokines in oral tolerance: abrogation of nonresponsiveness by treatment with anti-MCP-1. Ann NY Acad Sci 778:133–142

Kelsall BL, Strober W (1996) The role of dendritic cells in antigen processing in the Peyer's patch. Ann NY Acad Sci 778:47–54

Khoury SJ, Lider O, Al-Sabbagh A, Weiner HL (1990) Suppression of experimental autoimmune encephalomyelitis by oral administration of myelin basic protein. III. Synergistic effect of lipopolysaccharide. Cell Immunol 131:302–310

Khoury SJ, Hancock WW, Weiner HL (1992) Oral tolerance to myelin basic protein and natural recovery from experimental autoimmune encephalomyelitis are associated with downregulation of inflammatory cytokines and differential upregulation of TGF-β, IL-4 and PGE expression in the brain. J Exp Med 176:1355–1364

Launois P, Maillard I, Pingel S, Swilhart KG, Xenarios I, Acha Orbea H, Diggelmann H, Locksley RM, Mac Donald HR, Louis JA (1997) IL-4 rapidly produced by V beta 4 alpha 8 CD4+ T cells instructs the development and susceptibility to Leishmania major in Balb/c mice. Immunity 6:541–549

Liu LM, MacPerson GG (1993) Antigen acquisition by dendritic cells: intestinal dendritic cells acquire antigen administered orally and can prime naive T cells in vivo. J Exp Med 177:1299–1307

MacPherson GG, Jenkins CD, Stein MJ, Edwards C (1995) Endotoxin-mediated dendritic cell release from the intestine. Characterization of released dendritic cells and TNF dependence. J Immunol 154:1317–1322

Mayer L, So LP, Yio XY, Small G (1996) Antigen trafficking in the intestine. Ann NY Acad Sci 778: 28–35

Mc Sorley SJ, Rask C, Pichot R, Julia V, Czerkinsky C, Glaichenhaus N (1998) Selective tolerization of Th1-like cells after nasal administration of a cholera toxoid-LACK conjugate. Eur J Immunol 28:424–432

McKenzie SJ, Halsey JF (1984) Cholera toxin B subunit as a carrier protein to stimulate a mucosal immune response. J Immunol 133:1818–1824

McWilliam AS, Nelson D, Thomas JA, Holt PG (1994) Rapid dendritic cell recruitment is a hallmark of the acute inflammatory response at mucosal surfaces J Exp Med 179:1331–1336

Murray PD, McKenzie DT, Swain SL, Kagnoff MF (1987) Interleukin 5 and interleukin 4 produced by Peyer's patch T cells selectively enhance immunoglobulin A expression. J Immunol 139:2669–2674

Nagler-Anderson C, Bober LA, Robinson ME, Siskind GW, Thorbecke GJ (1986) Suppression of type II collagen-induced arthritis by intragastric administration of soluble type II collagen. Proc Natl Acad Sci U S A 83:7443–7446

Nedrud JG, Liang X, Hague N, Lamm ME (1987) Combined oral/nasal immunization protects mice from Sendai virus infection. J Immunol 139:3484–3492

Pavli P, Woodhams CE, Doe WF, Hume DA (1990) Isolation and characterization of antigen-presenting dendritic cells from the mouse intestinal lamina propria. Immunology 70:40–47

Rebien W, Puttonen E, Maasch HJ, Stix E, Wahn U (1982) Clinical and immunological response to oral and subcutaneous immunotherapy with grass pollen extracts. A prospective study. Eur J Pediatry 138:341–344

Rubin D, Weiner HL, Fields BN, Greene MI (1981) Immunological tolerance after oral administration of reovirus: requirement for two viral gene products for tolerance induction. J Immunol 127:1697–1701

Spalding DM, Williamson SI, Koopman WJ, McGhee JR (1983) Accessory cells in murine Peyer's patches. I. Identification and enrichment of a functional dendritic cell. J Exp Med 157:1646–1659

Staines NA, Harper N, Ward FJ, Thompson HSG, Bansal S (1996) Arthritis: animal models of oral tolerance. Ann NY Acad Sci 778:297–305

Strobel S, Mowat AM, Drummond HE, Pickering MG, Ferguson A (1983) Immunological responses to fed protein antigens in mice. II Oral tolerance for CMI is due to activation of cyclophosphamide-sensitive cells by gut-processed antigen. Immunology 49:451–456

Sun J-B, Holmgren J, Czerkinsky C (1994) Cholera toxin B subunit: an effective tansmucosal carrier delivery system for induction of peripheral immunological tolerance. Proc Natl Acad Sci U S A 91:10795–10799

Sun J-B, Rask C, Olsson T, Holmgren J, Czerkinsky C (1996) Treatment of experimental autoimmune encephalomyelitis by feeding myelin basic protein conjugated to cholera toxin B subunit. Proc Natl Acad Sci U S A 93:7196–7201

Tamura S, Hatori E, Tsuruhara T, Aizawa C, Kurata T (1997) Suppression of delayed-type hypersensitivity and IgE antibody responses to ovalbumin by intranasal administration of Escherichia coli heat-labile enterotoxin B subunit-conjugated ovalbumin. Vaccine 15:225–229

Thompson HSG, Staines NA (1986) Gastric administration of type II collagen delays the onset and severity of collagen-induced arthritis in rats. Clin Exp Immunol 64:581–586

van Vlasselaer P, Punnonen J, de Vries JE (1992) Transforming growth factor-β directs IgA switching in human B cells. J Immunol 148:2062–2067

Weiner HL (1997) Oral tolerance: immune mechanisms and treatment of autoimmune diseases. Immunol Today 18:335–343

Wells H (1911) Studies on the chemistry of anaphylaxis III. Experiments with isolated proteins, especially those of hen's egg. J Infect Dis 9:147–171

Whitacre C, Gienapp C, Orosz IE, Bitar D (1991) Oral tolerance in experimental autoimmune encephalomyelitis. III. Evidence for clonal anergy. J Immunol 147:2155–2163

Williams NA, Stasiuk LM, Nashar TO, Richards CM, Lang AK, Day MJ, Hirst TR (1997) Prevention of autoimmune disease due to lymphocyte modulation by the B-subunit of Escherichia coli heat-labile enterotoxin. Proc Natl Acad Sci U S A 94:5290–5295

Wortmann F (1977) Oral hyposensitization of children with pollinosis or house dust asthma. Allergol Immunopathol (Madr) 5:15–26

Zang ZJ, Lee CSY, Lider O, Weiner HL (1990) Suppression of adjuvant arthritis in Lewis rats by oral administration of type II collagen. J Immunol 145:2489–2493

Antibody-Mediated Protection of Mucosal Surfaces

B. Corthésy[1] and J.-P. Kraehenbuhl[2]

1 Introduction

Mucosal surfaces of the oral cavity, the digestive and urogenital tracts, and the airways are protected against environmental pathogens by innate and adaptive immune defense mechanisms. Innate defense involves physical, chemical and cellular factors. Entrapment of pathogens in mucus facilitates their clearance by peristalsis in the gut and ciliary movement in the airways. The longitudinal flow of fluids across the epithelial layer mediated by chloride channels helps to flush away microorganisms and prevent their attachment to the epithelial cell surface. The apical cell surface-associated glycocalyx, which consists of a dense network of

[1] Division of Immunology and Allergology, Centre Hospitalier Universitaire Vaudois, 1011 Lausanne, Switzerland
[2] Swiss Institute for Cancer Research and Institute of Biochemistry, University of Lausanne, 1066 Epalinges, Switzerland

glycoproteins, also limits the access of pathogens to the epithelial surface. Chemical factors (lysozyme, lactoferrin, peroxidase and defensins) secreted by specialized epithelial cells, such as the Paneth cells in the crypts of the small intestine, gastric acid and intestinal hydrolases (proteases, lipases, nucleases) constitute an efficient defense mechanism.

Host leukocytes present or recruited into mucosal tissues following microbial–epithelial cell interactions actively participate in innate defense. The recruited cells include monocytes, macrophages, neutrophils and eosinophils. These cells preferentially express Fcα receptors for immunoglobulin A (IgA) antibodies in both mice (Hayami et al. 1997) and humans (Monteiro et al. 1990) in mucosal tissues. These Fcα receptor-bearing cells may help to clear microorganisms, following opsonization by secretory IgA (sIgA) antibodies, facilitate uptake of antigen–antibody complexes by professional antigen-presenting cells and modulate immune functions by controlling the release of specific cytokines. Therefore innate immunity may not only provide rapid anti-microbial defense, but may also determine how pathogens activate the adaptive mucosal immune system and control the nature of the immune response (Fearon and Locksley 1996; Bendelac and Fearon 1997). Adaptive immunity requires that the antigens cross the epithelium to reach the underlying organized lymphoid tissues where antigen presentation and priming of T and B lymphocytes can occur.

2 Humoral Immune Responses in Mucosal Tissues

sIgA antibodies play a major role in the protection of mucosal surfaces which is reflected by the very large quantity of antibodies produced and transported into secretions each day (~10g). Following stimulation by antigens, naive B cells in organized mucosal lymphoid tissues (MALT) of the gut, the airways and the oropharyngeal cavity move to the germinal center where they proliferate clonally. During clonal expansion, B cells undergo affinity maturation, first, by somatic hypermutation which generates variability in B-cell receptors and, second, by selection of those with highest affinity for the antigen. Selection of cells bearing these mutated receptors by the antigen occurs on the surface of the follicular dendritic cells a process which rescues cells expressing high affinity Ig receptors from apoptosis (for review, see Liu et al. 1992).

In MALT germinal centers, B lymphocytes undergo isotype switch and differentiate further into B cells that express IgA receptors. (Cebra et al. 1991). MALT $CD4^+$ T cells have been shown to promote IgA isotype switch of IgM-bearing B cells (Kawanishi et al. 1983). Cytokines produced by activated Th2 helper $CD4^+$ T cells, including IL-5, IL-10 and transforming growth factor beta (TGF-β) play a major role in triggering switch, but the precise molecular mechanism that mediates the recombination event has not yet been fully elucidated (Strober and Ehrhardt 1994). Ligation of co-stimulatory molecules, such as

CD40, provides an important signal for switch induction (DEFRANCE et al. 1992). Bacterial lipopolysaccharide (LPS) has been shown to stimulate expression of the recombination machinery in pre-B lymphocytes (LI et al. 1996) and mucosal adjuvants, including cholera toxin and *E. coli* heat-labile toxin, are known to facilitate switch (LYCKE and STROBER 1989).

Subsequently, B lymphocytes differentiate into effector or memory cells, following contact with T helper lymphocytes and CD40–CD40 ligand interactions (LIU et al. 1991). In MALT, stimulated B and T cells acquire a mucosal homing program. The effector and memory lymphocytes lose their adhesion to stromal cells, leave organized MALT structures and enter the blood stream via the lymph. Depending on the mucosal sites at which priming takes place, different homing receptors will be expressed by B lymphocytes. Virtually all IgA- and even IgG-antibody-secreting cells detected after peroral and rectal immunization expressed $\alpha4\beta7$ integrin receptors, with only a minor fraction of these cells expressed the peripheral L-selectin receptor. In contrast, circulating B cells, induced by intranasal immunization, co-express L-selectin and $\alpha4\beta7$ receptors. This may explain the compartmentalization of mucosal immune responses initiated in the upper vs the lower aerodigestive tract (QUIDING-JARBRINK et al. 1997).

Effector and memory B cells are able to home to distant mucosal tissues or return to MALT structures (ROTT et al. 1997 and, for review, BUTCHER and PICKER 1996) (Fig. 1). The lymphocytes expressing mucosal $\alpha4\beta7$ homing receptors interact with flat post-capillary venule endothelial cells bearing mucosal addressins on their luminal surface (for review, see BUTCHER and PICKER 1996). Antigen receptors (surface Igs) do not participate in the selectivity of lymphocyte binding to the vascular bed. It has been suggested that antigen-specific plasmablasts become locally enriched in mucosal sites via retention at sites of antigen deposition (BUTCHER and PICKER 1996).

The mucosal addressin MadCam-1 (mucosal addressin cell adhesion molecule) is preferentially expressed in human and mouse intestinal flat post-capillary venules of the lamina propria and high endothelial venules of organized MALT, but not in other mucosal tissues (BRISKIN et al. 1997). MadCam-1, in mice, is also found associated with the endothelial cells of mammary-gland post-capillary venules. The vascular addressins mediating selective binding of lymphocytes in the airways and the genital tract have yet not been identified. Extravasation of lymphocytes requires the action of chemokines (for review, see BUTCHER and PICKER 1996). B lymphocytes express a specific chemokine receptor, which when knocked out prevents lymphocytes from reaching mucosal tissues (FÖRSTER et al. 1996). The chemokine that attracts B cells has recently been identified (GUNN et al. 1998; LEGLER et al. 1998), but it is not known whether B cell-specific chemokines are involved in the recruitment of B cells in mucosal tissues.

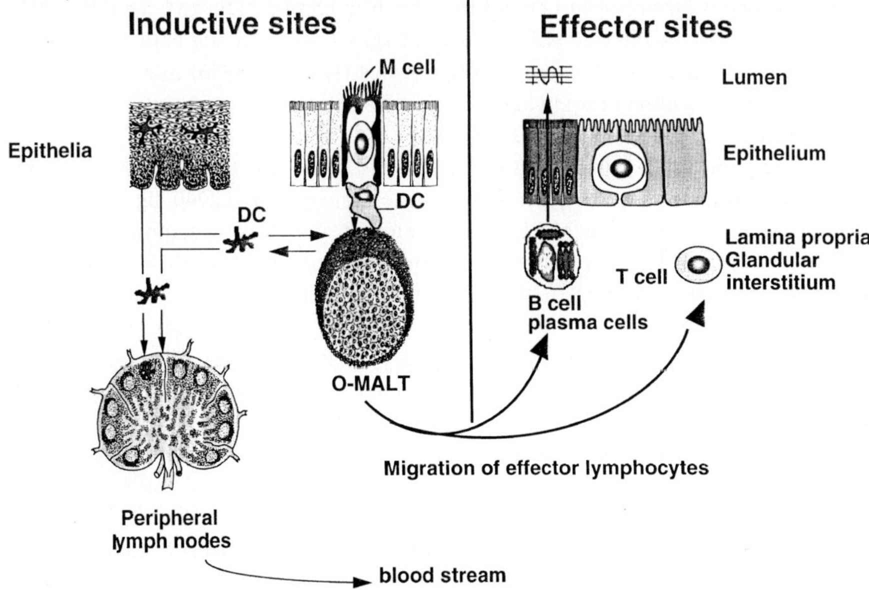

Fig. 1. In stratified epithelia (tonsil, vagina), antigen sampling is accomplished by migratory dendritic cells that transport antigens to distant lymph nodes or to local organized mucosa-associated lymphoid tissues (*O-MALT*). In simple epithelia (intestine, bronchi), membraneous (M) cells selectively take up the antigens and deliver them to underlying local *O*-MALT. Following antigen stimulation in lymphoid tissues, effector lymphocytes enter the submucosal lymphatics enter the blood circulation and establish themselves in distant MALT and glandular sites. Migration to effector sites depends on the specific interaction between homing receptor on the lymphocytes and addressin on the vascular bed bathing the epithelia. B-cell differentiation into antibody-producing cells leads to the local production of immunoglobulin A (IgA) which, upon association with the polymeric immunoglobulin receptor (pIgR), are transported across the epithelial layer and serve to neutralize pathogenic agents

3 Maturation of B Cells within Plasma Cells in Mucosal Tissues

Locally, in the lamina propria, effector B lymphocytes differentiate into antibody-secreting plasma cells, this process being regulated by T lymphocyte-, and epithelial, cell-derived cytokines (Fig. 2). In the intestinal mucosa, the number of plasma cells producing IgA exceeds those producing all other Ig isotypes. Maturation of IgA-bearing B cells into plasma cells is triggered by T cell-derived IL-5 (Matsumoto et al. 1989) and epithelial IL-6 (McGhee et al. 1991). In IL-6-deficient mice, a reduced number of IgA-producing plasma cells has been observed in the respiratory tract, and targeting IL-6 DNA into bronchial epithelial cells restored maturation of IgA B cells into plasma cells (Ramsay et al. 1994). This result, however, has not been confirmed within the digestive tract, suggesting that IL-6 is probably not the only cytokine involved in B cell maturation (Bromander et al. 1996). In the mucosal environment, all plasma cells, irrespective of their Ig isotype, express J chain, the small polypeptide required for IgA polymerization (Bjerke and Brandtzaeg 1990).

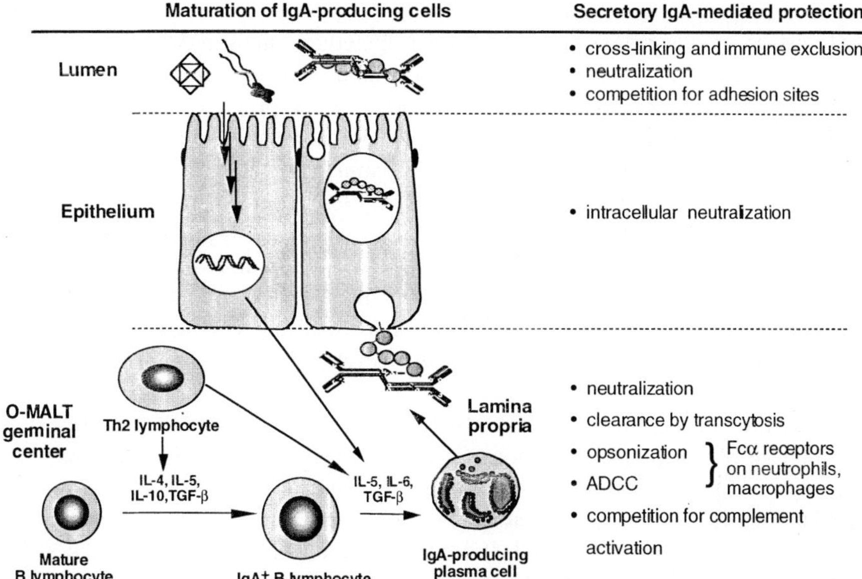

Fig. 2. Role of cytokines in the immunoglobulin A (IgA) response and biosynthesis. Mature (antigen responsive) B lymphocytes expressing membrane IgM switches to surface IgA production, under the control of Th2 cytokines, including transforming growth factor beta (TGF-β), interleukin (IL)-4, IL-5 and IL-10. This happens most likely in the germinal centers of organized mucosa-associated lymphoid tissues (*O*-MALT). IgA+ cells eventually leave the *O*-MALT and migrate to distant mucosal and glandular sites to undergo final differentiation. Enterocyte-derived IL-6 and Th2-released IL-5 and TGF-β induce secretion of soluble IgA oligomers through concomitant induction of J chain expression. Following binding to polymeric immunoglobulin receptor (pIgR), J chain-containing IgA polymers are selectively transported across the epithelium by transcytotic vesicles which, following fusion with the apical membrane, release sIgA complexes comprising IgA and the clipped form of pIgR called secretory component (*SC*). The possible sites of action of IgA/sIgA are listed on the right and discussed case by case within the text

4 Transepithelial Transport of Mucosal Immunoglobulins

The prevention of microbial infection requires that the humoral immune effectors gain access to the luminal compartment to block adhesion and invasion of environmental pathogens. Simple epithelia are sealed at their apex by tight junctions that prevent the lateral diffusion of antibodies between cells. Thus, translocation of antibodies from their site of synthesis in the lamina propria to the luminal compartments necessitates an efficient transepithelial transport machinery (Fig. 2). The pathway followed by polymeric Igs (IgA and IgM) across epithelial cells has been extensively studied (Solari and Kraehenbuhl 1984; Mostov and Deitcher 1986, Barroso and Sztul 1994; Natvig et al. 1997; Weimbs et al. 1997), and the receptor that mediates transcytosis has been well characterized (Eiffert et al. 1984; Krajci et al. 1989; Schaerer et al. 1991; Coyne et al. 1994).

The polymeric immunoglobulin receptor (pIgR) is expressed at the basolateral surface of a variety of epithelial cells, including small and large intestinal, nasal, tracheal and bronchial, and cervical and uterine epithelial cells of all mammals studied so far (BRANDTZAEG 1994). The receptor comprises five extracellular domains of 110–120 residues which share homology with Vκ and VH Ig domains, a 23-amino acid long membrane-spanning segment, and a cytoplasmic tail of about 100 amino acids (MOSTOV et al. 1984). Human, rat, mouse, rabbit and bovine receptors show extensive homology, particularly in the first domain and in the cytoplasmic tail. Following polymeric IgA binding and internalization into the recycling basolateral endosomal compartment, the complex is sorted into transcytotic vesicles that reach the apical endosomal compartment (SONG et al. 1994), where the ligand–receptor complex recycles transiently (APODACA et al. 1994).

The presence of the J chain in polymeric IgA accounts for the selective discrimination of polymeric vs monomeric IgA (BRANDTZAEG and PRYDZ 1984). At the apical surface, the poly-Ig receptor is proteolytically cleaved, and the extracellular fragment known as secretory component (SC) is released, whether bound or not to its polymeric IgA ligand. Rabbits express a spliced version of the poly-Ig receptor lacking domains 2 and 3 (DEITCHER and MOSTOV 1986) which is fully competent for transport (SOLARI and KRAEHENBUHL 1987) and generates SC noncovalently bound to IgA (FRUTIGER et al. 1987). The information necessary for the complex trafficking of the pIgR in epithelial cells resides within its cytoplasmic tail. The sequences that control basolateral sorting and endocytosis (MOSTOV et al. 1992), which are highly conserved among species (BANTING et al. 1989), are flanked by serine residues that undergo phosphorylation once the receptor is terminally glycosylated in the Golgi apparatus (HIRT et al. 1993). Transcytosis of the poly-Ig receptor is stimulated by binding of polymeric IgA (SONG et al. 1994). Endocytosis and transcytosis are regulated by the trimeric G proteins, protein kinase C and calmodulin (CHAPIN et al. 1996). In rat liver, the asialoglycoprotein receptor (ASGP-R) has also been shown to participate in IgA endocytosis (SCHIFF et al. 1986). In humans, transport of IgA from blood to bile is minimal, due to the lack of pIgRs in hepatocytes.

IgG antibodies are also found in secretions and it has been proposed that they cross the epithelial barrier by transudation. Recently, however, it has been shown that small-intestinal enterocytes express the neonatal Fc receptor (ISRAEL et al. 1997). In newborn rodents, the receptor mediates transport of maternal milk IgG antibodies from the lumen to the lamina propria. Binding of IgG antibodies in the luminal compartment is facilitated by low pH and release in the interstitium by neutral pH (RODEWALD and KRAEHENBUHL 1984). Transport in the opposite direction requires that IgG antibodies are able to reach an acidic compartment in order to bind to Fc receptors and escape to the lysosomal pathway. In the yolk sac, IgGs are taken up by fluid-phase endocytosis and accumulate in an acidic endosomal compartment where binding to neonatal Fc receptors takes place (ROBERTS et al. 1990). Whether the neonatal Fc-receptor mediates IgG transudation in other mucosal epithelia remains to be established.

5 Antibodies in Mucosal and Glandular Secretions

In human, roughly 60% of all Igs produced are IgAs, of which about half are selectively transported into external secretions (MESTECKY and MCGHEE 1987; KERR 1990). The majority of sIgA in humans is derived from local synthesis, and not from the circulation. Humans have two subclasses of monomeric and polymeric IgA: IgA1 and IgA2 (KAWAMURA et al. 1992), which differ in their hinge region and in their carbohydrate composition (WOLD et al. 1995). IgA1, but not IgA2, can be cleaved by IgA-specific proteases produced by a number of bacteria (PLAUT 1988). The protease recognizes a distinct sequence in the hinge region of IgA1 antibodies and renders the antibodies less efficient in cross-linking microorganisms. SC does not protect against cleavage (PLAUT et al. 1985). The proportion of IgA1 and IgA2 varies in individual secretions. IgA1-producing cells predominate in most mucosal tissues and glands, including the tonsils, the stomach, the duodenum, the mammary gland and the respiratory tract, while IgA2-secreting cells populate, preferentially, the large intestine and the female genital tract. Antibodies specific for protein antigens are found predominantly in the IgA1 subclass, whereas bacterial LPS and carbohydrate antigens lead preferentially to IgA2 antibody production. In IgA-deficient individuals and animals, a compensatory increase of sIgM in secretions has been reported (PLEBANI et al. 1983), and experimental IgA-deficient mice challenged with *H. felis* and cholera toxin have much higher titers of IgM antibodies than wild-type animals (NEDRUD et al. 1996).

IgE are also locally produced by plasma cells. They play a major role in the lamina propria where they can interact and activate local mast cells. IgE can mediate protection against certain parasites, as demonstrated in response to the feeding of larval ticks (MATSUDA et al. 1990). Recruitment of leukocytes at sites of IgE-dependent immune response by tumor necrosis factor alpha (TNF-α) released by mast cells could be a critical component of natural immunity to infection of bacteria and other pathogens. It has been reported that not only Th2 cells, but also mast cells and the B cells that mediate IgE responses express the same chemokine receptor, the eotaxin receptor CCR3 (SALLUSTO et al. 1997), which facilitates their recruitment at mucosal sites. Due to their strategic anatomical distribution near surfaces exposed to the external environment and their ability to release mediators of inflammation, mast cells are well suited to function as sentinels of mucosal immunity.

5.1 Molecular Structure of Immunoglobulin A (IgA)

Secretory sIgA has unique features that make this antibody particularly well suited to protect mucosal surfaces. Its polymeric structure enhances its avidity for antigens and pathogens, its distinct carbohydrate composition contributes to its binding to mucus components and also to its resistance to proteolysis (KERR 1990; MESTECKY et al. 1991). Secretory IgA consists of at least two monomeric IgA units and two additional polypeptide chains, a J chain and SC (Fig. 3). These four polypeptides are produced by two distinct cell types. The Ig heavy and the light chains and the J chain are synthesized and assembled by plasma cells, while SC, which corresponds to the five extracellular domains of the poly-Ig receptor, is contributed to by epithelial cells of mucus membranes and exocrine glands.

5.1.1 Polypeptide Composition of Secretory (s) IgA

Mammalian Ig α heavy chain consists of three constant-region domains (Cα1–Cα3), composed of seven β strands in a four–three configuration with intervening loops to create a β-barrel conformation (HUNKAPILLER and HOOD 1989). In humans, non-human primates and rabbits, multiple α-chain isotypes have been described (SPIEKER-POLET et al. 1993), in contrast to all other species, where only one IgA isotype is found. In human, two subclasses of IgA have been described: IgA1 and IgA2. The major difference between these two subclasses lies in the hinge region, lacking a 13-amino acid stretch in the IgA2 molecule. In addition, the IgA2

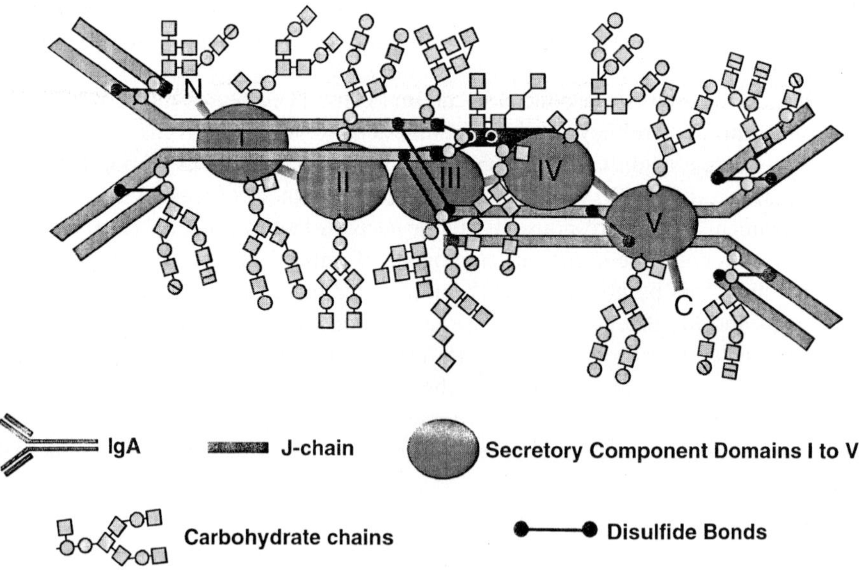

Fig. 3. Schematic representation of a dimeric secretory immunoglobulin A (sIgA). Two IgA monomers are depicted in a tail-to-tail arrangement, with J chain covalently linked to one monomer through two disulfide bridges. Secretory component (SC) is made of five Ig-like domains and corresponds to the extracellular portion of polymeric immunoglobulin receptor (pIgR). Domain I contains the information for initial anchoring to oligomeric IgA, while two cysteine residues situated in the IgA α chain and the domain 5 of SC form a covalent bond in several species. Sugar moieties most likely contributing to the stability and solubility of sIgA are also drawn, yet not on scale

subclass comprises two allotypes: IgA2 m(1) and IgA2 m(2), the major structural differences of which lie in the arrangement of the α and light chain interchain disulfide bridges. The Cα3 domain displays, at its C terminus, an 18-amino acid tail that, during biosynthesis of IgA, associates with J chain.

J chain is a 15-kDa glycoprotein, covalently linked to polymeric IgA or IgM before secretion by plasma cells. The role of J chain in the polymerization process remains controversial (CATTANEO and NEUBERGER 1987; DAVIS et al. 1989). IgA dimer formation occurs in J chain-deficient mice, but the IgA dimer/monomer ratio in serum is perturbed (HENDRICKSON et al. 1995): hepatic IgA transport is impaired, whereas intestinal, breast milk and nasal IgA levels compare with wild-type animals (HENDRICKSON et al. 1996). J chain-deficient IgA found in secretions, however, is not associated with SC. These data confirm the essential role of J chain in interaction with the poly-Ig receptor (BRANDTZAEG and PRYDZ 1984).

SC represents the extracellular portion of the poly-Ig receptor. The tridimensional structure of SC has not been elucidated, but molecular modeling of rabbit domains 1 and 2 indicates that both domains consist of the typical nine Igβ strands (A, B, C, C′, C, D, E, F, G) connected by loops of variable size (COYNE et al. 1994; CORTHÉSY et al. 1996). Each domain contains an internal disulfide bond, characteristic of Ig homology units, which apparently links strand B with strand F. In human sIgA, amino acids 14–38 in the *N*-terminus domain 1 of SC constitute the IgA binding epitope (BAKOS et al. 1991). Cysteine 467 in the distal fifth domain participates in disulfide bridge formation with cysteine 311 of one Cα2 domain of dimeric IgA. It is not known in which cell organelle covalent disulfide linkages are established during transcytosis, but the process is rapid and spontaneous, as reflected in *in vitro* reconstitution experiments (RINDISBACHER et al. 1995). SC delays IgA degradation by intestinal proteases, but does not alter the affinity for the antigen (LÜLLAU et al. 1996).

5.1.2 Glycosylation of sIgA

The glycosylation pattern of sIgA is complex and highly heterogeneous (ENDO et al. 1994; WOLD et al. 1994). The 13-amino acid stretch in the IgA1 hinge region contains five *O*-linked glycans, recognized by the lectin Jacalin (ROQUE-BARREIRA and CAMPOS-NETO 1985), which presumably enhances the flexibility between Fc and $F(ab')_2$ fragments. The presence of this extra sequence in IgA1 molecules is responsible for the sensitivity of the antibody to bacterial specific proteases as discussed above. Glycosylation of the α chain is essential for intracellular stability and normal secretion of IgA (TAYLOR and WALL 1988). In contrast, absence of glycosylation does not lead to loss of antigen binding activity (DONADEL et al. 1994). Human SC purified from milk is heavily glycosylated with 5–7 *N*-linked sugar side chains, accounting for over 20% of its molecular mass (MIZOGUCHI et al. 1982). IL-4 and IL-5 have recently been shown to alter the terminal glycosylation pattern of IgA (CHINTALACHARUVU and EMANCIPATOR 1997). These authors suggest that increased production of IL-4 and IL-5 by peripheral blood lymphocytes from IgA nephropathy patients might result in the production of abnormally glycosylated IgA which, in turn, may promote the disease by deposition of IgA in glomeruli.

5.1.3 Regulation of SC and α Chain Gene Expression

Poly-Ig receptor expression in mucosal epithelia is regulated by a complex interplay among lymphocytes, macrophages and epithelial cells. The presence of cytokine regulatory elements in the promotor of the poly-Ig receptor genes (Piskurich et al. 1997) allows expression of the receptor to be upregulated in response to both Th1 and Th2 cytokines. Upregulation of poly-Ig receptor expression stimulates transcytosis of IgA antibodies. Several interferon-stimulated response elements (ISRE) have been identified in the human poly-Ig receptor promoter and first exon (Piskurich et al. 1997; Verrijdt et al. 1997). In HT-29 human colon carcinoma cells, poly-Ig receptor is upregulated by interferon gamma (IFN-γ) (Piskurich et al. 1993), TNF-α (Kvale et al. 1988), and IL-4 (Phillips et al. 1990). In contrast to the promoter ISREs, the ISRE in the first exon binds interferon responsive factor (IRF)-1, a member of the interferon regulatory-factor family, only after stimulation of HT-29 cells with IFN-γ. Many microorganisms are able to stimulate cytokine or chemokine secretion directly by epithelial cells or indirectly via intraepithelial lymphocytes or inflammatory cells recruited into mucosal tissues (Ishikawa et al. 1993; Quiding et al. 1991). Upregulation of poly-Ig receptor by cytokines involves a protein tyrosine kinase-dependent signaling pathway (Denning 1996), in which the cytokine receptors recruit cytosolic tyrosine kinases (Janus kinases or JAKs) which, in turn, activate transcription factors of the STAT (signal transduction and activation of transcription) family (Ihle et al. 1994).

Several enhancer regions (Dariavach et al. 1991; Matthias and Baltimore 1993) and hypersensitive sites (Madisen and Groudine 1994) identified downstream of the murine Cα gene are involved in its developmental and transcriptional regulation. The best characterized enhancer element, 3′αE, contains motifs (octamer, κB, G-rich sequence) that mediate upregulation of transcription during maturation to plasma cells (Singh and Birshtein 1993; Meyer et al. 1995) as well as motifs (Elf-1, AP-1) for activation induced by B cell receptor cross-linking (Grant et al. 1995). 3′αE also comprises the hypersensitive sites HS1 and HS2, which share 90% identity with their human α1 and α2 gene counterparts (Mills et al. 1997). The HS1,2 core homology is likely to contain essential motifs important for the strong late B cell-specific enhancer characteristics of HS1,2 in mice and humans. Although the function of transcription binding sites in the human HS1,2 has not yet been demonstrated experimentally, this sequence element comprises Oct, AP-1, Ets, μE5 motifs (Mills et al. 1997), all of which are nearly identical to, and functional in, the murine HS1,2 enhancer.

5.2 Functions of sIgA Antibodies

It is usually accepted that sIgA in secretions bathing the mucosal surfaces protect these surfaces against environmental pathogens by cross-linking microorganisms or macromolecules, thus facilitating their elimination by persistalsis or mucociliary movement and preventing their contact with the surface of epithelial cells, a phenomenon called immune exclusion. The molecular mechanisms underlying sIgA antibody-mediated protection still remain poorly understood, and it is likely that this unique effector molecule plays multiple roles in addition to immune exclusion (Fig. 2).

The ability of secretory antibodies to recognize intact bacteria, viruses or parasites at mucosal surfaces is a prerequisite for protection. This is reflected by the observation that polyclonal sIgA responses to mucosal pathogens is dominated by antibodies recognizing microbial surface antigens. An antibody response is not restricted to pathogens, but commensals such as gram-negative microorganisms introduced into the gut of germ-free mice also produce an immune response (Shroff et al. 1995). This suggests that a successful sIgA response can attenuate chronic stimulation of germinal-center reactions in gut-associated lymphoid tissue in response to bacteria persisting in the gut. The role of sIgA in the establishment and maintenance of the gut flora will require further investigation.

It has been difficult to evaluate the relative importance of sIgA in protection or to assess whether "immune exclusion" by sIgA alone is able to prevent mucosal infection. Techniques for the production of monomeric IgA antibodies have been developed by several groups during the last few decades (Styles et al. 1984; Colwell et al. 1986; Rits et al. 1986; Mazanec et al. 1987; Weltzin et al. 1989), and production of recombinant sIgA has been achieved in mammalian cells (Chintalacharuvu and Morrison 1997; Berdoz et al. 1998) and in plants (Ma et al. 1995). This has allowed assessment of protection in vivo either by passive transfer of monoclonal antibodies specific for viruses (Mazanec et al. 1987: Renegar and Small 1991), bacteria (Michetti et al. 1992) or toxins (Apter et al. 1991; Apter et al. 1993b). The implantation of IgA-producing hybridoma cells into the backs of mice has also allowed efficient delivery of antibodies into mucosal sites (Winner et al. 1991). The development of methods (Czerkinsky et al. 1983; Haneberg et al. 1994) to measure antibodies in mucosal tissues has also contributed to a better understanding of their role in protection.

5.2.1 Immune Exclusion by Antigen Cross-Linking

Resistance to mucosal infection has been correlated with specific sIgA. The secretory antibodies provide an immunological barrier that prevents foreign antigens, including bacteria, viruses, parasites and toxins, from attaching to mucosal surfaces (for review see Neutra et al. 1991). Due to their multivalency, sIgA antibodies are ideally designed to cross-link target macromolecules or microorganisms in the mucosal environment. They do so by preventing their diffusion through the glycocalyx at the surface of the epithelial cells, by blocking their binding to epithelial surface receptors, by inhibiting their motility or by facilitating their entrapment in mucus. Prevention of viral attachment and subsequent internalization represents a major mechanism by which IgA-coated viruses are neutralized (Taylor and Dimmock 1985). In the absence of mucus and clearance mechanisms, sIgA antibodies are able to protect epithelial cell monolayers (Michetti et al. 1994). Aggregation of the pathogens, however, requires antibody concentrations which probably cannot be reached in the mucosal environment. This suggests that protection mediated by sIgA must involve other mechanisms.

5.2.2 Interaction with Mucus

Mucus is generally thought to protect epithelial cells by forming a diffusional barrier with pore size of about 100 nm, through which only small molecules including antibodies can pass. IgG, IgG fragments, IgA and IgM diffuse as rapidly in cervical mucus as in water, and it has been proposed that particles as large as viruses can similarly diffuse through human cervical mucus, provided that the particle forms no adhesive interactions with mucus glycoproteins (Saltzman et al. 1994). Cross-linking of viral particles by sIgA antibodies, however, significantly impedes their diffusion in cervical mucus. Whether mucus from other mucosal surfaces behave similarly remains to be established. Coating hydrophobic bacteria such as *Salmonella typhimuirum* or *Escherichia coli* with sIgA renders their surface more hydrophilic, thus facilitating their entry and movement into mucus gels (Magnusson and Stjernstroem 1982; McCormick et al. 1988). Thus, microorganisms coated with sIgA antibodies would be more readily retained in the moving stream of mucus. Secretory antibodies would bind to mucus only when tightly packed on the surface of the microorganisms which increases the avidity of the low affinity IgA–mucin interaction.

5.2.3 Neutralization and Opsonization

Most IgA antibodies do not neutralize microorganisms by standard mechanisms used by IgG. Generally, they do not opsonize via the macrophage Fcγ receptor in vitro, they do not activate the classical complement pathway and, thus, they do not lyse bacteria (Kilian et al. 1988). There is evidence that sIgA can directly block the microbial sites that mediate epithelial attachment, either by binding to their adhesins or by sterically hindering their interaction with the epithelial cell surface (Williams and Gibbons 1972; Svanborg-Eden and Svennerholm 1978; Weltzin et al. 1996). IgA antibodies which protect mucosal surfaces were shown to be inefficient when injected systemically (Michetti et al. 1992), suggesting that neutralization in a mucosal environment operates through different mechanisms than systemically (Neutra et al. 1994).

Several studies have demonstrated that protection of epithelial surfaces against viruses, bacteria or toxins can be accomplished with monoclonal IgA antibodies directed against microbial surface epitopes not necessarily involved in pathogenesis or adhesion (Apter et al. 1993a; Phalipon et al. 1995).

Secretory IgA antibodies are able to cooperate with the innate defense system and enhance, for instance, the bacteriostatic activity of lactoperoxidase or lactoferrin present in mucosal fluids (TENOVUO et al. 1982). Inflammatory cells (neutrophils, eosinophils and macrophages) and immune cells (lymphocytes and monocytes) express Fcα receptors in mucosal tissues (FANGER et al. 1983; MONTEIRO et al. 1990). Occupancy and aggregation of the Fcα receptor on monocytes release pro-inflammatory cytokines (PATRY et al. 1995). The role of IgA-dependent cellular cytotoxicity which has been reported for the clearance of *Salmonella typhi* in mice (TAGLIABUE et al. 1985) has to be further investigated. It is likely that cell-mediated phenomena play an important role in protection of the host against mucosal and systemic infection once the pathogen has crossed the epithelial barrier and entered mucosal tissues.

5.2.4 Intracellular Neutralization and Antigen Clearance

Microorganisms that are internalized by epithelial cells in mucosal tissues and accumulate in endosomes could be neutralized intracellularly in endosomal compartments provided that there is a meeting point between virus-containing endosomes and IgA-containing endosomal vesicles. This has been tested experimentally: IgA monoclonal antibodies against Sendai virus, a parainfluenza virus, co-localized with the viral hemagglutinin-neuraminidase protein within infected epithelial cells and reduced intracellular viral titers (MAZANEC et al. 1995).

A mouse model and "backpack tumor" transplantation (WINNER et al. 1991) were recently used to determine the protective effect of antibodies against rotavirus capsid proteins. Two non-neutralizing IgA antibodies to VP6 were capable of preventing virus shedding in infected mice (BURNS et al. 1996). These antibodies were not active when administered to the luminal side of the intestinal tract. The lack of virus shedding in the stools was taken as an indication that virus infection could be prevented inside the infected cells. However, the decrease of virus shedding could result from inhibition of virus assembly. It is difficult to imagine how VP6 specific antibodies that do not have access to surface-exposed epitopes on the virus could prevent infection. In addition, rotavirus infects villus epithelial cells, particularly those on the upper villus and IgA antibodies are transported mainly in crypt cells. Thus, it is not clear how IgA antibodies can neutralize the virus intracellularly. When prevention of diarrhea was assessed, rather than virus shedding, anti-VP6 antibodies were unable to prevent diarrhea in newborn mice (RUGGERI et al. 1998).

Soluble dimeric IgA–antigen immune complexes are transported from the basolateral to the apical surface of poly-Ig receptor-expressing epithelial monolayers and released into the apical compartment (KAETZEL et al. 1991; KAETZEL et al. 1994). This indicates that the pathways of epithelial transcytosis of free and complexed dimeric IgA are the same. Given the high population density of mucosal IgA plasma cells and the enormous surface area of poly-Ig receptor-expressing mucosal epithelium, it is likely that significant local transcytosis of IgA immune complexes occurs in vivo, providing a means to eliminate microorganisms that have gained access to mucosal tissues. Such a mechanism may thus provide an important defense function for IgA.

Acknowledgements. We are grateful to the current and former members of our laboratories who have contributed to the work summarized in this review. We would like to thank Marian R. Neutra for stimulating discussion, and for critically reading the manuscript. The authors are supported by the Swiss National Science Foundation (Grants 31–47110–96 and 31–47296.96) and the Swiss Research Against Cancer (AKT 622).

References

Apodaca G, Katz LA, Mostov KE (1994) Receptor-mediated transcytosis of IgA in MDCK cells is via apical recycling endosomes. J Cell Biol 125:67–86

Apter FM, Lencer WJ, Mekalanos JJ, Neutra MR (1991) Analysis of epithelial protection by monoclonal IgA antibodies directed against cholera toxin B subunits in vivo and in vitro. J Cell Biol 115:399a

Apter FM, Lencer WI, Finkelstein RA, Mekalanos JJ, Neutra MR (1993a) Monoclonal immunoglobulin A antibodies directed against cholera toxin prevent the toxin-induced chloride secretory response and block toxin binding to intestinal epithelial cells in vitro. Infect Immun 61:5271–5278

Apter FM, Michetti P, Winner LSI, Mack JA, Mekalanos JJ, Neutra MR (1993b) Analysis of the roles of antilipopolysaccharide and anti-cholera toxin immunoglobulin A (IgA) antibodies in protection against Vibrio cholerae and cholera toxin by use of monoclonal IgA antibodies in vivo. Infect Immun 61:5279–5285

Bakos MA, Kurosky A, and Goldblum RM (1991) Characterization of a critical binding site for human polymeric Ig on secretory component. J Immunol 147:3419–3426

Banting G, Brake B, Braghetta P, Luzio JP, Stanley KK (1989) Intracellular targeting signals of polymeric immunoglobulin receptors are highly conserved between species. FEBS Lett 254:177–183

Barroso M, Sztul ES (1994) Basolateral to apical transcytosis in polarized cells is indirect and involved BFA and trimeric G protein sensitive passage through the apical endosome. J Cell Biol 124:83–100

Bendelac A, Fearon DT (1997) Innate immunity – innate pathways that control acquired immunity. Curr Opin Immunol 9:1–3

Bjerke K, Brandtzaeg P (1990) Terminally differentiated human intestinal B cells. J chain expression of IgA and IgG subclass-producing immunocytes in the distal ileum compared with mesenteric and peripheral lymph nodes. Clin Exp Immunol 82:411–415

Brandtzaeg P (1994) Distribution and characterization of mucosal immunoglobulin-producing cells. In: PL Ogra, J Mestecky, ME Lamm, W Strober, JR McGhee, J Bienenstock (eds) Handbook of mucosal immunity, Academic Press, New York, pp 251–259

Brandtzaeg P, Prydz H (1984) Direct evidence for an integrated function of J chain and secretory component in epithelial transport of immunoglobulins. Nature 311:71–73

Briskin M, Winsorhines D, Shyjan A, Cochran N, Bloom S, Wilson J, McEvoy LM, Butcher EC, Kassam N, Mackay CR, Newman W, Ringler DJ (1997) Human mucosal addressin cell adhesion molecule-1 is preferentially expressed in intestinal tract and associated lymphoid tissue. Am J Pathol 151:97–110

Bromander AK, Ekman L, Kopf M, Nedrud JG, Lycke NY (1996) IL-6-deficient mice exhibit normal mucosal IgA responses to local immunizations and Helicobacter felis infection. J Immunol 156:429–297

Burns JW, Siadatpajouh M, Krishnaney AA, Greenberg HB (1996) Protective effect of rotavirus VP6-specific IgA monoclonal antibodies that lack neutralizing activity. Science 272:104–107

Butcher EC, Picker LJ (1996) Lymphocyte homing and homeostasis. Science 272:60–66

Cattaneo A, Neuberger MS (1987) Polymeric immunoglobulin M is secreted by transfectants of non-lymphoid cells in the absence of immunoglobulin J chain. EMBO J 6:27532–758

Cebra JJ, Logan AC, Weinstein PD (1991) The preference for switching to expression of the IgA isotype of antibody exhibited by B lymphocytes in Peyer's patches is likely due to intrinsic properties of their microenvironment. Immunol Res 10:393–395

Chapin SJ, Enrich C, Aroeti B, Havel RJ, Mostov KE (1996) Calmodulin binds to the basolateral targeting signal of the polymeric immunoglobulin receptor. J Biol Chem 271:133–342

Chintalacharuvu KR, Morrison SL (1997) Production of secretory immunoglobulin A by a single mammalian cell. Proc Natl Acad Sci U S A 94:636–368

Chintalacharuvu SR, Emancipator SN (1997) The glycosylation of IgA produced by murine B cells is altered by Th2 cytokines. J Immunol 159:232–333

Colwell DE, Michalek SM, McGhee JR (1986) Method for generating a high frequency of hybridomas producing monoclonal IgA antibodies. Methods Enzymol 121:42–51

Corthésy B, Kaufmann M, Phalipon A, Peitsch MC, Neutra MR, Kraehenbuhl J-P (1996) A pathogen-specific epitope inserted into recombinant secretory immunoglobulin A is immunogenic by the oral route. J Biol Chem 271:33670–33677

Coyne RS, Siebrecht M, Peitsch MC, Casanova JE (1994) Mutational analysis of polymeric immunoglobulin receptor/ligand interactions. J Biol Chem 269:31620–31625

Czerkinsky C, Nilsson LA, Nygren H, Ouchterlony O, Tarkowski A (1983) A solid-phase enzyme-linked immunospot (ELISPOT) assay for enumeration of specific antibody-secreting cells. J Immunol Methods 65:109–121

Dariavach P, Williams GT, Campbell K, Pettersson S, Neuberger MS (1991) The mouse IgH 3′-enhancer. Eur J Immunol 21:149–504

Davis AC, Roux KH, Pursey J, Shulman MJ (1989) Intermolecular disulfide bonding in IgM: effects of replacing cysteine residues in the mu heavy chain. EMBO J 8:251–526

Defrance T, Vanbervliet B, Briere F, Durand I, Rousset F, Banchereau J (1992) Interleukin 10 and transforming growth factor beta cooperate to induce anti-CD40-activated naive human B cells to secrete immunoglobulin A. J Exp Med 175:67–82

Deitcher DL, Mostov KE (1986) Alternate splicing of rabbit polymeric immunoglobulin receptor. Mol Cell Biol 6:2712–2715

Denning GM (1996) IL-4 and IFN-gamma synergistically increase total polymeric IgA receptor levels in human intestinal epithelial cells. Role of protein tyrosine kinases J Immunol 156:480–814

Donadel G, Calabro A, Sigounas G, Hascall VC, Notkins AL, Harindranath N (1994) Human polyreactive and monoreactive antibodies: effect of glycosylation on antigen binding. Glycobiology 4:49–96

Eiffert H, Quentin E, Decker J, Hillemeir S, Hufschmidt M, Klingmüller D, Weber MH, Hilschmann N (1984) Die Primärstruktur der menschlichen Sekretkomponente und die Anordnung der Disulfidbrücken. Hoppe-Seyler's Z Physiol Chem 365:1489–1495

Endo T, Mestecky J, Kulhavy R, Kobata A (1994) Carbohydrate heterogeneity of human myeloma proteins of the IgA1 and IgA2 subclasses. Mol Immunol 31:141–422

Fanger MW, Goldstine SN, Shen L (1983) The properties and role of receptors for IgA on human leukocytes. Ann N Y Acad Sci 409:552–563

Fearon DT, Locksley RM (1996) Elements of immunity – the instructive role of innate immunity in the acquired immune response. Science 272:50–54

Förster R, Mattis AE, Kremmer E, Wolf E, Brem G, Lipp M (1996) A putative chemokine receptor, BLR1, directs B cell migration to defined lymphoid organs and specific anatomic compartments of the spleen. Cell 87:1037–1047

Frutiger S, Hughes GJ, Fonck C, Jaton JC (1987) High and low molecular weight rabbit secretory components. Evidence for the deletion of the second and third domains in the smaller polypeptide. J Biol Chem 262:1712–1715

Grant PA, Thompson CB, Pettersson S (1995) IgM receptor-mediated transactivation of the IgH 3′ enhancer couples a novel Elf-1-AP-1 protein complex to the developmental control of enhancer function. EMBO J 14:4501–4513

Gunn MD, Ngo VN, Ansel KM, Ekland EH, Cyster JG, Williams LT (1998) A B-cell-homing chemokine made in lymphoid follicles activates Burkitts-lymphoma receptor-1. Nature 391:799–803

Haneberg B, Kendall D, Amerongen HM, Apter FM, Kraehenbuhl J-P, Neutra MR (1994) Induction of specific immunoglobulin A in the small intestine, colon-rectum, and vagina measured by a new method for collection of secretions from local mucosal surfaces. Infect Immun 62:15–23

Hayami K, Fukuta D, Nishikawa Y, Yamashita Y, Inui M, Ohyama Y, Hikida M, Ohmori H, Takai T (1997) Molecular cloning of a novel murine cell-surface glycoprotein homologous to killer cell inhibitory receptors. J Biol Chem 272:7320–7327

Hendrickson BA, Conner DA, Ladd DJ, Kendall D, Casanova JE, Corthésy B, Max EE, Neutra MR, Seidman CE, Seidman JG (1995) Altered hepatic transport of immunoglobulin A in mice lacking the J chain. J Exp Med 182:1905–1911

Hendrickson BA, Rindisbacher L, Corthésy B, Kendall D, Waltz DA, Neutra MR, Seidman JG (1996) Lack of association of secretory component with IgA in J chain-deficient mice. J Immunol 157:750–754

Hirt RP, Hughes GJ, Frutiger S, Michetti P, Perregaux C, Jeanguenat N, Neutra MR, Kraehenbuhl J-P (1993) Transcytosis of the polymeric Ig receptor requires phosphorylation of serine 664 in the absence but not the presence of dimeric IgA. Cell 74:245–255

Hunkapiller T, Hood L (1989) Diversity of the immunoglobulin gene superfamily. Adv Immunol 44:1–63

Ihle JN, Witthuhn BA, Quelle FW, Yamamoto K, Thierfelder WE, Kreider B, Silvennoinen O (1994) Signaling by the cytokine receptor superfamily: JAKs and STATs. Trends Biochem Sci 19:222–227

Ishikawa H, Li Y, Abeliovich A, Yamamoto S, Kaufmann SH, Tonegawa S (1993) Cytotoxic and interferon gamma-producing activities of gamma delta T cells in the mouse intestinal epithelium are strain dependent. Proc Natl Acad Sci U S A 90:820–208

Israel EJ, Taylor S, Wu Z, Mizoguchi E, Blumberg RS, Bhan A, Simister NE (1997) Expression of the neonatal Fc receptor, FcRn, on human intestinal epithelial cells. Immunology 92:69–74

Kaetzel CS, Robinson JK, Chintalacharuvu KR, Vaerman J-P, Lamm M (1991) The polymeric immunoglobulin receptor (secretory component) mediates transport of immune complexes across epithelial cells: a local defense function for IgA. Proc Natl Acad Sci U S A 88:8796–8800

Kaetzel CS, Robinson JK, Lamm ME (1994) Epithelial transcytosis of monomeric IgA and IgG cross-linked through antigen to polymeric IgA. A role for monomeric antibodies in the mucosal immune system. J Immunol 152:72–76

Kawamura S, Saitou N, Ueda S (1992) Concerted evolution of the primate immunoglobulin alpha-gene through gene conversion. J Biol Chem 267:735–367

Kawanishi H, Saltzman LE, Strober W (1983) Mechanisms regulating IgA class-specific immunoglobulin production in murine gut-associated lymphoid tissues. I. T cells derived from Peyer's patches that switch. J Exp Med 157:433–450

Kerr MA (1990) The structure and function of human IgA. Biochem J 271:285–296

Kilian M, Mestecky J, Russell MW (1988) Defense mechanisms involving Fc-dependent functions of immunoglobulin A and their subversion by bacterial immunoglobulin A proteases. Microbiol Rev 52:296–303

Krajci P, Solberg R, Sandberg M, Oyen O, Jahnsen T, Brandtzaeg P (1989) Molecular cloning of the human transmembrane secretory component (poly-Ig receptor) and its mRNA expression in human tissues. Biochem Biophys Res Com 158:783–789

Kvale D, Lovhaug D, Sollid LM, Brandtzaeg P (1988) Tumor necrosis factor-alpha upregulates expression of secretory component, the epithelial receptor for polymeric Ig. J Immunol 140:3086–3089

Legler DF, Loetscher M, Stuber Roos R, Clark-Lewis I, Baggiolini M, Moser B (1998) B cell-attracting chemokine 1,Ka human CXC chemokine expressed in lymphoid tissues, selectively attracts B lymphocytes via BRL1/CxCR5. J Exp Med 187:665–660

Li MJ, Peakman MC, Golub EI, Reddy G, Ward DC, Radding CM, Maizels N (1996) Rad51 expression and localization in B cells carrying out class switch recombination. Proc Natl Acad Sci USA 93:1022–0227

Liu YJ, Mason DY, Johnson GD, Abbot SD, Gregory CD, Hardie DL, Gordon J, MacLennan IC (1991) Germinal center cells express bcl-2 protein after activation by signals which prevent their entry into apoptosis. Eur J Immunol 21:1905–1910

Liu YJ, Johnson GD, Gordon J, MacLennan ICM (1992) Germinal centres in T-cell-dependent antibody responses. Immunol Today 13:17–21

Lüllau E, Heyse S, Vogel H, Marison I, von Stockar U, Kraehenbuhl J-P, Corthésy B (1996) Antigen binding properties of purified IgA and reconstituted secretory IgA antibodies. J Biol Chem 271:16300–16309

Lycke N, Strober W (1989) Cholera toxin promotes B cell isotype differentiation. J Immunol 142:3781–3787

Ma JKC, Hiatt A, Hein M, Vine ND, Wang F, Stabila P, Vandolleweerd C, Mostov K, Lehner T (1995) Generation and assembly of secretory antibodies in plants. Science 268:716–719

Madisen L, Groudine M (1994) Identification of a locus control region in the immunoglobulin heavy-chain locus that deregulates c-myc expression in plasmacytoma and Burkitt's lymphoma cells. Genes Dev 8:2212–2226

Magnusson KE, Stjernstroem I (1982) Mucosal barrier mechanisms. Interplay between secretory IgA (sIgA), IgG and mucins on the surface properties and association of salmonellae with intestine and granulocytes. Immunology 45:239–248

Matsuda H, Watanabe N, Kiso Y, Hirota S, Ushio H, Kannan Y, Azuma M, Koyama H, Kitamura Y (1990) Necessity of IgE antibodies and mast cells for manifestation of resistance against larval Haemaphysalis longicornis ticks in mice. J Immunol 144:25–62

Matsumoto R, Matsumoto M, Mita S, Hitoshi Y, Adno M, Araki S, Yamaguchi N, Tominaga A, Takatsu K (1989) Interleukin-5 induces maturation but not class switching of surface IgA-positive B cells into IgA-secreting cells. Immunology 66:32–41

Matthias P, Baltimore D (1993) The immunoglobulin heavy chain locus contains another B-cell-specific 3′ enhancer close to the alpha constant region. Mol Cell Biol 13:154–553

Mazanec MB, Nedrud JG, Lamm ME (1987) Immunoglobulin A monoclonal antibodies protect against Sendai virus IgA nephropathy: pathogenesis of the most common form of glomerulonephritis. J Virol 61:2624–2626

Mazanec MB, Kaetzel CS, Lamm ME, Fletcher D, Peterra J, Nedrud JG (1995) Intracellular neutralization of Sendai and influenza viruses by IgA monoclonal antibodies. Adv Exp Med Biol 371A:651–654

McCormick BA, Stocker BA, Laux DC, Cohen PS (1988) Roles of motility, chemotaxis, and penetration through and growth in intestinal mucus in the ability of an avirulent strain of Salmonella typhimurium to colonize the large intestine of streptomycin-treated mice. Infect Immun 56:220–217

McGhee JR, Fujihashi K, Beagley KW, Kiyono H (1991) Role of interleukin-6 in human and mouse mucosal IgA plasma cell responses. Immunol Res 10:418–422

Mestecky J, McGhee JR (1987) Immunoglobulin A (IgA): molecular and cellular interactions involved in IgA biosynthesis and immune response. Adv Immunol 40:153–245

Mestecky J, Lue C, Russell MW (1991) Selective transport of IgA: cellular and molecular aspects. Gastroenterol Clin North Am 20:441–471

Meyer KB, Skogberg M, Margenfeld C, Ireland J, Pettersson S (1995) Repression of the immunoglobulin heavy chain 3′ enhancer by helix-loop-helix protein Id3 via a functionally important E47/E12 binding site: implications for developmental control of enhancer function. Eur J Immunol 25:1770–1777

Michetti P, Mahan MJ, Slauch JM, Mekalanos JJ, Neutra MR (1992) Monoclonal secretory immunoglobulin A protects mice against oral challenge with the invasive pathogen Salmonella typhimurium. Infect Immun 60:1786–1792

Michetti P, Porta N, Mahan MJ, Slauch JM, Mekalanos JJ, Blum AL, Kraehenbuhl J-P, Neutra MR (1994) Monoclonal immunoglobulin A prevents invasion of polarized epithelial cell monolayers by Salmonella typhimurium. Gastroenterology 107:915–923

Mills FC, Harindranath N, Mitchell M, Max EE (1997) Enhancer complexes located downstream of both human immunoglobulin C alpha genes. J Exp Med 186:84–58

Mizoguchi A, Mizuochi T, Kobata A (1982) Structures of the carbohydrate moieties of secretory component purified from human milk. J Biol Chem 257:961–621

Monteiro RC, Kubagawa H, Cooper MD (1990) Cellular distribution, regulation, and biochemical nature of an Fc alpha receptor in humans. J Exp Med 171:597–613

Mostov KE, Deitcher DL (1986) Polymeric immunoglobulin receptor expressed in MDCK cells transcytoses IgA. Cell 46:613–621

Mostov KE, Friedlander M, Blobel G (1984) The receptor for transepithelial transport of IgA and IgM contains multiple immunoglobulin-like domains. Nature 308:37–43

Mostov KE, Apodaca G, Aroeti B, Okamoto C (1992) Plasma membrane protein sorting in polarized epithelial cells. J Cell Biol 116:577–583

Natvig IB, Johansen FE, Nardeng TW, Haranldsen G, Brandtzaeg P (1997) Mechanism for enhanced external transfer of dimeric IgA over pentameric IgM. J Immunol 159:4330–4340

Nedrud JG, Blanchard T, Czinn S, Harriman GR (1996) Orally immunized IgA deficient mice are protected against H. felis infection. Gut 39 [Suppl 2]:A-45

Neutra MR, Weltzin R, Winner L, Mack J, Michetti P, Morrison L, Fields BN, Mekalanos JJ, Kraehenbuhl J-P (1991) Identification and use of protective monoclonal IgA antibodies against viral and bacterial pathogens. Adv Exp Med Biol 310:179–182

Neutra MR, Michetti P, Kraehenbuhl J-P (1994) Secretory immunoglobulin A : induction, biogenesis and function. In: L R Johnson (ed) Physiology of the gastrointestinal tract. Raven, New York, pp 685–709

Patry C, Herbelin A, Lehuen A, Bach JF, Monteiro RC (1995) Fc alpha receptors mediate release of tumour necrosis factor-alpha and interleukin-6 by human monocytes following receptor aggregation. Immunology 86:1–5

Phalipon A, Kaufmann M, Michetti P, Cavaillon JM, Huerre M, Sansonetti PJ, Kraehenbuhl J-P (1995) Monoclonal IgA antibody directed against serotype-specific epitope of Shigella flexneri liposaccharide protects against murine experimental shigellosis. J Exp Med 182:769–778

Phillips JO, Everson MP, Moldoveanu Z, Lue C, Mestecky J (1990) Synergistic effect of IL-4 and IFN-gamma on the expression of polymeric Ig receptor (secretory component) and IgA binding by human epithelial cells. J Immunol 145:174–744

Piskurich JF, France JA, Tamer CM, Willmer CA, Kaetzel CS, Kaetzel DM (1993) Interferon-gamma induces polymeric immunoglobulin receptor mRNA in human intestinal epithelial cells by a protein synthesis dependent mechanism. Mol Immunol 30:413–421

Piskurich JF, Yougman KR, Phillips KM, Hempen PM, Blanchard MH, France JA, Kaetzel CS (1997) Transcriptional regulation of the human polymeric immunoglobin receptor gene by interferon-gamma. Mol Immunol 34:75–91

Plaut AG (1988) Production and isolation of neissereal IgA proteases. Methods Enzymol 165:11–20

Plaut AG, Gilbert JV, Leger G, Blumenstein M (1985) IgA1 protease cleaves heavy chains independently in dimeric human IgA1. Mol Immunol 22:821–826

Plebani A, Mira E, Mevio E, Monafo V, Notarangelo LD, Avanzini A, Ugazio AG (1983) IgM and IgD concentrations in the serum and secretions of children with selective IgA deficiency. Clin Exp Immunol 53:689–696

Quiding M, Nordstrom I, Kilander A, Andersson G, Hanson LA, Holmgren J, Czerkinsky C (1991) Intestinal immune responses in humans. Oral cholera vaccination induces strong intestinal antibody responses and interferon-gamma production and evokes local immunological memory. J Clin Invest 88:143–148

Quiding-Jarbrink M, Nordstrom I, Granstrom G, Kilander A, Jertborn M, Butcher EC, Lazarovits AI, Holmgren J, Czerkinsky C (1997) Differential expression of tissue-specific adhesion molecules on human circulating antibody-forming cells after systemic, enteric, and nasal immunizations. A molecular basis for the compartmentalization of effector B cell responses. J Clin Invest 99:128–286

Ramsay AJ, Husband AJ, Ramshaw IA, Bao S, Matthaei KI, Koehler G, Kopf M (1994) The role of interleukin-6 in mucosal IgA antibody responses in vivo. Science 264:561–563

Renegar KB, Small PA (1991) Passive transfer of local immunity to influenza virus infection by IgA antibody. J Immunol 146:1972–1978

Rindisbacher L, Cottet S, Wittek R, Kraehenbuhl J-P, Corthésy B (1995) Production of human secretory component with dimeric IgA binding capacity using viral expression systems. J Biol Chem 270:14220–14228

Rits M, Cormont F, Bazin H, Maykens R, Vaerman J-P (1986) Rat monoclonal antibodies. VI. Production of IgA secreting hybridomas with specificity for the 2,4-dinitrophenyl (DNP) hapten. J Immunol Methods 89:81–87

Roberts DM, Guenthert M, Rodewald R (1990) Isolation and characterization of the Fc receptor from the fetal yolk sac of the rat. J Cell Biol 111:186–876

Rodewald RD, Kraehenbuhl J-P (1984) Receptor-mediated transport of IgG. J Cell Biol 99:159–164

Roque-Barreira MC, Campos-Neto A (1985) Jacalin : an IgA-binding lectin. J Immunol 134:1740–1743

Rott LS, Rose JR, Bass D, Williams MB, Greenberg HB, Butcher EC (1997) Expression of mucosal homing receptor alpha-4-beta-7 by circulating Cd4(+) cells with memory for intestinal rotavirus. J Clin Invest 100:1204–1208

Ruggeri FM, Johansen K, Basile G, Kraehenbuhl J-P, Svensson L (1998) Rotavirus IgA neutralize virus in vitro after transcytosis through epithelial cells and protect infant mice from diarrhea. J Virol (in press)

Sallusto F, Mackay CR, Lanzavecchia A (1997) Selective expression of the eotaxin receptor CCR3 by human T helper 2 cells. Science 277:2005–2007

Saltzman WM, Radomsky ML, Whaley KJ, Cone RA (1994) Antibody diffusion in human cervical mucus. Biophysical J 66:508–515

Schaerer E, Neutra MR, Kraehenbuhl J-P (1991) Molecular and cellular mechanisms involved in transepithelial transport. J Membr Biol 123:93–103

Schiff JM, Fisher MM, Jones AL, Underdown BJ (1986) Human IgA as a heterovalent ligand : switching from the asialoglycoprotein receptor to secretory component during transport across the rat hepatocyte. J Cell Biol 102:920–931

Shroff KE, Meslin K, Cebra JJ (1995) Commensal enteric bacteria engender a self-limiting humoral mucosal immune response while permanently colonizing the gut. Infect Immun 63:3904–3913

Singh M, Birshtein BK (1993) NF-HB (BSAP) is a repressor of the murine immunoglobulin heavy-chain 3′ alpha enhancer at early stages of B-cell differentiation. Mol Cell Biol 13:3611–3622

Solari R, Kraehenbuhl J-P (1984) Biosynthesis of the IgA antibody receptor: a model for the transepithelial sorting of a membrane glycoprotein. Cell 36:61–71

Solari R, Kraehenbuhl J-P (1987) Receptor-mediated transepithelial transport of polymeric immunoglobulins In: MC Neville, CW Daniel (eds) The mammary gland, Plenum, New York, pp 269–298
Song W, Apodaca G, Mostov K (1994) Transcytosis of the polymeric immunoglobulin receptor is regulated in multiple intracellular compartments. J Biol Chem 269:29474–29480
Spieker-Polet H, Yam PC, Knight KL (1993) Differential expression of 13 IgA-heavy chain genes in rabbit lymphoid tissues. J Immunol 150:5457-5465
Strober W, Ehrhardt OR (1994) Regulation of IgA B cell development. In: PL Ogra, J Mestecky, ML Lamm, W Strober, JR McGhee, J Bienenstock (eds) Handbook of mucosal immunology, Academic Press, New York, pp 159–176
Styles JM, Dean CJ, Gyure A, Hobbs SM, Halél JG (1984) The production of hybridomas from the gut-associated lymphoid tissue of tumor-bearing rats. II. Peripheral intestinal lymph as a source of IgA producing cells. Clin Exp Immunol 57:365–370
Svanborg-Eden C, Svennerholm AM (1978) Secretory immunoglobulin A and G antibodies prevent adhesion of Escherichia coli to human urinary tract epithelial cells. Infect Immun 22:790–797
Tagliabue A, Villa L, Boraschi D, Peri G, de Gori V, Nencioni L (1985) Natural anti-bacterial activity against Salmonella typhi by human T4+ lymphocytes armed with IgA antibodies. J Immunol 135:4178–4182
Taylor AK, Wall R (1988) Selective removal of alpha heavy-chain glycosylation sites causes immunoglobulin A degradation and reduced secretion. Mol Cell Biol 8:419–203
Taylor HP, Dimmock NJ (1985) Mechanism of neutralization of influenza virus by secretory IgA is different from that of monomeric IgA or IgG. J Exp Med 161:198–209
Tenovuo J, Moldoveanu Z, Mestecky J, Pruitt KM, Rahemtulla BM (1982) Interaction of specific and innate factors of immunity: IgA enhances the antimicrobial effect of the lactoperoxidase system against Streptococcus mutans. J Immunol 128:726–731
Verrijdt G, Swinnen J, Peeters B, Verhoeven G, Rombauts W, Claessens F (1997) Characterization of the human secretory component gene promoter. Biochim Biophys Acta 1350:147–154
Weimbs T, Low SH, Chapin SJ, Mostov KE (1997) Apical targeting in polarized epithelial cells – there's more afloat than rafts. Trends Cell Biol 7:393–399
Weltzin RA, Lucia Jandris P, Michetti P, Fields BN, Kraehenbuhl J-P, Neutra MR (1989) Binding and transepithelial transport of immunoglobulins by intestinal M cells: demonstration using monoclonal IgA antibodies against enteric viral proteins. J Cell Biol 108:1673–1685
Weltzin R, Traina-Dorge V, Soike K, Zhang JY, Mack P, Soman G, Drabik G, Monath TP (1996) Intranasal monoclonal IgA antibody to respiratory syncytial virus protects rhesus monkeys against upper and lower respiratory tract infection. J Infect Dis 174:256–261
Williams RC, Gibbons RJ (1972) Inhibition of bacterial adherence by secretory immunoglobulin A: a mechanism of antigen disposal. Science 177:697–699
Winner LSI, Mack J, Weltzin RA, Mekalanos JJ, Kraehenbuhl J-P, Neutra MR (1991) New model for analysis of mucosal immunity: intestinal secretion of specific monoclonal immunoglobulin A from hybridoma tumors protects against Vibrio Cholerae infection. Infect Immun 59:977–982
Wold AE, Motas C, Svanborg C, Mestecky J (1994) Lectin receptors on IgA isotypes. Scand J Immunol 39:195–201
Wold AE, Motas C, Svanborg C, Hanson LA, Mestecky J (1995) Characterization of IgA1, IgA2 and secretory IgA carbohydrate chains using plant lectins Adv Exp Med Biol 371A:585–589

Effector and Regulatory Lymphoid Cells and Cytokines in Mucosal Sites

T.T. MacDonald

Department of Paediatric Gastroenterology, St Bartholomews and the Royal London School of Medicine and Dentistry, St Bartholomews Hospital, London EC1A 7BE, UK

1 Introduction

The key feature of the mucosal immune system is the dual ability to respond promptly and effectively to invasive and lumen-dwelling pathogens, while retaining the ability to be unresponsive to non-pathogenic agents such as the autochthonous flora and foods. To specifically ignore some antigens in the gut (i.e. foods) and respond to others (i.e. cholera vibrios) is not an option, since there is no way in which the immune system can predict the nature of a pathogen. The enormous diversity of potential T-cell receptors, therefore, means that all foreign peptides in the gut are likely to be recognised. One of the major goals of modern immunology is to ascertain the determinants which can predict whether recognition results in active immunity or unresponsiveness, and nowhere is this more important than in the gastrointestinal tract.

In the last few years there has been an enormous advance in our understanding of immunity and tolerance at mucosal surfaces and of the molecules that mediate these responses. The literature divides itself into two main areas which, until recently, had little overlap. Animal literature that is largely experimental exists, describing studies in which various antigens with different vectors and with different adjuvants have been given orally, nasally, intravaginally, rectally etc, and the cytokine profiles of mucosal-associated and systemic lymphoid cells has been studied. There are virtually no equivalent studies in humans because of the difficulty of obtaining cells from sites other than the blood, and of justifying such an approach ethically. However, there is much literature, based mostly on human studies, in which the cytokine profiles of mucosal (mostly intestinal lamina propria) cells have been investigated in the context of chronic inflammatory bowel disease (IBD). With the description of a large number of animal models of IBD, there is now an appreciation that the lessons learned from physiological responses to mucosally administered antigens have application to mucosal pathologies.

An additional feature, unique to the gastrointestinal tract, is that there is a dense lymphoid infiltrate in normal individuals, presumably in response to normal gut antigens. Moreover, attempts to elicit mucosal immune responses experimentally take place on a physiological background of low-level inflammation. With all this in mind, in this chapter, I will review the lymphoid cells and cytokines seen in normal patients and animals; the types of cells and immune responses elicited by delivery of specific antigens at mucosal surfaces; the effect of feeding antigens or giving antigens intranasally on systemic immunity (oral tolerance) and the mechanisms involved; and the cells and cytokines thought to be important in mucosal injury and how they can be downregulated for therapeutic benefit. The majority of the examples will be based on the gut rather than the lung or reproductive tract, since this is where most data are available and, wherever possible, I will try to emphasise findings in humans rather than animals.

2 Phenotypic Features of T Lymphocytes at Mucosal Surfaces

2.1 Peyer's Patches (PP)

The Peyer's patches (PP) are the site at which luminal antigens are transported by M cells across the epithelium where they can diffuse into the dome region of the patch. Unsuriprisingly, some CD3+ T cells in the dome are CD45R0+, L-selectin^{lo} and $\alpha 4\beta 7^{hi}$, indicating antigen activation, and some cells with a similar phenotype can be seen in sections of PP efferent lymphatics, probably migrating to the lamina propria via the blood (Farstad et al. 1996; Farstad et al. 1997a). PP are on the major route of small lymphocyte recirculation and therefore, as expected, in the interfollicular zones around the high endothelial venules, there are abundant small naive CD45RA+, L-selectin^{hi} cells (Farstad et al. 1997b). The vast majority of T cells in PP are integrin-β1+ (Farstad et al. 1997b). Normal human PP also contain large reactive follicle centres. All of the features of PP in humans are consistent with the tissue being a secondary lymphoid organ, where there is ongoing responses to enteric antigens.

2.2 Lamina Propria

Lamina propria T cells are predominantly CD4+ and, since they are probably derived from PP T blasts, they are all α4β7+, although about one-third also express αEβ7 (Farstad et al. 1996). γδ cells are uncommon. They have other features of activation such as being L-selectin^{lo}, CD45R0+, CD25+, and Fas+ (Berg et al. 1991; Halstensen et al. 1990; Schieferdecker et al. 1992; Boirivant et al. 1996; De Maria et al. 1996). A sub-population are also FasL+. Fas cross-linking on lamina propria T cells rapidly leads to apoptosis (De Maria et al. 1996). Lamina propria CD8+ cells are also CD45R0+ and express high levels of αEβ7 (Farstad et al. 1996). Since it has been established for many years that T blasts from the mesenteric lymph nodes preferentially lodge in the lamina propria and do not leave, in order to avoid the lamina propria filling up with T cells, there must be extensive cell death. Activated T cells are programmed to die unless rescued by antigen or a common γ-chain cytokine (Akbar et al. 1996), and the expression of Fas and FasL on lamina propria T cells also suggests that most lamina propria T cells die in situ. Interleukin-7 (IL-7) has been reported to be made by epithelial cells and lamina propria T cells have functional IL-7 receptors (Watanabe et al. 1995). The interactions between survival factors, such as IL-7, and death factors, such as Fas, have not been well explored in the gut lamina propria.

2.3 Epithelial T Cells

There are major species differences between rodents and humans. In rodents, especially young animals, γδ T cells predominate, whereas in humans, except in celiac disease, γδ T cells are uncommon (<5%). In mice, there is largely confusing literature on the thymus-dependence of intraepithelial lymphocytes (IEL) and it is beyond the remit of this review to discuss this topic. Interested readers are referred to a recent review on this matter (LEFRANCOIS and PUDDINGTON 1995). However, it now seems that the original proposition which stated that γδ IEL T cells are thymus-independent, whereas, the vast majority of IEL αβ T cells are thymus-dependent (VINEY et al. 1989), has turned out to be correct (LEFRANCOIS and OLSON 1997).

In humans this discussion is largely irrelevant since IEL contain few γδ T cells and athymic individuals are rare. Likewise, thymus-independent IEL in mice express CD8 αα-homodimers on their surface (GUY-GRAND et al. 1991), but cells with this phenotype are rare in human IEL (JARRY et al. 1990). Very limited evidence for extrathymic maturation of T cells in gut epithelium has been published in postnatal intestine (LYNCH et al. 1995) and it is probable that this is not an important pathway in humans. Most studies have focused on small intestinal IEL. IEL are present in the colon, although at a markedly reduced density. CD8+ cells still dominate, but there is a higher frequency of CD4+ IEL and γδ IEL in the colon than small bowel (BRANDTZAEG et al. 1989).

In humans, however, as in mice, virtually all IEL are CD8+, although there are small populations of CD4+ cells, CD3+, 4-, 8- and CD3+, 4+, 8+ cells (SPENCER et al. 1991). There are also non-T cells with a CD3-, 7+ phenotype at the tips of the villi (SPENCER et al. 1991). The vast majority are L-selectin^{lo}, CD45RO+, and αEβ7^{hi} (JARRY et al. 1990, LEFRANCOIS 1987). There is very little evidence that once in the epithelium, IEL migrate back into the lamina propria, and rodent studies indicate that IEL can be long-lived (POUSSIER et al. 1992, PENNEY et al. 1995). T cells, therefore, persist in the epithelium for extended periods as the enterocyte layer is renewed around them.

Although perhaps more related to function than phenotype, one of the most curious features of epithelial αβ and γδ T cells is that they are oligoclonal (BALK et al. 1991, HOLTMEIER et al. 1995). The αβ population is dominated by several clones, as is the γδ population. The oligoclonal T-cell receptors are highly complex with extensive *N*-region insertions, suggesting that they have been selected by specific ligands. Consistent with this, the γδ T cells in new-born humans are very diverse and polyclonal (HOLTMEIER et al. 1997), and in the human foetus the αβ T cells are also polyclonal (KONIGSBERGER et al. 1997), suggesting again that some clones are selected post-natally by enteric antigens. Dominant clones vary among individuals, but seem to be quite stable with time and along the length of the gut.

3 Function of T Cells at Mucosal Surfaces

3.1 PP Cells

The study of T cells in PP is an area dominated by animal studies. There are no published reports of antigen-specific responses of PP cells in human. In mice PP, there are numerous CD4+ T cells that spontaneously secrete either IL-5 or interferon gamma (IFN-γ) (TAGUCHI et al. 1990). These are not present in the spleen, suggesting that the cells in PP are responding to enteric antigens with both a Th1-type and Th2-type cytokine pattern. The vast majority of studies in which the antigen-specific cytokine pattern of PP T cells to fed antigens have been studied have been carried out in mice that were fed antigens with cholera toxin (CT) as an adjuvant. Without CT, usually only minimal responses are generated. However, it has been reported that feeding sheep erythrocytes results in PP CD4+ T cells that preferentially secrete IL-5, rather than IFN-γ, when re-stimulated in vitro or cloned (XU-AMANU et al. 1992). Interestingly, this difference is not seen when PP T cells are polyclonally activated, although there is a report that PP CD8+ T cells secrete IL-5 and IL-10 when activated with anti-CD3 (LAGOO et al. 1994).

The predilection of PP T cells to secrete Th2-type cytokines in response to orally administered antigens has been demonstrated in a large series of papers from Kiyono, McGhee and co-workers who have almost invariably used CT as an adjuvant. In this work, it has been shown that CT functions as an oral adjuvant because of its ability to enhance IL-4 production by PP T cells and hence boost immunoglobulin A (IgA) responses (XU-AMANU et al. 1993; MARINARO et al. 1995). However, it has been reported by others that CT can function as an oral adjuvant to boost IL-2 and IFN-γ production and that the adjuvant effect is not so much Th2-directed (HORNQUIST and LYCKE 1993). Finally, salmonella are oral pathogens which require cell-mediated Th1 responses for acquired immunity. When antigens are given orally, expressed in salmonella vectors, PP Th1 responses are generated along with Th2-cells secreting IL-10 (VAN COTT et al. 1996).

While it is difficult to summarise this work at present, the balance of evidence suggests that in PP there may be some bias towards generating Th2-type responses. The situation is, however, not so clear-cut, even in mice, and the data generated using toxins, although of interest to vaccinologists, is of little relevance to what happens when PP T cells interact with food antigens in healthy individuals.

3.2 Lamina Propria T Cells

In contrast to PP, there have been a number of studies on the cytokine profile of human lamina propria lymphocytes (LPL), usually as a control for the alterations seen in IBD. In mice, however, using ELISPOTs it was first shown that LPL had a very high number of CD4+ T cells spontaneously secreting IL-5, about twice as many as secreted IFN-γ (TAGUCHI et al. 1990). This would again support the notion that the gut is a Th2 dominated site. T cells from the lamina propria of mice fed with keyhole limpet hemocyanin (KLH) and CT also preferentially secrete IL-4, IL-5 and IL-6 when challenged in vitro (Wilson et al. 1991). However, in a TcR transgenic model, it has been shown that lamina propria T cells predominantly make IFN-γ, and IL-10-, IL-4- and IL-5-secreting cells are less abundant (Saparov et al. 1997). IFN-γ transcripts have also been seen in mouse lamina propria, by in situ hybridisation, at the bases of the villi (Bao et al. 1993).

In humans, the situation with freshly isolated or polyclonally activated lamina propria T cells is clear. Freshly isolated human LPL contain a very high frequency of cytokine-secreting cells compared with blood T cells from the same individual, and the response is dominated by IFN-γ (Hauer et al. 1997, Carol et al. 1998). There are about 10-fold fewer cells spontaneously secreting IL-4. IL-5 and IL-10 secreting cells are uncommon. Many studies have also shown that IL-4 and IL-5 transcripts are barely detectable in normal human gut (Niessner and Volk 1995; Autschbach et al. 1995). Human LPL do not respond well to CD3 ligation, but stimulation with anti-CD2, with or without CD28 co-stimulation, elicits a strong proliferative and cytokine response. When the cytokines produced by LPL are analysed, it is again clear that the response is dominated by IFN-γ (Targan et al. 1995; Fuss et al. 1996). In the work of Fuss, CD2/CD28-activated normal human LPL secreted 50,000 pg/ml IFN-γ compared with just over 100 pg/ml IL-4 and less than 50 pg/ml IL-5. The use of IL-4 and IL-5 as prototypic Th2 cytokines may, however, be misleading in the human gut. Although human LPL make large amounts of IFN-γ, it has recently been shown that they can secrete several thousand picograms of IL-10 when activated via CD2 (Braunstein et al. 1997).

3.3 Intraepithelial Lymphocytes

Intraepithelial lymphocytes are the only population of CD8+ cells in the body that are constitutively activated. In re-directed cytotoxicity assays using anti-T cell-receptor antibodies in mice, both αβ and γδ TcR+ IEL from normal animals are potent cytotoxic effectors at low effector/target ratios (Viney et al. 1990). γδ IEL and the few thymus-independent αβ IEL express natural killer (NK) receptors and kill target cells by perforin secretion and/or the Fas/FasL pathway (Guy-Grand et al. 1996). Likewise in humans, IEL activated via CD3 are also cytotoxic effectors (Lundqvist et al. 1996). Mice infected with reovirus also develop antigen-specific αβ TcR+ cytotoxic effectors in their IEL compartment (Cuff et al. 1993). Mice orally infected with *Toxoplasma gondii* develop CD8+, αβ TcR+ IEL which can lyse infected enterocytes (Chardes et al. 1994) and which can provide passive protection to naive recipients (Buzoni-Gatel et al. 1997). Passive protection in this system is inhibited by anti-IFN-γ, and the CD8+ IEL express IFN-γ messenger RNA. In the author's view, the main function of CD8+ IEL is as cytotoxic effectors, capable of killing virally-infected or parasitised epithelial cells.

Nonetheless, IEL are also capable of secreting cytokines. When activated with concanavalin A (ConA), whole IEL populations secrete low amounts of IFN-γ and

IL-3 (Dillon et al. 1986). IEL from mice infected with the parasite *Trichinella spiralis* also secrete IL-3 when re-stimulated with worm antigens in vitro (Dillon et al. 1986). Freshly isolated mouse αβ and γδ IEL also contain high numbers of cells that spontaneously secrete IL-5 or IFN-γ (Taguchi et al. 1990). Mice fed with sheep red blood cells (SRBC), without adjuvants, are reported to contain CD4+, and CD4+, 8+, αβ TcR+ T cells which produce IL-4 and IL-5, but reduced IFN-γ when challenged in vitro and which can provide help for IgA responses (Fujihashi et al. 1993). It has also been reported that γδ T cells from SRBC-fed mice can abrogate oral tolerance when transfused systemically into tolerant recipients (Fujihashi et al. 1992). It is difficult, however, to see the significance of this observation, given that there is no evidence of γδ T cells ever leaving the mucosa. γδ IEL in mice are also reported to be a potent source of the chemokine, lymphotactin (Boismenu et al. 1996). Mice given *Listeria monocytogenes* orally show an increase in the number of IFN-γ-secreting αβ and γδ IEL (Yamamoto et al. 1993). β2-Microglobulin knockout mice show reduced IFN-γ-secreting αβ and γδ TcR+ IEL, but when infected with *L. monocytogenes*, there is a dramatic increase in the frequency of IFN-γ-secreting γδ IEL (Emoto et al. 1996).

Studies in human beings are limited. Both freshly isolated and polyclonally activated IEL in humans have been reported to contain transcripts for, and secrete IFN-γ, and IL-4 is detectable in only a minority of samples (Lundqvist et al. 1996). Using ELISPOTs, it has also been reported that freshly isolated human IEL predominantly secrete IFN-γ rather than IL-4 (Carol et al. 1998).

4 Summary of Cytokines in Human and Mouse Gut Associated with Lymphoid Tissue

Murine studies where CT has been used as an adjuvant indicate that the T cell response in PP is skewed towards the Th2-type, although perhaps not as polarised as was originally considered. The failure of IL-4 knockout mice, fed with antigen and CT, to generate a secretory IgA response supports this notion (Vajdy et al. 1995). With LPL and IEL in rodents, the situation is more contentious. There is little solid data on LPL, which are difficult to isolate and hard to work with. With IEL it seems clear that CD8+ cells, both αβ and γδ TcR+, are cytotoxic effectors and can also secrete IFN-γ. The minority CD4+, IEL population may be more Th2-type, but the author wonders how they can influence the terminal differentiation of IgA plasmablasts in the lamina propria.

In humans, however, there is no data on PP cells, therefore their cytokine profile is unknown. However, with IEL and LPL the situation seems clear, the response of freshly isolated or polyclonally activated cells is dominated by IFN-γ. So far, no antigen-specific T cell clones have been isolated from the intestine of healthy individuals, but gluten-specific T-cell clones and lines isolated from the lamina propria of celiac-disease patients have a Th1 or Th0 cytokine profile (Nilsen et al. 1995).

5 Oral Tolerance

Although oral tolerance was originally thought of as a way to prevent harmful reactions to dietary proteins in the gut, this has only rarely been investigated because of the difficulty in generating a T cell-mediated tissue damaging reaction to a nominal antigen in the gut. Instead, the standard protocol has been to feed with antigens and, then, after a few weeks, challenge the animals systemically to elicit immunity or disease and try to determine whether prior feeding modulates the response. Therefore, strictly speaking, most studies have examined mucosally induced systemic tolerance. Generally, tolerance involves either active suppression, anergy or clonal deletion. All have been demonstrated to occur after feeding with antigens and the topic is well-reviewed elsewhere (WEINER 1997). For the purpose of this chapter, I will concentrate on active suppression by T cells capable of secreting immunosuppressive cytokines.

5.1 Dose of Antigen

Generally, immunoregulatory cells are elicited by prolonged low-level exposure to oral antigen, most probably identical to the physiological presentation of antigens in the gut (CHEN et al. 1996). Clonal deletion after oral administration of antigens is best demonstrated in transgenic mice bearing a single TcR and seems to require extremely large amounts of antigens (CHEN et al. 1995).

5.2 Regulatory Cells Produced by Feeding with Antigens

It was demonstrated over 20 years ago that oral tolerance was infectious, i.e. that spleen cells or cells from the PP or mesenteric lymph nodes (MLN) of fed mice were able to transfer systemic unresponsiveness to a naive animal. Depletion of CD8+ cells eliminated this effect in most cases, suggesting that the cell responsible was a CD8+ suppressor T cell. Delayed-type hypersensitivity responses appear to be more readily tolerised than antibody responses. This earlier work is well summarised by MOWAT (1987). Oral tolerance has also been demonstrated in humans (HUSBY et al. 1994). Although a considerable amount of work has been done on feeding with nominal antigens, the bulk of recent work and most progress has focused on immunoregulatory cells produced by feeding self-antigens – a process known to inhibit the development of experimental autoimmune disease in rodents.

Rats fed with guinea-pig myelin basic protein are highly resistant to the development of experimental autoimmune encephalitis induced by systemic immunisation of this protein mixed with complete Freund's adjuvant (CFA). A similar phenomenon is seen in rats fed with collagen that are resistant to collagen arthritis; diabetic prone-mice fed with insulin are resistant to disease; and rats fed with retinal proteins are resistant to experimental uveitis (summarised by WEINER 1997).

All of the three potential mechanisms of peripheral tolerance (apoptosis, anergy and active suppression) have been demonstrated in oral tolerance, but their relative importance has not been determined and may vary from system to system.

As expected, CD8+ cells from rats fed with myelin basic protein (MBP) can transfer tolerance to naive rats. The strongest evidence that tolerance is mediated by cytokines comes from in vivo experiments. Rats made tolerant to experimental autoimmune encephalomyelitis (EAE) by prior feeding develop disease if injected with anti-TGFβ antibodies (MILLER et al. 1992). The lesions in the nervous system of rats with EAE contain lots of cells containing Th1-type cytokines and, as the lesion resolves, these become replaced by cells containing TGFβ, PGE_2 and other Th2 type cytokines. Prior feeding with MBP changes the cytokine profiles in the brain early in disease from a Th1 to a Th2 type (KHOURY et al. 1992). In a conceptually important experiment, rats fed with ovalbumin and then immunised with ovalbumin and MBP in CFA did not develop disease-showing non-specific bystander suppression of the encephalogenic T cells by immunoregulatory T cells reacting to ovalbumin (MILLER et al. 1991). This demonstrates that the immunoregulatory cell need not be specific for the encephalogenic peptides as long as it can recognise a peptide in the immediate vicinity. Bystander tolerance has been demonstrated now in a number of systems. In addition, it has also been demonstrated that regulatory CD4 cells are induced after oral immunisation (CHEN et al. 1995), showing that the effect is not restricted to CD8 cells.

Detailed analysis of immunoregulatory T cells was next carried out in SJL mice, a strain susceptible to EAE. After feeding mouse MBP to SJL mice, it proved possible to clone immunoregulatory CD4+ cells from the mesenteric lymph nodes (CHEN et al. 1994). These cells secreted IL-10, IL-4 and TGFβ and when passively transferred into normal mice could prevent EAE. Similar kinds of results were found in multiple sclerosis patients fed with low dose MBP. Short term cultures of blood lymphocytes resulted in an increase in the frequency of T cell lines that secreted TGFβ1 in the MBP-fed patients compared with controls (FUKAURA et al. 1996). In intriguing studies, it was also demonstrated that blocking either B7.1 or B7.2 could alter the cytokine profile of immunoregulatory T cell induced by feeding with antigens (KUCHROO et al. 1995), but it is not clear whether this is a general rule in immunoregulation.

The real impetus for this work, however, came from two preliminary clinical trials in which it was found that rheumatoid arthritic patients fed with chicken collagen did significantly better than those on placebo, and a tantalising trial which showed some benefit of feeding bovine myelin to multiple sclerosis patients (WEINER et al. 1993, TRENTHAM et al. 1993). However in a recent 515 patient double-blind placebo controlled multicenter phase III trial, a single dose of bovine myelin was ineffective (WEINER 1997). More encouraging results were obtained with a 280 phase II double blind dose ranging trial of chicken type II collagen in rheumatoid arthritis (BARNET et al. 1998) where an effect was seen at the lowest dose (20 μg).

In a very recent study, HANEDA and colleagues (1997) have also shown that oral tolerance can downregulate the Th2-dependent airway eosinophilia observed when ovalbumin TcR transgenic mice have ovalbumin instilled into the trachea. Suppression is infectious, i.e. it can be transferred into naive animals and, moreover, it can be ablated by anti-TGFβ treatment. Oral tolerance may therefore be a potential way to downregulate bronchial asthma. There is also evidence in other systems that oral tolerance can downregulate Th2 responses (Garside et al. 1995), though what distinguishes this type of response from the Th2-type responses elicited by oral antigen and described by McGhee and colleagues (Xu-Amanu et al. 1993; Marinaro et al. 1995) is still unclear. This somewhat confusing picture, i.e. that Th2 cells can be made tolerant can be explained by envisaging that Th1 cells in the same animals may be made tolerant by anergy or apoptosis – thus, the unresponsiveness. Alternatively, it is more likely that it is merely a nomenclature problem. If one defines a Th2 as a cell that makes IL-4 and little IFN-γ, then it may be possible to inhibit IL-4 responses. But, unless one measures an array of other cytokines, i.e. IL-10, IL-13 and TGFβ, it is premature to say that one has made Th2 cells tolerant.

6 Nasal Immunisation and Tolerance

6.1 Nasal Associated Lymphoid Tissue

Rodents have a pair of organised cylindrical lymphoid structures, parallel to the oral septum, and overlain by a ciliated epithelium. Rodent gut-associated lymphoid tissue (GALT) has lymphoid follicles and T and B cell areas. It is not known whether the epithelium contains M-cells (Kuper et al. 1992). These structures are not present in man and the main nasal-associated lymphoid tissue (NALT) is the tonsils, adenoids and the lateral pharyngeal bands. The main difference between NALT and GALT is that the former is exposed to antigens entering through the nose, without prior modification, whereas the antigens exposed to GALT have been acidified and digested. Given the easy access to the airways through the nose, many recent studies have investigated the effects of nasal immunisation. Responses can be generated locally in the NALT or in the draining cervical lymph nodes.

6.2 Nasal Immunisation

It seems clear that to generate good serum and mucosal responses to non-replicating antigens following intranasal immunisation, an adjuvant should be used. CT and recombinant CTB have been widely used, as have liposomes. It has recently been reported that a non-toxic mutant of CT, given intranasally with nominal antigens, has a potent adjuvant effect and elicits Th2 responses in lung and spleen (Yamamoto et al. 1997). Local IgA responses can be generated in NALT and, for some unknown reason, nasal immunisation is quite effective at producing IgA in the female reproductive tract (BERQUIST et al. 1997). Serum IgG antibodies are also markedly increased following intranasal immunisation and there have been reports that CTL can be generated (summarised by STAATS 1997). Therefore, there is little doubt that nasal immunisation can induce strong local and systemic immunity, and clinical trials for efficacy are underway at a number of centres against a variety of diseases of the upper respiratory tract (TAMURA and KURATA 1997).

6.3 Nasal Tolerance

Intranasal administration of peptides is an extremely potent way to make animals tolerant, especially in the context of autoimmunity. Prior nasal immunisation with an altered MBP peptide prevents experimental autoimmune encephalitis in mice (METZLER and WRAITH 1993); glutamic acid decarboxylase (GAD) and insulin peptides given intranasally protect against diabetes in nonobese diabetic mice (DANIEL and WEGMANN 1996; TIAN et al. 1996); and collagen peptides decrease the severity of collagen arthritis (STAINES et al. 1996). Nasal immunisation with house dust-mite peptides also specifically make mice tolerant to this important environmental allergen (HOYNE et al. 1993). There has been much less work on the cytokines induced following nasally-induced tolerance. As with oral tolerance, it has been reported that intranasal immunisation with nicotinic acetyl-choline receptor biases subsequent responses to the same antigen given parenterally towards a T cell response in which TGFβ is prominent and IFN-γ is downregulated (MA et al. 1996).

7 Cytokines in Disease

7.1 The Relationship Between Immunity and Gut Injury

The overlap between protective immunity and tissue injury in the gut was first highlighted in the 1970s by Anne Ferguson (FERGUSON and JARRETT 1975) who showed that the villous atrophy that accompanies expulsion of the parasite *Nippostrongylus brasiliensis* from the gut of rats was thymus-dependent. Tissue injury such as villous atrophy, increased epithelial permeability, epithelial cell death and overt ulceration can be tolerated by the host if the transient loss of gut function results in expulsion of the pathogen. However, if the pathogen persists, chronic diarrhoea, which can be life-threatening, occurs, particularly in children. A different problem occurs in the developed world where enteric pathogens are less common. In this case, an inappropriately directed immune response to food antigens (i.e. wheat gluten in celiac disease) or to unknown antigens (probably gut bacteria) in Crohn's disease also leads to chronic tissue injury. In both these diseases, there is compelling evidence that cytokines are important in tissue injury.

7.2 Cytokine Profiles in Human Gut Disease

7.2.1 Inflammatory Bowel Disease

It is self-evident that in chronic IBD the number of inflammatory cells are increased in the mucosa. There is no specificity to this. In both Crohn's and ulcerative colitis (UC), there is a massive increase in plasma cells of all types, especially IgG. Although there are statistical differences between IgG and IgA subclasses in Crohn's and UC, definitive findings are hard to make (Brandtzaeg et al. 1989). However, there is no doubt that in UC, the lesion is more neutrophilic, and in Crohn's it is more mononuclear, but there is tremendous overlap. There have been many studies of cytokines in IBD using in situ hybridisation, immunohistochemistry, organ culture of biopsies, cytokine profiles of freshly isolated cells, and in vitro activated T cells and macrophages. All of these techniques have relative advantages and disadvantages and the results from any single study should not be over-interpreted. However, sufficient concordance has been found among different studies for some kind of consensus to exist.

The difference between Crohn's and UC was first highlighted by Mullin and colleagues (1992) who used quantitative reverse-transcriptase polymerase chain reaction (RT-PCR) to show that IL-2 transcripts were markedly elevated in Crohn's tissue but not UC. This result was confirmed and extended by the author who demonstrated that freshly isolated lamina propria T cells from Crohn's disease patients contained a high proportion that secreted IFN-γ or IL-2 (Breese et al. 1993). These results have been confirmed by an in situ hybridisation technique (Autschbach et al. 1995). Slightly different results have been obtained with polyclonally activated isolated cells that make less IL-2 than controls (Kusugami et al. 1991). Nonetheless, lamina propria CD4+ T cells from Crohn's patients activated in vitro with anti-CD2 and anti-CD28 make enhanced amounts of IFN-γ and less IL-4 compared with controls. In UC, more IL-5 is made than with either Crohn's or control patients (Fuss et al. 1996). IL-4 has been reported to be reduced in active Crohn's disease by several groups (Karttunen et al. 1994; Fuss et al. 1996; West et al. 1996), but a recent report found that IL-4 transcripts were increased in the mucosa in the early lesions of Crohn's disease (Desreumaux et al. 1997). Polyclonally activated CD4+ T cells from Crohn's patients have a Th1-type profile (Parronchi et al. 1997). It, therefore, appears clear that in Crohn's disease the activated T cells in the lamina propria show a Th1-type profile. It would be premature to say that in UC they show a Th2 cytokine pattern.

A possible explanation for the skewed Th1 response in Crohn's disease is that there is also over expression of IL-12 in Crohn's but not in UC (Monteleone et al. 1997). The IL-12 is made by macrophages. As would be expected from chronically inflamed tissue, in both UC and Crohn's there is a dense infiltrate of macrophages, recently extravasated from the blood (Rugtveidt et al. 1997). These are probably a major source of the elevated IL-1, IL-6, IL-8 and TNFα found in diseased mucosa. Epithelial cells can also make a number of pro-inflammatory cytokines (Eckman et al. 1993; Jung et al. 1995). It should also be remembered that in UC, particularly, the mucosa is replete with neutrophils and they also can release pro-inflammatory cytokines. Inflamed mucosa also contains natural inhibitors of cytokines, such as the IL-1 receptor antagonist and soluble TNFα receptors (CASINI-RAGGI et al. 1995). The biological effect of the cytokines will therefore reflect the relative ratio of the cytokine and its antagonist within the tissue.

7.2.2 Food-Sensitive Enteropathies

Celiac disease has a great advantage over the study of IBD in that the antigen is known. If celiac biopsies are challenged ex vivo with gliadin, there is a rapid increase in the number of CD25+ T cells in the lamina propria (HALSTENSEN et al. 1993). Cloning of these T cells shows that many are gliadin specific and also restricted by DQ2 (LUNDIN et al. 1993). The cytokine profile of the T cell clones and lines is of a Th1 or Th0 type with IFN-γ dominant (NILSEN et al. 1995). It has also been reported that, following gluten challenge in vivo, there is a rapid increase in cells making IFN-γ transcripts (KONTAKOU et al. 1995). IL-4 and IL-10 transcripts are not elevated in active celiac disease (BECKETT et al. 1996). The evidence therefore suggests that in celiac disease, there is also an exaggerated Th1-type response in the mucosa. Freshly isolated lamina propria T cells from celiac patients have increased frequencies of IL-2- and IFN-γ-secreting cells (BREESE et al. 1994a).

The only other disease in which cytokines have been studied is cows-milk sensitive enteropathy/post-enteritis syndrome. This is a relatively uncommon disease of children who are intolerant of cows milk either because of immune-mediated hypersensitivity or lactose intolerance following gut infection. Jejunal biopsies from these children contain a dramatically increased frequency of CD4+ T cells secreting IFN-γ. IL-4 secreting cells are also raised (HAUER et al. 1997).

7.2.3 *Helicobacter pylori*

Helicobacter pylori (Hp) is the most common pathogen of the human race and it is obviously important to ascertain the type of T cell responses ongoing in the gastric mucosa. T cells freshly isolated from patients with gastritis show a raised frequency of cells secreting IFN-γ and decreased numbers secreting IL-4 (KARTTUNEN et al. 1995). This is not related to Hp status since higher numbers of IFN-γ-secreting cells were found in non-Hp gastritis than Hp gastritis. T-cell clones specific for Hp antigens have been isolated, and the majority express a Th1 cytokine profile (D'ELIOS et al. 1997a). In a comparison of T-cell clones isolated from patients with either chronic gastritis or chronic gastritis and peptic ulcer, it was found that the latter group contained a higher frequency of clones with a Th1 cytokine profile (high IFN-γ, low IL-4). In the former group, the cytokine pattern was more Th0 (equivalent amounts of IFN-γ and IL-4). The authors speculate that the polarised Th1 response may be related to the genesis of the peptic ulcer (D'ELIOS et al. 1997b).

7.3 Cytokines in Rodent Models of Gut Inflammation

7.3.1 Mouse Models

The main advance in the study of gut inflammation has been the development of a large number of small-animal models of IBD (reviewed by ELSON et al. 1995). Models include spontaneous colitis in TcRα knockout mice, IL-2 knockout mice, IL-10 knockout mice, $G\alpha_{i2}$ knockout mice, bone-marrow reconstituted human CD3ε transgenic mice, C3H/HeJBir mice, and CD4+CD45RB^{hi} T cells → SCID (severe combined immunodeficient) mice. Together with trinitrobenzene sulphonic acid (TNBS) colitis, there is now an armamentarium of animal models with which to dissect the disease processes in chronic gut inflammation. Details of these models are well reviewed elsewhere (ELSON et al. 1995; MACDONALD 1997), but all of the models show two common features. First, in diseased intestine, there is a large increase in cells with a Th1-type cytokine pattern and, second, disease is not seen in germ-free mice, strongly suggesting that the aberrant response is to the normal gut flora.

The source of the IFN-γ need not be αβ T cells. In mice lacking conventional αβ T cells with colitis, there are large numbers of NK cells and CD4+, TcR α-β^{lo}+ cells secreting IFN-γ (MIZOGUCHI et al. 1996). In the CD3ε transgenic mice reconstituted with bone marrow cells from TcRα or TcRβ knockout mice, colitis still develops. In this case, the diseased colon becomes infiltrated with IFN-γ-secreting γδ T cells (SIMPSON et al. 1997). Thus IFN-γ and TNFα protein and transcripts are dramatically increased in diseased gut from all of these mice. Neutralisation of these two cytokines with antibodies is of therapeutic benefit in IBD models where this approach has been tested (POWRIE et al. 1994b; BERG et al. 1996; NEURATH et al. 1997). Likewise, neutralisation of IL-12, a monokine that promotes Th1 responses, is highly effective in attenuating TNBS colitis in mice (NEURATH et al. 1995). It is not only in response to the normal flora that excess IFN-γ in the mucosa is deleterious. B6 mice infected orally with the protozoan *Toxoplasma gondii* die of massive small-bowel necrosis (LIESENFELD et al. 1996). This is dependent on CD4+, αβ TcR+ cells and, importantly, is inhibited by anti-IFN-γ antibody treatment. In many intestinal parasite infections, there is pathology and a great deal of work has been carried out on Th1 and Th2 cytokines in immunity in these models. In contrast, there is virtually no information on the cytokines expressed by T cells in the mucosa of these models.

An important lesson to be taken from these models is that potentially damaging immune responses to the normal flora are tightly regulated. This is best shown in the CD4+CD45RB^{hi} → SCID model where disease is prevented by the simultaneous infusion of CD4+ CD45RB^{lo} cells (POWRIE et al. 1994a). The protective effect of the RB^{lo} cells is inhibited by anti-TGFβ, but not anti-IL-4 (POWRIE et al. 1996). Likewise, in TNBS colitis in mice, where intracolonic infusion of the hapten causes a T cell-mediated inflammatory response to both haptenated self-proteins and gut bacteria, neutralisation of endogenous IL-12 and IL-10 therapy prevents disease and sensitisation to the flora (DUCHMANN et al. 1996).

7.3.2 Oral Tolerance as a Therapy to Prevent Chronic Colitis

If mice are rendered tolerant to trinitrophenol (TNP) by feeding with haptenated self-proteins (oral tolerance), disease induced by subsequent intracolonic TNBS challenge is attenuated (ELSON et al. 1996; NEURATH et al. 1996). The mice made tolerant orally show increased production of TGFβ by PP and lamina propria T cells and decreased IFN-γ. By immunohistochemistry, TGFβ is markedly upregulated in the colonic mucosa and IL-12 is downregulated. Disease is exacerbated in tolerant mice by treatment with recombinant IL-12 or anti-TGFβ. These results fit well with the scheme proposed by WEINER (1997) that oral immunisation preferentially activates a population of CD4+ Th3 cells which secrete TGFβ (FUKAURA et al. 1996). These cells may reside normally in the $CD45RB^{lo}$ population which can protect against colitis in the $CD45RB^{hi} \rightarrow$ SCID model of colitis (POWRIE et al. 1994a), and whose effect is neutralised by anti-TGFβ (POWRIE et al. 1996).

Slightly different results have been obtained in some very recent studies where GROUX and colleagues (1997) identified a regulatory T-cell population that inhibits colitis by secreting IL-10. In this elegant study, naive CD4+ T cells from mice transgenic for a TcR which recognises an ovalbumin peptide, were grown in vitro with IL-10 and ovalbumin. Some of the clones produced by this protocol secreted large amounts of IL-10 and remained ovalbumin specific. When transfused into SCID mice reconstituted with $CD45RB^{hi}$ CD4+ T cells, the T regulatory cells on their own had no effect. However, if the mice were fed ovalbumin, disease was prevented. This is an extremely good example of bystander tolerance in the gut, preventing T cell tissue-damaging responses to unrelated antigens. Other data to support a crucial role for IL-10 in tolerance in the gut comes from Kronenberg's group who have shown that, if you use $CD45RB^{hi}$ cells from a mouse overexpressing IL-10 as donors in the SCID model of colitis, onset of disease is prevented (HAGENBAUGH et al. 1997).

7.3.3 How do Cytokines Damage the Gut?

There are two ways in which an elevated concentration of pro-inflammatory cytokines in the mucosa could cause gut injury. Either the cytokines directly damage the tissue or they activate other pathways of injury. While there is no doubt that cytokines such as TNFα and IL-1β can have profound and well-documented effects on epithelial cells and endothelial cells, for example (HEYMAN et al. 1994; KAISER and POLK 1997), we believe that this plays only a minor role. Instead, we consider that the altered cytokine milieu in the gut results in the production of endogenously secreted molecules which then alter and damage the mucosa.

Normally in the gut, lamina propria stromal cells secrete low amounts of the matrix metalloproteinases (MMP) – a family of zinc-containing neutral endopeptidases which slowly replace the matrix. They also secrete tissue inhibitors of MMPs (TIMPs) which, together with inhibitors such as α2-macroglobulin, prevent excess enzyme activity. Cytokines rapidly upregulate MMP production without altering TIMP production. We first became aware of the potential importance of this pathway in the gut when studying an ex vivo model of T cell-mediated gut inflammation in which lamina propria T cells in explant cultures of human small intestine were polyclonally activated with pokeweed mitogen (PWM) (PENDER et al. 1997).

The T-cell response in the fetal explants is strongly Th1-biased and within 3–4 days there is severe tissue injury, with destruction of the mucosa. Organ culture supernatants of PWM-treated explants contain markedly raised concentrations of MMP-1 (collagenase) and MMP-3 (stromelysin-1). Importantly, MMP inhibitors prevent injury without preventing T-cell activation. Cultures of isolated stromal cells secrete extremely high amounts of MMP-1 and MMP-3 when activated with TNFα and IL-1β. A p55 TNFα-receptor human IgG fusion protein prevents PWM-induced injury and also inhibits MMP-3 production (PENDER et al. 1998). Stromelysin-1 is extremely potent at degrading gut mucosa. It has a relatively broad specificity, functioning as a type-IV collagenase and, thus, can destroy basement membranes. In addition, it is very potent at cleaving proteoglycan core protein and can destroy the lamina propria extracellular matrix and cell surface proteoglycans involved in cell-signalling and growth factor presentation. MMP-3 message is abundant near ulcers in the gut, whether it be a gastric ulcer caused by Hp or a Crohn's ulcer in the colon (SAARIALHO-KERE et al. 1996). We would like to suggest that at least one of the ways by which the excess TNFα in IBD (BREESE et al. 1994b) causes tissue injury in Crohn's disease is through activation of endogenous MMPs by stromal cells. Anti-TNFα therapy is efficacious in Crohn's disease (TARGAN et al. 1997) because TNF neutralisation interrupts this pathway.

A second way in which cytokines could indirectly produce changes in the gut is through their effects on stromal cell production of epithelial growth factors, such as members of the fibroblast growth factor (FGF) family. One of these molecules, keratinocyte growth factor (KGF-FGF-7) is made by stromal cells and its production is also up-regulated by cytokines. It is also over-expressed in IBD (FINCH et al. 1996; BRAUCHLE et al. 1996; BAJAJ-ELLIOT et al. 1997). Epithelial cells make their own mitogens and motogens, such as TGFα, EGF, and members of the trefoil factor family which are increased during injury (reviewed by GOKE and PODOLSKY 1996). It is highly likely that the production of these molecules and their receptors can be altered by cytokines.

8 Summary and Conclusions

In this review, I hope to have highlighted that cytokines are of crucial importance in the normal homeostasis of the gut immune system, the interactions of the gut immune system with enteric antigens and also in tissue injury associated with IBD. There is evidence from a number of different systems that the response to nominal non-replicating antigens, administered nasally or orally, is skewed towards a non-Th1 type of response. To say that the response is Th2, Th3 or Tr is premature.

IL-10 and TGFβ seem to be important in downregulating potentially tissue-damaging Th1 responses to the normal flora and possibly food antigens. However, it needs to be seen whether the mouse results also apply to humans. A consistent pattern in disease states, whether it be human or mouse, is an exaggerated Th1 type response with excess local production of IFN-γ and TNFα, and its association with tissue injury. An important question to address is whether this represents a switch from the Th2, Th3, or Tr pathway towards a Th1 pathway, or whether the Th1 pathway is in fact always present in the gut, but is kept in check and non-pathogenic by regulatory cells. Equally important is the need to discover where regulation occurs: is it in the PP or the lamina propria? Intriguing results from Kronenberg and colleagues have shown that SCID mice reconstituted with $CD45RB^{hi}$ or $CD45RB^{lo}$ cells show no difference in the re-population of the gut prior to disease (Aranda et al. 1997). The reason for colitis developing in those mice reconstituted with $CD45RB^{hi}$ cells is therefore more complex than merely differential re-population kinetics. No matter what the outcome is, these and other related questions dealing with the induction and expression of mucosal T-cell responses are going to produce some surprises in the next few years.

References

Akbar AN, Borthwick NJ, Wickremasinghe RG, Panayoitidis P, Pilling D, Bofill M, Krajewski S, Reed JC, Salmon M (1996) Interleukin-2 receptor common chain signalling cytokines regulate activated T cell apoptosis in response to growth factor withdrawal: selective induction of anti-apoptotic (bcl-2,bcl-xL) but not pro-apoptotic (bax, bcl-xS) gene expression. Eur J Immunol 26:294–29

Aranda B, Sydora BC, McAllister PL, Binder SW, Yang HY, Targan SR, Kronenberg M (1997) Analysis of intestinal lymphocytes in mouse colitis mediated by transfer of CD4+, $CD45RB^{hi}$ T cells to SCID recipients. J Immunol 158:3464–73

Autschbach F, Schurmann G, Qiao L, Merz H, Wallich R, Meuer SC (1995) Cytokine messenger RNA expression and proliferation status of intestinal mononuclear cells in noninflamed gut and Crohn's disease. Virchows Arch 426:51–60

Bajaj-Elliot M, Breese E, Poulsom R, Fairclough PD, Pender SF, MacDonald TT (1997) Keratinocyte growth factor in inflammatory bowel disease: increased mRNA transcripts in ulcerative colitis compared to Crohn's disease in biopsies and isolated stromal cells. Am J Pathol 151:1469–1476

Balk SP, Ebert EC, Blumenthal RL, McDermott FV, Wucherpfennig KW, Landau SB, Blumberg RS (1991) Oligoclonal expansion and CD1 recognition by human intestinal intraepithelial lymphocytes. Science 253:1411–1415

Bao S, Goldstone S, Husband AJ (1993) Localisation of IFN-gamma and IL-6 mRNA in murine intestine by in situ hybridisation. Immunology 80:666–670

Barnet ML, Kremer JM, St Clair EW, Clegg DO, Furst D, Weisman M, Fletcher MJ, Chasan-Taber S, Finger E, Morales A, Le CH, Trentham DE (1998) Treatment of rheumatoid arthritis with oral type II collagen. Arthritis Rheum 41:290–297

Beckett CG, Dell'Olio D, Kontakou M, Przemioslo RT, Rosen-Bronson S, Ciclitira PJ (1996) Analysis of interleukin-4 and interleukin-10 and their association with the lymphocytic infiltrate in the small intestine of patients with coeliac disease. Gut 39:818–23

Berg M, Murakawa Y, Camerini D, James SP (1991) Lamina propria lymphocytes are derived from circulating cells that lack the Leu-8 lymph node homing receptor. J Immunol 101:90–99

Berg DJ, Davidson N, Kuhn R, Muller W, Menon S, Holland G, Thompson-Snipes L, Leach MW, Rennick D (1996) Enterocolitis and colon cancer in interleukin-10-deficient mice are associated with aberrant cytokine production and CD4+ Th1-like responses. J Clin Invest 98:1010–1020

Berquist C, Johansson EL, Lagergard T, Holmgren J, Rudin A (1997) Intranasal vaccination of humans with recombinant cholera toxin B subunit induces systemic and local antibody responses in the upper respiratory tract and vagina. Infect Immun 65:2676–2684

Boirivant M, Pica R, DeMaria R, Testi R, Pallone F, Strober W (1996) Stimulated human lamina propria T cells manifest enhanced Fas-mediated apoptosis. J Clin Invest 98:2616–2622

Boismenu R, Feng L, Xia YY, Chang JCC, Havran WL (1996) Chemokine expression by intraepithelial T cells. J Immunol 157:985–992

Brandtzaeg P, Hastensen TS, Kett K, Krajki P, Kvale D, Rognum TO, Scott H, Sollid LM (1989) Immunobiology and immunopathology of human gut mucosa: humoral immunity and intraepithelial lymphocytes. Gastroenterology 97:1562–1584

Brauchle M, Madlener M, Wagner AD, Angermeyer K, Lauer U, Hofschneider PH, Gregor M, Werner S (1996) Keratinocyte growth factor is highly overexpressed in inflammatory bowel disease. Am J Pathol 149:521–529

Braunstein J, Qiao L, Autschbach F, Schurmann G, Meuer S (1997) T cells of the human intestinal lamina propria are high producers of interleukin-10. Gut 41:215–220

Breese E, Braegger CP, Corrigan CR, Walker-Smith JA, MacDonald TT (1993) Interleukin-2 and interferon-γ secreting T cells in normal and diseased human intestinal mucosa. Immunology 78:127–131

Breese EJ, Farthing MJG, Kumar P, MacDonald TT (1994a) Interleukin-2 and interferon-gamma producing cells in the lamina propria in coeliac disease. Dig Dis Sci 39:2243

Breese EJ, Michie CA, Nicholls SW, Murch SH, Williams CB, Domizio P, Walker-Smith JA, MacDonald TT (1994b) Tumour necrosis factor-alpha producing cells in the intestinal mucosa of children with inflammatory bowel disease. Gastroenterology 106:1455–1466

Buzoni-Gatel D, Lepage AC, Dimier-Poisson IH, Bout DT, Kasper LH (1997) Adoptive transfer of gut intraepithelial lymphocytes protects against murine infection with Toxoplasma gondii. J Immunol 158:5883–5889

Carol M, Lambrechts A, Van Gossum A, Libin M, Goldman M, Mascart-Lemone F (1998) Spontaneous secretion of interferon-gamma and interleukin-4 by human intraepithelial and lamina propria gut lymphocytes. Gut 42:643–9

Casini-Raggi V, Kam L, Chong YJT, Ficchi C, Pizzaro TT, Cominelli F (1995) Mucosal imbalance of IL-1 and IL-1 receptor antagonist in inflammatory bowel disease. J Immunol 154:2434–2440

Chardes T, Buzoni-Gatel D, Lepage A, Bernard F, Bout D (1994) Toxoplasma gondii oral infection induces specific cytotoxic CD8αβ-Th1+ gut intraepithelial lymphocytes, lytic for parasite-infected enterocytes. J Immunol 153:4596–603

Chen Y, Kuchroo VK, Inobe J, Hafler DA, Weiner HL (1994) Regulatory T cell clones induced by oral tolerance: suppression of autoimmune encephalomyelitis. Science 265:1237–1240

Chen Y, Inobe J, Marks R, Gonnella P, Kuchroo VK, Weiner HL (1995) Peripheral deletion of antigen-reactive T cells in oral tolerance. Nature 376:177–180

Chen Y, Inobe J, Weiner HL (1995) Induction of oral tolerance to myelin basic protein in CD8-depleted mice: both CD4+ and CD8+ cells mediated active suppression. J Immunol 155:910–916

Chen Y, Inobe J, Kuchroo VK, Baron JL, Janeway CA, Weiner HL (1996) Oral tolerance in myelin basic protein T-cell receptor transgenic mice: suppression of autoimmune encephalomyelitis and dose-dependent induction of regulatory cells. Proc Natl Acad Sci USA 93:388–391

Cuff CF, Cebra CK, Rubin DH, Cebra JJ (1993) Developmental relationship between cytotoxic T cell receptor αβ-positive intraepithelial lymphocytes and Peyer's patch lymphocytes. Eur J Immunol 23:1333–1339

Daniel R, Wegmann DR (1996) Protection of nonobese diabetic mice from diabetes by intranasal or subcutaneous administration of insulin peptide B-(9–23). Proc Natl Acad Sci USA 93:956–960

D'Elios MM, Manghetti M, Almerigogna F, Amedei A, Costa F, Burroni D, Baldari CT, Romagnani S, Telford JL, Del Prete G (1997a) Different cytokine profile and antigen-specificity repertoire in Helicobacter pylori-specific T cell clones from the antrum of chronic gastritis patients with and without peptic ulcer. Eur J Immunol 27:1751–1755

D'Elios MM, Manghetti M, DeCarli M, Costa F, Baldari CT, Burroni D, Telford JL, Romagnani S, Del Prete G (1997b) T helper 1 effector cells specific for Helicobacter pylori in the gastric antrum of patients with peptic ulcer disease. J Immunol 158:962–967

De Maria M, Boirivant M, Cifone MG, Roncaioli P, Hahne M, Tschopp J, Pallone F, Santoni A, Testi R (1996) Functional expression of Fas and Fas ligand on human gut lamina propria T lymphocytes: a potential role for acidic sphingomyelinase pathway in normal immunoregulation. J Clin Invest 97:316–322

Desreumaux P, Brandt E, Gambiez L, Emilie D, Geboes K, Klein O, Ectors N, Cortot A, Capron M, Colombel JF (1997) Distinct cytokine patterns in early and chronic ileal lesions of Crohn's disease. Gastroenterology 113:118–126

Dillon SB, Dalton BJ, MacDonald TT (1986) Lymphokine production by antigen and mitogen activated murine intraepithelial lymphocytes. Cell Immunol 103:326–338

Duchmann R, Schmitt E, Knolle P, Meyer zum Buschenfelde KH, Neurath M (1996) Tolerance towards resident intestinal flora in mice is abrogated in experimental colitis and restored by treatment with interleukin-10 or antibodies to interleukin-12. Eur J Immunol 26:934–8

Eckman L, Jung HC, Schuerer-Maly CC, Panja A, Morzycka-Wroblewska E, Kagnoff MF (1993) Differential cytokine expression by human intestinal epithelial cell lines: regulated expression of IL-8. Gastroenterology 105:1689–1697

Elson CO, Sartor RB, Tennyson GS, Riddell RH (1995) Experimental models of inflammatory bowel disease. Gastroenterology 109:1344–1367

Elson CO, Beagley KW, Sharmanov AT, Fujihashi K, Kiyono H, Tennyson GS, Cong Y, Black CA, Ridwan BW, McGhee JR (1996) Hapten-induced model of murine inflammatory bowel disease: mucosa immune responses and protection by tolerance. J Immunol 157:2174–2185

Emoto M, Neuhaus O, Emoto Y, Kaufmann SH (1996) Influence of β2-microglobulin expression on gamma-interferon secretion and target cell lysis by intraepithelial lymphocytes during intestinal Listeria monocytogenes infection. Infect Immun 64:569–575

Farstad IN, Halstensen TS, Lien B, Kilshaw PJ, Lazarovitz AI, Brandtzaeg P (1996) Distribution of β7 integrins in human intestinal mucosa and organised gut-associated lymphoid tissue. Immunology 89:227–237

Farstad IN, Norstein J, Brandtzaeg P (1997a) Phenotypes of B and T cells in human intestinal and mesenteric lymph. Gastroenterology 112:163–173

Farstad IN, Halstensen TS, Kvale D, Fausa O, Brandtzaeg P (1997a) Topographic distribution of homing receptors on B and T cells in human gut associated lymphoid tissue. Am J Pathol 150:187–199

Ferguson A, Jarrett EEE (1975) Hypersensitivity reactions in small intestine. 1. Thymus-dependence of experimental "partial villous atrophy". Gut 16:114–117

Finch PW, Pricolo V, Wu A, Finkelstein SD (1996) Increased expression of keratinocyte growth factor messenger RNA associated with inflammatory bowel disease. Gastroenterology 110:441–451

Fukaura H, Kent SC, Pietrusewicz MJ, Khoury SJ, Weiner HL, Hafler DA (1996) Induction of circulating myelin basic protein and proteolipid protein-specific transforming growth factor-β1-secreting Th3 T cells by oral administration of myelin in multiple sclerosis patients. J Clin Invest 98:70–77

Fujihashi K, Taguchi T, Aicher WK, McGhee JR, Bluestone JA, Eldridge JH, Kiyono H (1992) Immunoregulatory functions for murine intraepithelial lymphocytes: γδ T cell receptor-positive (TCR+) T cells abrogate oral tolerance, while αβ TCR+ T cells provide B cell help. J Exp Med 175:695–707

Fujihashi K, Yamamoto M, McGhee JR, Kiyono H (1993) αβ T cell receptor-positive intraepithelial lymphocytes with CD4+, 8- and CD4+, CD8+ phenotypes from orally immunized mice provide Th2-like function for B cells responses. J Immunol 151:6681–6691

Fuss IJ, Neurath M, Boirivant M, Klein JS, de la Motte C, Strong SA, Fiocchi C, Strober W (1996) Disparate CD4+ lamina propria (LP) lymphokine secretion profiles in inflammatory bowel disease. J Immunol 157:1261–1270

Garside P, Steel M, Worthey EA, Satoskar A, Alexander J, Bluethmann H, Liew FY, Mowat AM (1995) T helper 2 cells are subject to high dose oral tolerance induction. J Immunol 154:5649–5655

Goke M, Podolsky DK (1996) Regulation of the mucosal epithelial barrier. Baillieres Clin Gastroenterol 10:393–405

Groux H, O'Garra A, Bigler M, Rouleau M, Antonenko S, DeVries JE, Roncarolo MG (1997) A CD4+ T-cell subset inhibits antigen-specific T-cell responses and prevents colitis. Nature 389:737–742

Guy-Grand D, Cerf-Bensussan N, Malissen B, Malassis-Seris M, Briottet C, Vassalli P (1991) Two gut intraepithelial CD8+ lymphocyte populations with different T cell receptors: a role for the gut epithelium in T cell differentiation. J Exp Med 173:471–481

Guy-Grand D, Cuenod-Jabri B, Malassis-Seris M, Selz F, Vassalli P (1996) Complexity of the mouse gut T cell immune system:identification of two distinct natural killer T cell intraepithelial lineages. Eur J Immunol 26:2248–2256

Hagenbaugh A, Sharma S, Dubinett SM, Wei SH-Y, Aranda R, Cheroutre H, Fowell DJ, Binder S, Tsao B, Locksley RM, Moore KW, Kronenberg M (1997) Altered immune responses in interleukin 10 transgenic mice. J Exp Med 185:2101–2110

Halstensen TS, Farstad IN, Scott H, Fausa O, Brandtzaeg P (1990) Intraepithelial TcR αβ+ lymphocytes express CD45R0 more often than the TcR γδ counterparts in celiac disease. Immunology 71:460–466

Halstensen TS, Scott H, Fausa O, Brandtzaeg P (1993) Gluten stimulation of coeliac mucosa in vitro induces activation (CD25) of lamina propria CD4+ T cells and macrophages but no crypt-cell hyperplasia. Scand J Immunol 38:581–590

Haneda K, Sano K, Tamura G, Sato T, Habu S, Shirato K (1997) TGFβ induced by oral tolerance ameliorates experimental tracheal eosinophilia. J Immunol 159:4484–4490

Hauer AC, Breese EJ, Walker-Smith JA, MacDonald TT (1997) The frequency of cells secreting interferon-γ, IL-4, IL-5 and IL-10 in the blood and duodenal mucosa of children with cows' milk hypersensitivity. Pediatr Res 42:1–10

Heyman M, Darmon N, Dupont C, Dugas B, Hirribaren A, Blaton MA, Desjeux JF (1994) Mononuclear cells from infants allergic to cow's milk secrete tumor necrosis factor α, altering intestinal function. Gastroenterology 106:1514–1523

Holtmeier W, Chowers Y, Lumeng A, Morzycka-Wroblewska E, Kagnoff MF (1995) The δT cell receptor repertoire in human colon and peripheral blood is oligoclonal irrespective of V region usage. J Clin Invest 96:1108–1117

Holtmeier W, Witthoft T, Hennemann A, Winter HS, Kagnoff MF (1997) The TCRδ repertoire in human intestine undergoes characteristic changes during fetal to adult development. J Immunol 158:5632–5641

Hornquist E, Lycke N (1993) Cholera toxin adjuvant greatly promotes antigen priming of T cells. Eur J Immunol 23:2136–2143

Hoyne GF, O'Hehir RE, Wraith DC, Thomas WR, Lamb JR (1993) Inhibition of T cell and antibody responses to house dust mite allergen by inhalation of the dominant T cell epitope in na and sensitised mice. J Exp Med 178:1783–1788

Husby S, Mestecky J, Moldoveanu Z, Holland S, Elson CO (1994) Oral tolerance in humans. J Immunol 152:4663–4670

Jarry A, Cerf-Bensussan N, Brousse N, Selz F, Guy-Grand D (1990) Subsets of CD3+ (T cell receptor αβ or γδ) and CD3- lymphocytes isolated from normal human gut epithelium display phenotypic features different from their counterparts in peripheral blood. Eur J Immunol 20:1097–2103

Jung HC, Eckmann L, Yang *S*-K, Panja A, Fierer J, Morzycka-Wroblewska E, Kagnoff MF (1995) A distinct array of pro-inflammatory cytokines is expressed in human colon epithelial cells in response to bacterial invasion. J Clin Invest 95:55–65

Kaiser GC, Polk DB (1997) Tumor necrosis factor α regulates proliferation in a mouse intestinal cell line. Gastroenterology 112:1231–1240

Karttunen R, Breese E, MacDonald TT (1994) Loss of interleukin-4 secreting cells in inflammatory bowel disease. J Clin Path 11:1015–1018

Karttunen R, Karttunnen T, Ekre H-P T, MacDonald TT (1995) Interferon-gamma and interleukin-4 secreting cells in the gastric antrum in Helicobacter pylori associated gastritis. Gut 36:341–345

Khoury SJ, Hancock WW, Weiner HL (1992) Oral tolerance to myelin basic recovery and natural recovery from experimental autoimmune encephalomyelitis are associated with downregulation of inflammatory cytokines and differential upregulation of TGFβ, interleukin-4, and prostaglandin E expression in the brain. J Exp Med 176:1355–1364

Koningsberger JC, Chott A, Logtenberg T, Wiegman LJ, Blumberg RS, van Berge Henegouwen GP, Balk SP (1997) TCR expression in human fetal intestine and identification of an early T cell receptor β-chain transcript. J Immunol 159:1775–1782

Kontakou M, Przemioslo RT, Sturgess RP, Limb GA, Ellis HJ, Day P, Ciclitira PJ (1995) Cytokine mRNA expression in the mucosa of treated coeliac patients after wheat peptide challenge. Gut 37:52–57

Kuchroo VK, Das MP, Brown JA, Ranger AM, Zamvil SS, Sobel RA, Weiner HL, Nabavi N, Glimcher LH (1995) B7–1 and B7–2 co-stimulatory molecules activate differentially the TH1/Th2 developmental pathways: application to autoimmune disease therapy. Cell 80:707–718

Kuper CF, Koornstra PJ, Hameleers DM, Biewenga J, Spit DJ, Duijvestijn AM, van Breda Vriesman PJ, Sminia T (1992) The role of nasopharyngeal lymphoid tissue. Immunol Today 13:219–224

Kusugami K, Matsuura T, West GA, Youngman KR, Rachmilewitz D, Fiocchi C (1991) Loss of interleukin-2-producing intestinal CD4+ T cells in inflammatory bowel disease. Gastroenterology. 101:1594–1605

Lagoo AS, Eldridge JH, Lagoo-Deenadaylan S, Black CA, Ridwan BU, Hardy KJ, McGhee JR, Beagley KW (1994) Peyer's patch CD8+ memory T cells secrete T helper type 1 and type 2 cytokines and provide help for immunoglobulin secretion. Eur J Immunol 24:3087–3092

Lefrancois L (1987) Carbohydrate differentiation antigens of murine T cells: expression on intestinal lymphocytes and intestinal epithelium. J Immunol 138:3375–3384

Lefrancois L, Olson S (1997) Reconstitution of the extrathymic intestinal T cell compartment in the absence of irradiation. J Immunol 159:538–541

Lefrancois L, Puddington L (1995) Extrathymic T-cell development: virtual reality? Immunol Today 16:16–21

Liesenfeld O, Kosek J, Remington JS, Suzuki Y (1996) Association of CD4+ T cell-dependent, interferon-γ-mediated necrosis of the small intestine with genetic susceptibility of mice to peroral infection with Toxoplasma gondii. J Exp Med 184:597–607

Lundin KE, Scott H, Hansen T, Paulsen G, Halstensen TS, Fausa O, Thorsby E, Sollid LM (1993) Gliadin specific, HLA-DQ ((*0501*0201) restricted T cells isolated from the small intestinal mucosa of celiac disease patients. J Exp Med 178:187–196

Lundqvist C, Melgar S, Yeung M-W, Hammarstrom S, Hammarstrom M-L (1996) Intraepithelial lymphocytes in human gut have lytic potential and a cytokine profile that suggests T helper 1 and cytotoxic functions. J Immunol 157:1926–1934

Lynch S, Kelleher D, McManus R, O'Farrelly C (1995) RAG1 and RAG2 expression in human intestinal epithelium:evidence of extrathymic T cell differentiation. Eur J Immunol 25:1143–1147

Ma CG, Zhang GX, Xiao BG, Link H (1996) Cellular mRNA expression of interferon-gamma (IFN-gamma), IL-4 and transforming growth factor-beta (TGF-beta) in rats nasally tolerised against experimental autoimmune myasthenia gravis (EAMG). Clin Exp Immunol 104:509–516

MacDonald TT (1997) Cytokine gene deleted mice in the study of gastrointestinal inflammation. Eur J Gastoenterol Hepatol 9:1051–1055

Marinaro M, Staats HF, Hiroi T, Jackson RJ, Coste M, Boyaka PN, Okahashi N, Yamamoto M, Kiyono H, Bluethmann H, Fujihashi K, McGhee JR (1995) Mucosal adjuvant effect of cholera toxin in mice results from induction of T helper 2 (Th2) cells and IL-4. J Immunol 155:4621–4629

Metzler B, Wraith DC (1993) Inhibition of experimental autoimmune encephalomyelitis by inhalation but not oral administration of encephalitogenic peptide: influence of MHC binding affinity. Int Immunol 5:1159–1165

Miller A, Lider O, Weiner HL (1991) Antigen-driven bystander suppression after oral administration of antigens. J Exp Med 174:791–798

Miller A, Lider O, Roberts AB, Sporn MB, Weiner HL (1992) Suppressor T cell generated by oral tolerization to myelin basic protein suppress both in vitro and in vivo immune responses by the release of transforming growth factor β after antigen-specific triggering. Proc Natl Acad Sci USA 89:421–425

Mizoguchi A, Mizoguchi E, Chiba C, Spiekermann GM, Tonegawa S, Nagler-Anderson C, Bhan AK (1996) Cytokine imbalance and autoantibody production in T cell receptor-alpha mutant mice with inflammatory bowel disease. J Exp Med 183:847–856

Monteleone G, Biancone L, Marasco R, Morrone G, Marasco O, Luzza F, Pallone F (1997) Interleukin 12 is expressed and actively released by Crohn's disease intestinal lamina propria mononuclear cells. Gastroenterology 112:1169–1178

Mowat AM (1987) The regulation of immune responses to dietary protein antigens. Immunol Today 1987 8:93–98

Mullin GE, Lazenby AJ, Harris ML, Bayless TM, James SP (1992) Increased interleukin-2 messenger RNA in the intestinal mucosal lesions of Crohn's disease but not ulcerative colitis. Gastroenterology 102:1620–1627

Neurath MF, Fuss I, Kelsall BL, Stuber E, Strober W (1995) Antibodies to interleukin 12 abrogate established experimental colitis in mice. J Exp Med 182:1281–1290

Neurath MF, Fuss I, Kelsall BL, Presky DH, Waegell W, Strober W (1996) Experimental granulomatous colitis in mice is abrogated by induction of TGFβ-mediated oral tolerance. J Exp Med 183:2605–2616

Neurath MF, Fuss I, Pasparakis M, Alexapoulou L, Haralambous S, Meyer zum Buschenfelde KH, Strober W, Kollias G (1997) Predominant pathogenic role of tumor necrosis factor in experimental colitis. Eur J Immunol 27:1743–1750

Niessner M, Volk BA (1995) Altered Th1/Th2 cytokine profiles in the intestinal mucosa of patients with inflammatory bowel disease as assessed by quantitative reversed transcribed polymerase chain reaction (RT-PCR). Clin Exp Immunol 101:428–435

Nilsen EM, Lundin KEA, Krajci P, Scott H, Sollid LM, Brandtzaeg P (1995) Gluten-specific, HLA-DQ restricted T cells from coeliac mucosa produce cytokines with Th1 or Th0 profile dominated by interferon-γ. Gut 37:766–776

Parronchi P, Romagnani P, Annunziato F, Sampognaro S, Becchio A, Giannarini L, Maggi E, Pupilli C, Tonelli F, Romagnani S (1997) Type 1 T-helper cell predominance and interleukin-12 expression in the gut of patients with Crohn's disease. Am J Pathol 150:823–832

Pender SLF, Tickle SP, Docherty AJP, Howie D, Wathen NC, MacDonald TT (1997) A major role for matrix metalloproteinases in T cell injury in the gut. J Immunol 158:1582–1590

Pender SLF, Fell JMC, Chamow SM, Ashkenazi A, MacDonald TT (1998) A p55 tumor necrosis factor (TNF) receptor immunoadhesin prevents T cell-mediated intestinal injury by inhibiting matrix metalloproteinase production. J Immunol 160:4098–103

Penney LM, Kilshaw PJ, MacDonald TT (1995) Regional variation in the proliferative rate and lifespan of alpha beta TCR+ and gamma delta TCR+ intraepithelial lymphocytes in the murine small intestine. Immunology 86:212–218

Poussier P, Edouard P, Lee C, Binnie M, Julius M (1992) Thymus-independent development and negative selection of T cells expressing T cell receptor alpha/beta in the intestinal epithelium: evidence for distinct circulation patterns of gut-and thymus-derived T lymphocytes. J Exp Med 176:187–199

Powrie F, Correa-Oliveira R, Mauze S, Coffman RL (1994a) Regulatory interactions between CD45RB^{high} and CD45RB^{low} CD4+ T cells are important for the balance between protective and pathogenic cell-mediated immunity. J Exp Med 179:589–600

Powrie F, Leach MW, Mauze S, Menon S, Caddle LB, Coffman RL (1994b) Inhibition of Th1 responses prevents inflammatory bowel disease in scid mice reconstituted with CD45RBhi CD4+ T cells. Immunity 1:553–562

Powrie F, Carlino J, Leach MW, Mauze S, Coffman RL (1996) A critical role for transforming growth factor-beta but not interleukin 4 in the suppression of T helper type 1-mediated colitis by CD45RB(low) CD4+ T cells. J Exp Med 183:2669–2674

Rugtveidt J, Nilsen EM, Bakka A, Carlsen H, Brandtzaeg P, Scott H (1997) Cytokine profiles differ in newly recruited and resident subsets of mucosal macrophages from inflammatory bowel disease. Gastroenterology 112:1493–1505

Saarialho-Kere UK, Vaalamo M, Puolakkainen P, Airola K, Parks WC, Karjalainen-Lindsberg ML (1996) Enhanced expression of matrilysin, collagenase, and stromelysin-1 in gastrointestinal ulcers. Am J Pathol 148:519–526

Saparov A, Elson CO, Devore-Carter D, Bucy RP, Weaver CT (1997) Single cell analyses of CD4+ T cells from αβ T cell receptor transgenic mice: a distinct mucosal cytokine phenotype in the absence of transgene-specific antigen. Eur J Immunol 27:1774–1781

Schieferdecker HL, Ullrich R, Hirseland H, Zeitz M (1992) T cell differentiation antigens on lymphocytes in the human intestinal lamina propria. J Immunol 149:2816–2822

Simpson SJ, Hollander GA, Mizoguchi E, Allen D, Bhan AK, Wang B, Terhorst C (1997) Expression of pro-inflammatory cytokines by TCR αβ+ and TCR γδ+ T cells in an experimental model of colitis. Eur J Immunol 27:17–25

Spencer J, Isaacson P, MacDonald TT, Thomas AJ, Walker-Smith JA (1991) Gamma/delta T cells and the diagnosis of coeliac disease. Clin Exp Immunol 85:109–113

Staines NA, Harper N, Ward FJ, Malmstrom V, Holmdahl R, Bansal S (1996) Mucosal tolerance and suppression of collagen-induced arthritis (CIA) induced by nasal inhalation of synthetic peptide 184–198 of bovine type II collagen (CII) expressing a dominant T cell epitope. Clin Exp Immunol 103:368–375

Staats HF (1997) Intranasal immunisation with carrier-free synthetic peptides. Mucosal Immunol Update 5:16–20

Taguchi T, McGhee JR, Coffman RL, Beagley KW, Eldridge JH, Takatsu K, Kiyono H (1990) Analysis of Th1 and Th2 cells in murine gut-associated tissues. Frequencies of CD4+ and CD8+ T cells that secrete IFN-γ and IL-5. J Immunol 145:68–77

Tamura S, Kurata T (1997) Intranasal immunization for human application. Mucosal Immunol Update 5:8–10

Targan SR, Deem DL, Liu M, Wang S, Nel A (1995) Definition of a lamina propria T cell responsive state. Enhanced cytokine responsiveness of T cells stimulated through the CD2 pathway. J Immunol 154:664–675

Targan SR, Hanauer SB, VanDeventer SJH, Mayer L, Present DH, Braakman T, DeWoody KL, Schaible TF, Rutgeerts PJ (1997) A short-term study of chimeric monoclonal antibody cA2 to tumor necrosis factor α (for Crohn's disease. New Engl J Med 337:1029–1035

Tian J, Atkinson MA, Clare-Salzer M, Herschenfeld A, Forsthuber T, Lehmann PV, Kaufman DL (1996) Nasal administration of glutamate decarboxylase (GAD65) peptides induces Th2 responses and prevents murine insulin-dependent diabetes. J Exp Med 183:1561–1567

Trentham DE, Dynesius-Trentham RA, Orav EJ, Combitchi D, Lorenzo C, Sewell KL, Hafler DA, Weiner HL (1993) Effects of oral administration of type II collagen on rheumatoid arthritis. Science 261:1727–1730

Vajdy M, Kosco-Vilbois MH, Kopf M, Kohler G, Lycke N (1995) Inpaired mucosal immune responses in interleukin-4 targeted mice. J Exp Med 181:41–53

Van Cott JL, Staats HF, Pasual DW, Roberts M, Chatfield SN, Yamamoto M, Coste M, Carter PB, Kiyono H, McGhee JR (1996) Regulation of mucosal and systemic antibody responses by T helper subsets, macrophages, and derived cytokines following oral immunization with live recombinant Salmonella. J Immunol 156:1504–1514

Viney J, MacDonald TT, Kilshaw PJ (1989) T cell receptor expression in intestinal intraepithelial lymphocyte subpopulations of normal and athymic mice. Immunology 66:583–587

Viney JL, Kilshaw PK, MacDonald TT (1990) Cytotoxic αβ and γδ T cells in murine intestinal epithelium. Eur J Immunol 20:1623–1626

Watanabe M, Ueno Y, Yajima T, Iwao Y, Tsuchiya M, Ishikawa H, Aiso S, Hibi T, Ishii H (1995) Interleukin 7 is produced by human intestinal epithelial cells and regulates the proliferation of intestinal mucosal lymphocytes. J Clin Invest 95:2495–2953

Weiner HL (1997) Oral tolerance: immune mechanisms and treatment of autoimmune disease. Immunol Today 19:335–343

Weiner HL, Mackin GA, Matsui M, Orav EJ, Khoury SJ, Dawson DM, Hafler DA (1993) Double-blind pilot trial of oral tolerization with myelin antigens in multiple sclerosis. Science 259:1321–1324

West GA, Matsuura T, Levine AD, Klein JS, Fiocchi C (1996) Interleukin 4 in inflammatory bowel disease and mucosal immune reactivity. Gastroenterology 110:1683–1695

Wilson AD, Bailey M, Williams NA, Stokes CR (1991) The in vitro production of cytokines by mucosal lymphocytes immunized by oral administration of keyhole limpet hemocyanin using cholera toxin as an adjuvant. Eur J Immunol 21:2333–2339

Xu-Amanu J, Aicher WK, Taguchi T, Kiyono H, McGhee JR (1992) Selective induction of Th2 cells in murine Peyer's patches by oral immunisation. Int Immunol 4:433–445

Xu-Amanu J, Kiyono H, Jackson RJ, Staats HF, Fujihashi K, Burrows PD, Elson CO, Pillai S, McGhee JR (1993) Helper T cell subsets for immunoglobulin A responses: oral immunization with tetanus toxoid and cholera toxin as adjuvant selectively induces Th2 cells in mucosa associated tissues. J Exp Med 178:1309–1320

Yamamoto S, Russ F, Teixeira HC, Conradt P, Kaufmann SH (1993) Listeria monocytogenes-induced γ-interferon secretion by intestinal intraepithelial lymphocytes. Infect Immun 61:2154–2161

Yamamoto S, Kiyono H, Yamamoto M, Imaoka K, Fujihashi K, Van Ginkel FW, Noda M, Takeda Y, McGhee JR (1997) A nontoxic mutant of cholera toxin elicits Th2-type responses for enhanced mucosal immunity. Proc Natl Acad Sci USA 94:5267–5272

Bacterial Epithelial Cell Cross Talk

B. Raupach[1,2], J. Mecsas[1], U. Heczko[3], S. Falkow[1,4], and B.B. Finlay[3]

[1] Department of Microbiology and Immunology, Stanford University, Stanford, CA 94305, USA
[2] Department of Immunology, Max-Planck Institut für Infektionsbiologie, 10117 Berlin, Germany
[3] Biotechnology Laboratory, University of British Columbia, #237-6174 University Blvd, Vancouver, B.C., Canada, V6T-1Z3
[4] Rocky Mountain Laboratory, Hamilton, MT 59840, USA

1 Introduction

1.1 Membraneous (M) Cells: A Gateway for Both the Mucosal Immune System and Pathogens

The ability to bind to epithelial cells is an important determinant of virulence for many pathogenic bacteria. The interaction of enteropathogens with host cells of the intestinal epithelia can lead to either extracellular adhesion or internalization of the microorganisms. Generally, binding to intestinal host cells is essential for the bacteria to resist the fluid flow of the luminal contents. Once bound to the epithelial surface, the bacteria may flourish and establish a micro-colony. However, adhesion also serves as a prerequisite for subsequent invasion. Increasingly, evidence suggests that many enteropathogenic bacteria target specialized antigen-transporting membraneous cells (M cells) within the follicle-associated epithelium (FAE) of the Peyer's patches (SIEBERS and FINLAY 1996). This is a remarkable and somewhat ironical adaptation by the pathogens, because not only are M cells a minority in the epithelial population within the intestine, but they also represent a key element of the gastrointestinal immune system. M cells serve as samplers of the intestinal content by constantly transporting potential antigens from the gut lumen to the underlying lymph tissues. Therefore, it is surprising that several enteropathogens have chosen M cells as their conduit to deeper tissues.

In this review, we discuss the present understanding of the cross talk between enteropathogenic bacteria and M cells. Our understanding of this is based on the study of the interactions between M cells and microbes in animal models and is also extrapolated from infection experiments of established tissue-culture models. Only recently has a system for in vitro differentiation of M cells been described (KERNEIS et al. 1997). We will focus on the different strategies that four enteric microbial pathogens utilize to interact with M cells and establish infection resulting in diarrhea: enteropathogenic *Escherichia coli*, *Shigella flexneri*, *Salmonella typhimurium*, and enteropathogenic *Yersinia* spp.

1.2 Characteristics of M Cells

It has become clear that a variety of microbes that cross the epithelial barrier of the gut commonly target the Peyer's patches and particularly a specialized epithelial cell, the M cell. Since M cells constantly endocytose, researchers were initially unsure whether this constitutive sampling mechanism was sufficient for transepithelial movement of microorganisms, or whether specific adhesion of the bacteria to M cells was required for invasion of pathogens.

M cells have characteristic glycosylation patterns which vary between species and tissue location. Since cell surface glycoconjugates have been shown to serve as receptors for some bacterial adhesins, the identification of glycoconjugates characteristic of M cell surfaces suggest that the interaction between bacteria and M cells can be specific and does not exclusively rely on passive internalization via M-cell endocytosis. Furthermore, the regional variation in the glycosylation state displayed by different intestinal M cells implies that bacterial adhesins and M cell-specific surface glycoconjugates may decide which portion of the intestine is colonized or infected by individual enterics.

1.3 How Enteropathogens Establish Infection/Disease

Some enteric and systemic pathogens have subverted the well-designed M-cell antigen delivery system to become the "Achilles heel" of the host. In fact, many common themes in the pathogenesis of enteropathogenic *E. coli* (EPEC), *Shigella flexneri*, *Salmonella typhimurium* and enteropathogenic *Yersinia* spp. exist with respect to how these pathogens interact with epithelial cells. However, despite the mechanistic similarities, which will be discussed later in this chapter, the course of infection and the outcome of disease differs significantly among the four enteropathogens.

The rabbit pathogen *Escherichia coli* RDEC-1 adheres specifically to M cells and causes characteristic attaching/effacing lesions (A/E lesions), but resists internalization by M cells. This enteroadherent phenotype allows the bacteria to colonize the intestinal epithelium without being delivered to the submucosa and without eliciting a mucosal immune response (Fig. 1).

Other enteropathogens exploit facilitated transepithelial movement through M cells, using it as an invasion route to enter the host. Under regular circumstances, this M-cell assisted uptake would lead to the elimination of the pathogens after

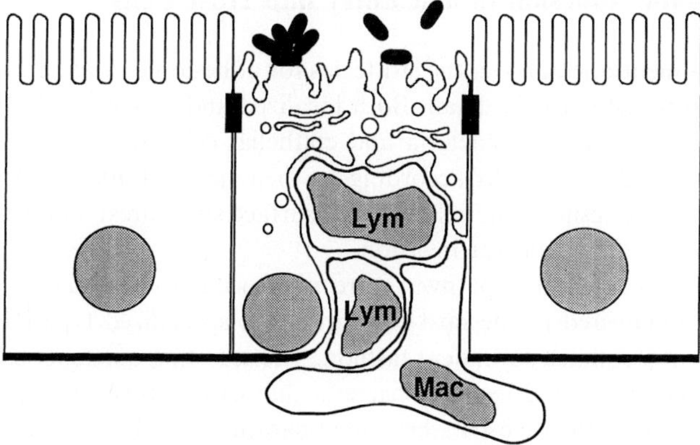

Fig. 1. Enteropathogenic *E. coli* (EPEC) interactions with membraneous (M) cells. EPEC adheres to the apical surface causing effacement of microvilli and pedestal formation beneath adherent bacteria. *Lym*, lyphoid; *Mac*, macrophage. Reproduced with permission from Siebers and Finlay (1996)

initiation of a successful host immune response, as it does for a *Vibrio cholerae* infection. However, certain microorganisms have developed clever strategies that counteract mucosal immunity and enable the bacteria to resist host defense mechanisms. For instance, *S. flexneri*, *S. typhimurium* and enteropathogenic *Yersinia* spp. can induce apoptosis in macrophages which are the host effector cells that the bacteria encounter immediately after M-cell passage. However, the subsequent steps in the pathogenesis of the three pathogens diverge, since *S. flexneri* establishes localized infection, whereas *S. typhimurium* and enteropathogenic *Yersinia* spp. can cause systemic disease as intracellular or extracellular pathogens, respectively.

To avoid contact with immune effector cells, *Shigella flexneri* escapes from the M-cell vacuole rapidly after it is endocytosed, spreads laterally from cell to cell, replicates intracellularly or extracellularly and can invade adjacent epithelial cells from the basolateral side. In the submucosa, the bacteria can trigger an acute inflammatory reaction, which is responsible for tissue damage and rapid spreading of the infection within the epithelium. The intracellular localization provides a replicative niche protected from the host immune response.

To survive the attack by subepithelial phagocytes after M cell transcytosis and establish systemic infection, *S. typhimurium* and the enteropathogenic species of *Yersinia* have developed mechanisms that allow them to cope with the antimicrobial properties of macrophages. *S. typhimurium* resides within a specialized intracellular compartment in which the bacteria can replicate and persist in the reticuloendothelial system, whereas *Yersinia* species have evolved a strategy to prevent internalization by macrophages and nonphagocytic host cells, and therefore replicate extracellularly. However, both *Salmonella* and *Yersinia* species exploit circulating cells to spread to sites distant from the initial port of entry and establish a successful systemic infection.

1.4 Common Themes for Adhesion to and Entry into Host Cells

For the enteropathogens mentioned above, the strategies for adhesion to and entry into host cells have been investigated in detail. Both localized and systemic infections result from a cross talk between bacteria and epithelial cells in which the microorganisms manipulate the host cells according to their needs. Initially, the specific interaction between adhesins (generally fimbrial surface structures) and the epithelial surface selects the target cells for invasion.

Subsequently, eukaryotic signaling pathways are influenced by bacterial effector molecules that are delivered to the host cell through a specialized type-III secretion system (TTSS) in a contact-dependent fashion. The secretion machineries are multi-component structures found in many bacteria that cause disease in animals and plants; as an example, the type-III secretion apparatus from *Yersinia* will be described. For the enteropathogens mentioned in this review, the TTSS are functionally conserved and located on pathogenicity islands. As a result of secretion and translocation of virulence proteins, stimulation of signal transduction cascades, including kinases, phosphatases and small GTPases, and the formation of second messenger molecules, such as calcium ions and inositol phospholipids, are frequently observed. In each case, structural reorganization of the actin cytoskeleton is a central requirement for both invasion and the formation of A/E lesions.

It is remarkable that various enteropathogens employ strikingly similar mechanisms to establish cross talk between the bacterium and its host, but yet are able to initiate distinctive responses from mammalian host cells that best serve their pathogenic strategy. The host-cell interactions that set the individual pathogens apart are discussed in detail in this review.

1.5 How Bacteria Interact with Surface Structures on Target Cells

Based on the difference in morphology of the eukaryotic cell surface during the interaction and in the identity of both bacterial ligand and (eukaryotic) receptor, several strategies for the interaction between the bacterium and the host-cell surface can be distinguished.

In the case of pathogens, like EPEC, that cause A/E lesions, bacterial attachment is generally not followed by uptake. The pedestals underneath the bacteria form after intimate binding of a ligand on the bacterial surface to a receptor protein in the host-cell plasma membrane. Remarkably, the microorganisms supply both the bacterial ligand and the receptor on the host-cell surface. The outer membrane protein intimin binds to a 90 kDa tyrosine phosphorylated protein (Tir) in the plasma membrane of the target cell which is produced by the bacterium and delivered to the host cell membrane by a TTSS.

For *S. typhimurium* and *S. flexneri* the entry process into the host cell is based on bacterial–epithelial cell cross talk in which a multi-component structure on the bacterial surface transmits a signal to the host cell, resulting in a reshaping of the host-cell surface that allows bacterial entry via macropinocytosis. Localized rearrangement of the mammalian plasma membrane at the site of bacterial contact is characteristic of the "splashes" caused by *S. typhimurium* immediately after interaction with host cells. A surface organelle called an invasome, secreted by the bacteria, is potentially responsible for the induction of membrane ruffles on target cells. However, neither a *Salmonella*-encoded ligand comparable to an invasin has been identified, nor is it clear whether a specific adhesion event is necessary for *Salmonella* invasion. Entry of *S. flexneri* into host cells requires secreted Ipa complexes that are believed to trigger the formation of splash-like structures characterized by the polymerization of actin and the formation of membrane projections at the sites of bacterium–host cell interaction. In addition, the Ipa complexes were shown to bind to alpha5beta1 integrins on Chinese hamster's ovary (CHO) cells and entry of *Shigella flexneri* into enterocytes is thought to rely on the binding of the bacteria to integrins located on the basolateral side of the cell surface.

The entry of enteropathogenic *Yersinia* spp. requires that appropriate receptors on the mammalian cell surface bind a defined bacterial protein that mimics a natural ligand, as is the case after interaction of *Yersinia* invasin and beta1 integrins. The individual bacterium establishes close contact with the epithelium, is subsequently engulfed by the host cell membrane in a zipper-like fashion and promotes its own entry.

2 Enteropathogenic *E. Coli* and Other Attaching and Effacing Pathogens

The EPEC family comprises a group of related enteric pathogens that adhere specifically to intestinal mucosal surfaces, especially the M cells of Peyer's patches, yet do not penetrate further into the underlying lymph system. Pathogens belonging to this group include EPEC; enterohemorrhagic *E. coli* (EHEC); several EPEC-like animal pathogens that cause disease in rabbits (REPEC, including the strain RDEC-1), dogs, pigs (PEPEC), etc; *Citrobacter rodentium*; *Hafnia alvei*; and possibly even *Helicobacter pylori*. These organisms cause cytoskeletal rearrangement and pedestal formation on relevant host epithelial cells (Donnenberg et al. 1997). EHEC, which causes enteric colitis ("hamburger disease") can also cause hemolytic uremic syndrome in approximately 10% of cases. EHEC possess all of the EPEC virulence factors needed for pedestal formation, but has an additional shiga toxin which contributes to its increased pathogenesis.

The binding of EPEC organisms to host cells is a two-step process. The initial loose contact and selection of target cells is established by fimbriae (bundle-forming pili). Subsequently, the bacteria closely associate with the host-cell surface and form a characteristic histological lesion, the A/E lesion (Fig. 1), which results in production of diarrhea. A/E lesions are marked by dissolution of the intestinal brush border surface and loss of epithelial microvilli (effacement) at the sites of bacterial attachment. Once bound, these organisms reside upon a cup-like projection or pedestal made from a reorganization of the host-cell cytoskeleton containing several cytoskeletal components, including actin, alpha-actinin, ezrin, talin, and myosin light chain (Finlay et al. 1992). Formation of the A/E lesion appears to be responsible for fluid secretion and diarrhea, although mechanistically this remains to be proven. It has been suggested that disruption of the brush border and microvilli may be responsible for diarrhea. Although these pathogens can enter (invade) tissue-culture cells, they do not normally cause invasive disease and rarely penetrate the intestinal barrier.

2.1 Initial Adherence

Pathogens that cause A/E lesions appear to have specialized adhesins that target them to specific locations of their host's intestinal surface. For example, EPEC colonizes the small bowel in humans, yet EHEC targets the large-intestinal surface.

Initial adherence in EPEC is mediated by the plasmid encoded bundle-forming pilus (BFP), found on a 55–70 MDa plasmid that is common to EPEC strains. Mutants in EPEC that arc defective in this pilus produce fewer A/E lesions on epithelial cells, but the lesions are indistinguishable from those caused by parental EPEC. Their virulence is also attenuated in human volunteers.

Adherence of the rabbit EPEC strain RDEC-1 has been characterized, and it has been shown that this pathogen targets, specifically, the M cells of Peyer's patches and this tissue tropism is mediated by adhesive factor/rabbit 1 (AF/R1) (Berendson et al. 1983). Upon adherence to mucosal M cells, the characteristic A/E lesion is formed on their surface (Fig. 2). Other REPEC strains encode other adhesins, such as AF/R2, which mediate similar targeting to M cells within Peyer's patches (Fiederling et al. 1997). Because of the similarity in A/E lesions, it is thought, but not proven, that EHEC and EPEC target to Peyer's patches in their human hosts.

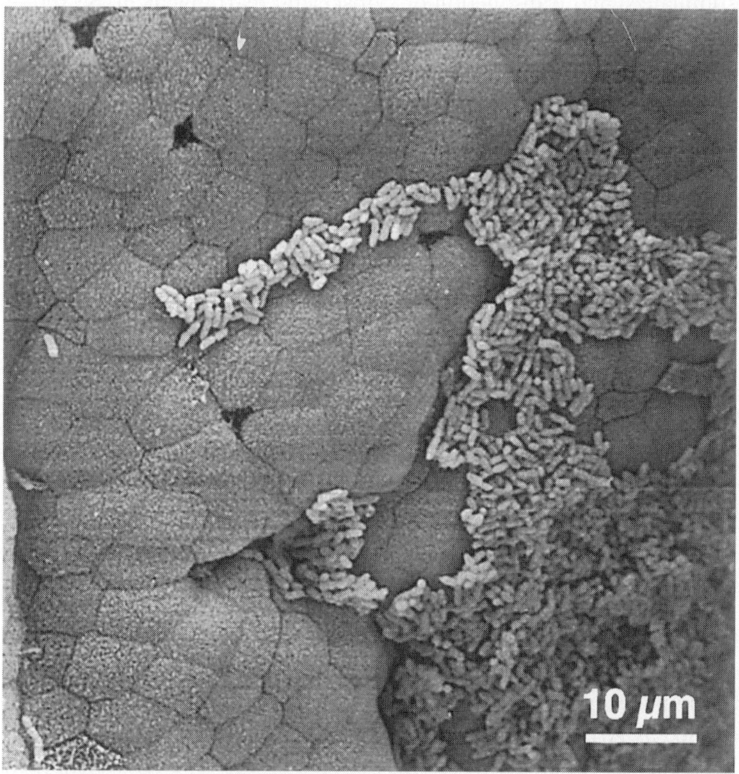

Fig. 2. Scanning electron micrograph of rabbit enteropathogenic *E. coli* (EPEC) O103 adhering to the intestinal epithelium of rabbits

2.2 Signal Transduction

When EPEC and other A/E pathogens interact with cultured epithelial cells, several signal transduction pathways are activated in the epithelial cells, including the release of the eukaryotic secondary messengers IP_3 and intracellular calcium, and phosphorylation of myosin light chain (DYTOC et al. 1994; FOUBISTER et al. 1994). Additionally, EPEC and REPEC (but not EHEC) binding to cultured epithelial cells also causes tyrosine phosphorylation of a protein in the host-cell plasma membrane called Hp90 (see Sect. 2.3 below) (ROSENSHINE et al. 1992).

All of the genes known to be involved in A/E formation (except the plasmid encoded regulator, *per*) are found within a unique contiguous region in the EPEC chromosome called LEE (*l*ocus of *e*nterocyte *e*ffacement) (MCDANIEL et al. 1995). The LEE region is a pathogenicity island inserted in the *E. coli* chromosome and is conserved in all the A/E pathogens, although its site of chromosomal insertion varies between pathogens. Several bacterial loci have been identified that are involved in activating epithelial signal transduction and are encoded within the LEE region (DONNENBERG et al. 1997). Strains containing mutations in the secreted proteins encoded by *espA* (*E. coli* *s*ecreted *p*rotein A), *espB* (formerly *eaeB*), and *espD* do not stimulate signal transduction or cytoskeletal rearrangement. Secretion of these proteins are mediated by a type-III secretion system, encoded by approximately 20 genes. Collectively, this information indicates that EPEC secretes at least three molecules (EspA, EspB, and EspD) that are critical for activating signal transduction and cytoskeletal rearrangement in epithelial cells.

2.3 Intimate Adherence

INTIMIN is the product of a bacterial LEE locus, *eaeA*, and is a 94 kDa outer membrane protein that is key for intimate adherence and A/E lesions (JERSE et al. 1990). *eaeA* mutants form immature A/E lesions and do not organize phosphotyrosine proteins and cytoskeletal components beneath adherent bacteria, although epithelial signal transduction is still activated. Intimin appears to participate in reorganization of the underlying host cytoskeleton after other bacterial factors stimulate epithelial signal transduction and insert the intimin receptor into the host cell (ROSENSHINE et al. 1996).

A 90 kDa tyrosine-phosphorylated host membrane-localized protein (originally called Hp90) is found beneath adherent organisms at the tip of extended pseudopods. This protein interacts with intimin and is the intimin receptor (ROSENSHINE et al. 1996). Recently, the identity of this protein has been established (KENNY et al. 1997). Surprisingly, it is a bacterial protein that is encoded by *tir* (translocated intimin receptor) in the LEE region, upstream of the intimin gene, *eaeA*. Tir is secreted by A/E pathogens into growth supernatant as a 78 kDa protein under specialized growth conditions. Proteins secreted by the type-III system (i.e., EspA, B, and D) are needed to deliver Tir into host cell membranes, where it is phosphorylated and its mobility shifts to 90 kDa. Upon intimin binding to Tir in the host cell membranes, a second series of signals occurs, including activation of phospholipase C gamma and cytoskeletal condensation under the adherent organisms (Kenny and Finlay 1997). It appears that Tir is responsible for organizing the underlying cytoskeletal rearrangements, directing the final condensation and organizing the host cytoskeleton upon association with intimin in the bacterial membrane. Injection of Tir, followed by intimate bacterial binding and pedestal formation appears critical for disease, as mutations in intimin or the secreted proteins are attenuated for disease in various animal models. Such binding also presumably initiates a host inflammatory response which is seen in response to disease.

3 *Shigella*

Bacillary dysentery, or shigellosis, is caused by penetration of *Shigella* spp. into the mucosa of the colon, where localized destruction of the intestinal epithelium and a strong inflammatory response is indicative of infection sites. The molecular mechanisms underlying the interactions between *Shigella* and their host cells have been studied using mainly *S. flexneri*, but most conclusions drawn from these investigations probably also apply to other *Shigella* species and to the closely related enteroinvasive *Escherichia coli* (EIEC).

3.1 Binding Colonic Versus Ileal M Cells

While *Shigella* appears to have an affinity for M cells, it is only found in the M cells of the colon and not the ileum. As discussed above, the initial attachment to M cells is often mediated by fimbrial adhesins. Despite much study, a *Shigella* adhesin has not yet been identified. Thus, the mechanism that *Shigella* uses to attach to M cells remains unclear. However, the observation that *Shigella* associates with M cells of the large intestine rather than the small intestine might mean that *Shigella* "finds" M cells via a passive mechanism (without the aid of an adhesin). The bacteria are flushed through the small intestine by the continuous peristalsis, whereas in the colon they have time to settle on M cells. Alternatively, as yet unidentified components of *Shigella* may specifically bind to surface structures on the M cells in the large intestine, since it has long been recognized that adhesion of wild-type *Shigella flexneri* to epithelial cells occurs more efficiently than *Shigella* mutants that lack the virulence plasmid.

3.2 Entry From the Basolateral Side

Surprisingly, the apical pole of polarized epithelial cells is only poorly invaded by *Shigella flexneri*, whereas the bacteria efficiently invade the basolateral pole (Fig. 3)

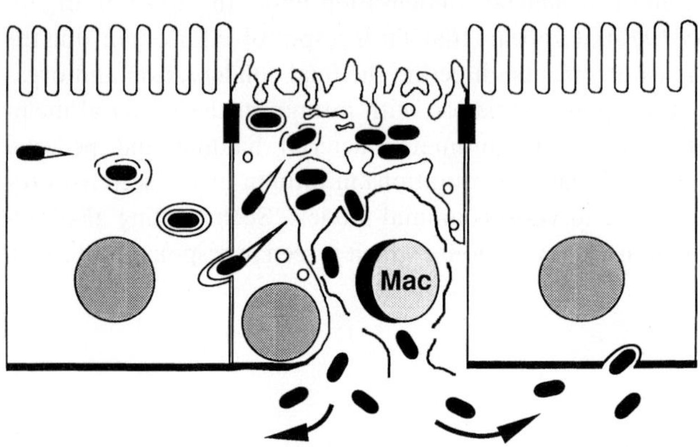

Fig. 3. *S. flexneri* interactions with membraneous (M) cells. Bacteria are taken up by M cells, released from the vacuole, and spread from cell to cell. They also cause apoptosis in underlying macrophages (*Mac*). Reproduced with permission from Siebers and Finlay, (1996)

(Mounier et al. 1992). This ability of *S. flexneri* (and also other organisms such as enteropathogenic *Yersinia* species) causes a dilemma for these microbes, since, after oral ingestion, it is the apical surface of the epithelium that the bacteria have access to from the intestinal lumen. The observation that *Shigella* uses M cells as the gateway to underlying tissues resolves this conundrum. Translocation through M cells guarantees rapid access for *S. flexneri* to the basolateral surface of epithelial cells. Moreover, *Shigella* has been shown to provoke a powerful chemotactic response leading to transmigration of polymorphonuclear leukocytes (PMN) to the apical surface. The bacteria can invade PMNs at the intercellular junctions between colonocytes and gain access to the basolateral surface of the epithelium this way (Perdomo et al. 1994a; Perdomo et al. 1994b).

3.3 Genetic Basis of *S. flexneri* Entry into Host Cells

All virulent *Shigella* isolates harbor a 220-kb plasmid that is also found in EIEC. A 31-kb fragment of the plasmid is necessary and apparently sufficient for entry into epithelial cells (Menard et al. 1996a). The DNA fragment contains 32 genes clustered into two operons that are transcribed in opposite orientation. One operon comprises the *mxi-spa* locus which encodes a specialized type-III secretion system; the other operon encodes the proteins translocated by this secretion system, IpaB–D and their chaperone IpgC. IpaB, IpaC, and IpaD are the secreted proteins that have been identified so far as being necessary for bacterial entry. After secretion, IpaB and IpaC can form a soluble complex, called the Ipa complex which is essential for invasion and can mediate uptake of latex beads by epithelial cells (Menard et al. 1996b). In vivo, however, it remains unclear whether the Ipa complex is used for uptake in M cells, PMNs and/or entry into epithelial cells from the basolateral side.

3.4 *Shigella*-Induced Host-Cell Changes

Shigella invasion into epithelial cells has been studied in vitro using HeLa cells. Bacterial uptake is accompanied by localized polymerization of actin and membrane ruffling (CLERC and SANSONETTI 1987). Actin-containing plasma-membrane projections are formed at the site of bacterial–host cell interaction, in which the actin filaments are cross linked by the actin-bundling protein plastin (ADAM et al. 1995). The engulfment of surface-associated bacteria by these projections leads to internalization of the organisms.

Additional processes required for *Shigella* invasion are the activation of protein tyrosine kinases (PTK) and the small GTPase Rho. However, a wide array of other host-cell components are recruited to the site of *Shigella* entry. Rho, which plays a role in the formation of stress fibers and focal adhesion plaques, has recently been shown to be required for actin polymerization associated with *Shigella* invasion in HeLa and CHO cells (ADAM et al. 1996; WATARAI et al. 1997). The invasive capacity of *Shigella flexneri* is greatly reduced when host cells are treated with the Rho-specific inhibitor C3 transferase from *Clostridium botulinum*. The proto-oncoprotein pp60c-src is also recruited to the *Shigella*-induced membrane projections. The protein tyrosine kinase pp60c-src specifically phosphorylates cortactin, an actin-binding protein, which represents the major PTK substrate at the sites of *Shigella* entry (DEHIO et al. 1995).

In addition to being the key components in the *Shigella*-induced cytoskeleton remodeling process, both pp60c-src and Rho are essential players in the signaling cascade that leads to the formation of focal adhesion plaques. Therefore, it has been suggested that the structures formed on contact of *Shigella flexneri* with epithelial cells resemble the interactions between focal adhesion plaques and extracellular matrix components. In addition, several other proteins recruited to the projections formed at sites of *Shigella* entry into epithelial cells are well-established constituents of adhesion plaques, including actin, alpha-actinin, talin, paxillin, pp125FAK, and vinculin (TRAN VAN NHIEU et al. 1997). Moreover, the binding of released Ipa proteins from *Shigella flexneri* to alpha5beta1 integrins on CHO cells, in a manner similar to that of the tissue form of fibronectin, has been recently reported (WATARAI et al. 1996). The interaction leads to the tyrosine phosphorylation of the integrin-regulated focal adhesion kinase pp125FAK and paxillin. Presumably, the activation of these proteins by phosphorylation is required for *S. flexneri* entry, since the pretreatment of cells with the tyrosine kinase inhibitor genistein reduces the invasive capacity of the bacteria significantly. However, the association of Ipa with alpha5beta1 integrins does not present a high-affinity receptor-ligand interaction as in the case of *Yersinia* invasin, binding to beta1 integrins (see Sect. 5.3.1 below).

Recently, a two-step model has been proposed for *Shigella*-induced entry into epithelial cells (Tran Van Nhieu et al. 1997). This model is based on the findings that IpaA, a *Shigella* protein secreted into host cells upon contact, rapidly associates with vinculin during bacterial entry. An *ipaA* mutant does not recruit vinculin and alpha-actinin and, thus, fails to induce the formation of focal adhesion-like structures to the site of bacterial entry, but is still capable of inducing foci of actin polymerization. In addition, *IpaA* deficient organisms possess only reduced invasive capacity compared with noninvasive *ipaB*, *ipaC* or *ipaD* mutants. Consequently, at least two events can be distinguished in *Shigella*-induced entry: an initial event which requires IpaB-D and leads to localized actin polymerization and inefficient *Shigella* internalization, and a subsequent phase that depends on IpaA, which induces the formation of focal adhesion-like structures required for efficient uptake.

4 *Salmonella typhimurium*

Host-adapted pathogenic *Salmonella* species can enter and grow within the reticuloendothelial system of the host and cause systemic enteric fevers (typhoid). In 1974, Carter and Collins analyzed the interactions of *Salmonella enteritidis* with intestinal tissue after oral infection of normal mice. They observed that early in infection, the bacteria are found in close association with the Peyer's patches of the terminal ileum rather than the epithelium of the intestinal wall (Carter and Collins 1974). Subsequent microscopic examinations revealed that both *S. typhi* (Kohbata et al. 1986) and *S. typhimurium* (Clark et al. 1994; Jones et al. 1994) can adhere to and invade the M cells of murine Peyer's patches (Fig. 4).

4.1 Specific Binding to M Cells

In tissue-culture models, *S. typhimurium* has been shown to invade all host cell lines tested thus far. Therefore, it is surprising that, in vivo, the bacterium displays an extraordinary host-cell selectivity. Why is it that in the FAE of ileal Peyer's patches, *S. typhimurium* specifically invades M cells and not neighboring enterocytes?

Many members of the family *Enterobacteriaceae* express fimbriae, or pili, which function as adhesions and mediate attachment to host surfaces. In fact, *S. typhimurium* contains three different fimbrial operons called *pef*, *lpf* and *fim*. Pef fimbriae and lpf fimbriae mediate selective binding to different host cells in the intestinal epithelium, whereas the presence or absence of type-1 fimbriae encoded by the *S. typhimurium fim* operon does not affect colonization in mice (Lockman and Curtiss 1992).

Recent experiments conducted by Bäumler et al. revealed that distinct fimbrial adhesions determine the epithelial target cell invaded by *S. typhimurium* in the

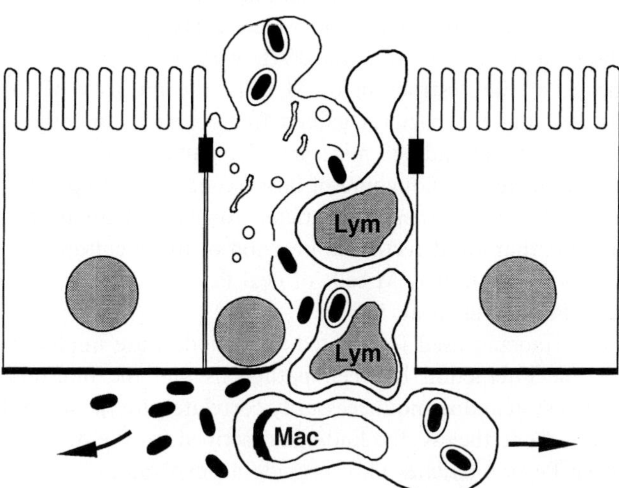

Fig. 4. *S. typhimurium* interactions with membraneous (M) cells. Bacteria invade the apical surface and remain within a membrane bound vacuole. They are then taken up by underlying macrophages (*Mac*), which can result in apoptosis, and lymphoid cells (*lym*). Reproduced with permission from Siebers and Finlay (1996)

small intestine (Baumler et al. 1997). The plasmid-encoded fimbriae from the *pef* operon mediate adhesion to the villous epithelium of the murine small intestine and are necessary for fluid accumulation in infant mice intestine (together with additional factors encoded on the *Salmonella* chromosome). In contrast, the *S. typhimurium* fimbrial operon termed *lpfABCDE* was shown to be responsible for specific adhesion of the microorganisms to the M cells of murine Peyer's patches in an intestinal organ-culture model (Baumler et al. 1996). Consequently, *S. typhimurium lpfC* mutants that fail to display selective binding to M cells should also fail to penetrate the intestinal mucosa and, therefore, lose virulence for mice by the oral route of infection. However, both mutations in genes that affect *S. typhimurium* invasion or colonization of Peyer's patches show only slightly increased values for intragastric LD_{50} indicating that additional factors are required for establishing infection and that initial interactions are important, but not essential, for the subsequent events.

4.2 Destruction of M Cells

To study interactions of bacteria with M cells, an intestinal ligated-loop model system is typically utilized, in which a portion of the intestine of an anesthetized animal is tied off and injected with pathogens. Experiments using a murine ligated ileal-loop model demonstrate that *S. typhimurium* preferentially binds to and enters M cells within 30 min post infection (Clark et al. 1994; Jones et al. 1994). Entry into the M cell is accompanied by membrane ruffling and actin reorganization as observed when *S. typhimurium* enters tissue-culture cell lines. At 60 min following injection, *S. typhimurium* had a cytotoxic effect on M cells. Dying and disintegrating M cells were observed to detach from the FAE with a *S. typhimurium* infection. As a consequence of the destruction of M cells, the integrity of the intestinal epithelium was lost and the bacteria gained free access to underlying tissues. At later time points (120 min and 180 min), massive damage to the intestinal epithelium was detected, including sloughing of large sections of enterocytes. The destructive potential of *S. typhimurium* is restricted to invasive organisms. Thus, *S. typhimurium* mutants that are defective in invasion of tissue-culture cells also lose their ability to invade and destroy M cells of ileal Peyer's patches in this murine ligated-loop model (Penheiter et al. 1997).

In general, the bacterial inocula used for ligated-loop models are unphysiologically high. Therefore, the interaction between pathogens and the intestinal epithelium observed in such a system does not necessarily correlate with the natural infection process completely. Nevertheless, the findings described for the *S. typhimurium* infection of murine Peyer's patches may provide an explanation for the pathology observed in many typhoid patients. The destruction of the FAE by invasive *S. typhi* can progress to intestinal ulceration and perforation, the major cause of mortality in typhoid fever.

4.3 Host Cell Determinants Contributing to *S. typhimurium* Invasion

The cross talk between *S. typhimurium* and the FAE has been reproduced in vitro using a variety of polarized and nonpolarized epithelial cell lines (Finlay and Falkow 1990; Finlay et al. 1988). Immediately after contact with the host-cell surface, *S. typhimurium* triggers a dramatic response from the target cell (Fig. 5). The bacteria are rapidly internalized in a process that causes localized membrane ruffling, reorganization of the host-cell cytoskeleton and macropinocytosis (Alpuche-Aranda et al. 1994; Francis et al. 1993; Garcia-del Portillo and Finlay 1994). The membrane structure associated with entering bacteria has been termed "splash," based on the resemblance to a stone being dropped in water. Uptake of *S. typhimurium* by host cells is a highly dynamic process, since the structural changes in host-cell morphology linked to *S. typhimurium* invasion are only transiently observed and the cytoskeletal-associated proteins, such as actin, alpha-actinin, talin, and ezrin, associated with the site of entry have been reported (Finlay et al. 1991).

S. typhimurium invasion activates a variety of signal transduction pathways within the host cell. Changes in intracellular calcium concentration and the production of inositol phospholipids and arachidonic acid metabolites have been described in response to *S. typhimurium* entry (Pace et al. 1993; Ruschkowski et al. 1992). However, it is unclear how these different second messengers mediate bacterial uptake. Furthermore, it is difficult to dissect which changes in the infected host cell are primary, i.e., induced by the entering bacteria, or secondary, i.e.,

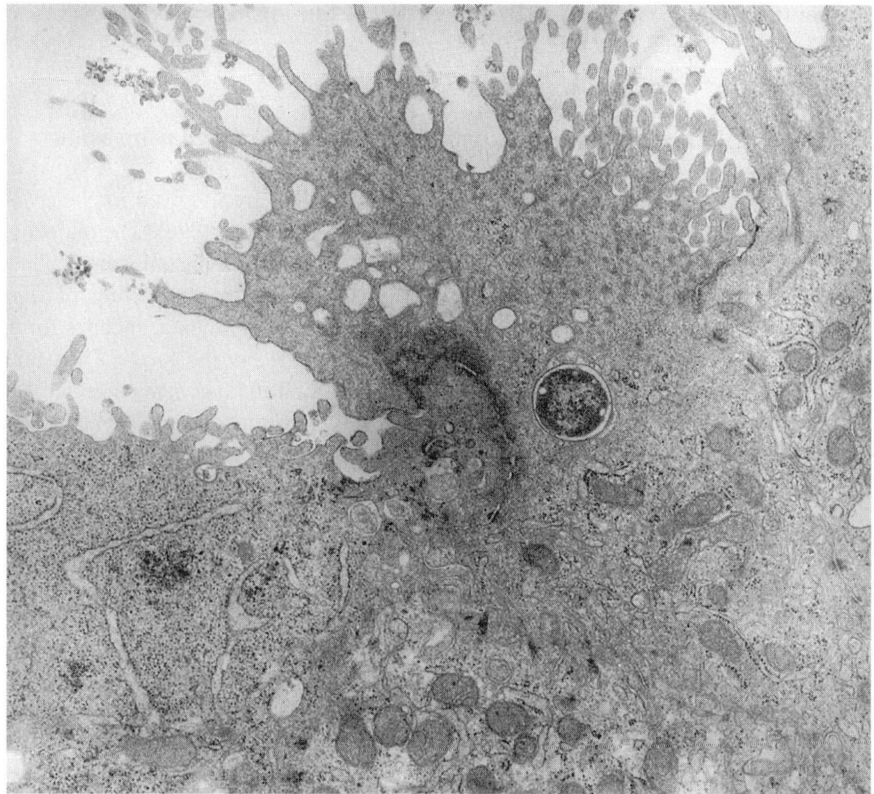

Fig. 5. Transmission electron micrograph of *S. typhimurium* interacting with the apical surface of polarized Caco-2 cells. Note the extensive localized membrane ruffling and microvilli rearrangements

induced by the host in response to bacterial invasion. The majority of information regarding the signal transduction pathways triggered by entry into the host cell is derived from experiments that describe the activation of the epidermal growth factor receptor (EGFR) by *S. typhimurium* in Henle-407 cells (GALAN et al. 1992). The reorganization of the host-cell surface associated with *Salmonella* uptake resembles the membrane ruffling induced by growth factor-mediated activation of certain oncogenes. However, based on observations that *S. typhimurium* is internalized by a variety of cells that lack EGFR (FRANCIS et al. 1993; MCNEIL et al. 1995; ROSENSHINE et al. 1994), it seems unlikely that the EGFR is directly involved in the *S. typhimurium* invasion process. Moreover, recent evidence suggests that *S. typhimurium* induces membrane ruffling by a growth factor-receptor-independent mechanism (JONES et al. 1993).

The contribution of actin-organizing small GTP-binding proteins to *S. typhimurium* internalization was assessed using cells transfected or microinjected with mutant forms of the various GTPases and exoenzyme C3, which ADP-ribosylates Rho. The observed results indicate that *S. typhimurium* entry into host cells occurs

Rac and Rho independent of (JONES et al. 1993), but involves the small GTPase Cdc42 in COS-1 cells (CHEN et al. 1996).

4.4 Bacterial Determinants Contributing to *S. typhimurium* Invasion

A large number of genetic loci have been identified that are involved in *S. typhimurium* entry into host cells (reviewed in FINLAY 1994; GALAN 1996). The majority of the genes cluster in a region at minute 63 of the *Salmonella* chromosome called *Salmonella* pathogenicity island 1 (SPI1) (LEE 1996). The main portion of this region, which comprises both regulatory and structural components, encodes for a contact-dependent type-III secretion system (TTSS), the *inv/spa* locus. Thus far, five different proteins, SptP and SipA-D, have been identified as being secreted by *S. typhimurium* via the inv/spa system; however, only mutations in *sipB-D* affect the ability of *S. typhimurium* to invade host cells (HUECK et al. 1995).

The Sips (*Salmonella i*nvasion *p*rotein*s*, also called Ssps for *Salmonella s*ecreted *p*rotein*s*) display high homology to the effectors of *Shigella* invasion, the Ipa proteins. Furthermore, some components of the two bacterial species are interchangeable, suggesting that *Salmonella* and *Shigella* may share common mechanisms for invasion (GINOCCHIO and GALAN 1995; HERMANT et al. 1995; KANIGA et al. 1995a; KANIGA et al. 1995b). One current model is that SipB and SipC are the effectors of *S. typhimurium* invasion and act as a soluble complex, based on observations obtained with the SipB and SipC homologues from *Shigella flexneri*, IpaB and IpaC. It has been reported recently that both SipB and SipC are translocated into the host-cell cytoplasm and that translocation depends on the presence of SipB, SipC and SipD (COLLAZO and GALAN 1997).

An alternative model is based on reports by WOOD et al. (1996), in which the authors describe the production of a set of novel secreted proteins termed Sops (*Salmonella o*uter *p*roteins) by a *Salmonella dublin* double *fliM*/polar *sipB* mutant. In addition, they show that SopE is translocated into the eukaryotic target cell via a *sip*-dependent mechanism. These data suggest that SipB-D are part of a translocation apparatus that delivers bacterial effectors of *Salmonella* invasion to the cytoplasm of the mammalian target cell.

Irrespective of which hypothesis is proven to be true, precipitates containing SipB, SipC, their *Shigella* homologues and other secreted proteins, including SopE, have been detected in culture supernatants. It remains unclear whether these precipitates result from unphysiological aggregation of hydrophobic proteins or rather represent the TTSS-dependent surface appendages that were observed to transiently coat the *S. typhimurium* surface after contact with polarized epithelial cells (GINOCCHIO et al. 1994).

5 Enteropathogenic *Yersinia* spp

Besides the famed *Yersinia pestis*, the causative agent of the bubonic plague, the genus *Yersinia* contains two additional species that are pathogenic for humans. The closely related species *Yersinia enterocolitica* and *Yersinia pseudotuberculosis* can cause gastroenteritis and mesenteric lymphadenitis. Disease is initiated by oral ingestion of contaminated food or water. The bacteria penetrate the epithelial barrier of the terminal ileum of the small intestine and the Peyer's patches, which are their primary site of infection. From there, the enteropathogenic *Yersiniae* drain to the mesenteric lymph nodes and it appears that, in human infections, disease terminates at this stage. Only in the mouse model, and occasionally in humans, can the bacteria cause systemic infections and spread to deeper tissues such as the liver and the spleen. Eventually, replication at these sites leads to death of the animal.

One striking feature of the disease is that the *Yersiniae* are located exclusively extracellularly during the complete course of infection, except for the initial passage across the intestinal epithelium (Fig. 6). The microorganisms replicate outside of host cells and manage to actively prevent uptake by phagocytic cells (Simonet and Falkow 1992). The factors responsible for this antiphagocytic behavior are encoded on the 70-kb virulence plasmid pYV. Physiological temperatures and Ca^{2+} limiting conditions are required for optimal expression of the plasmid-encoded, virulence-associated proteins (Straley et al. 1993). The presence of pYV is essential for the pathogenesis of enteropathogenic *Yersinia* spp., although strains

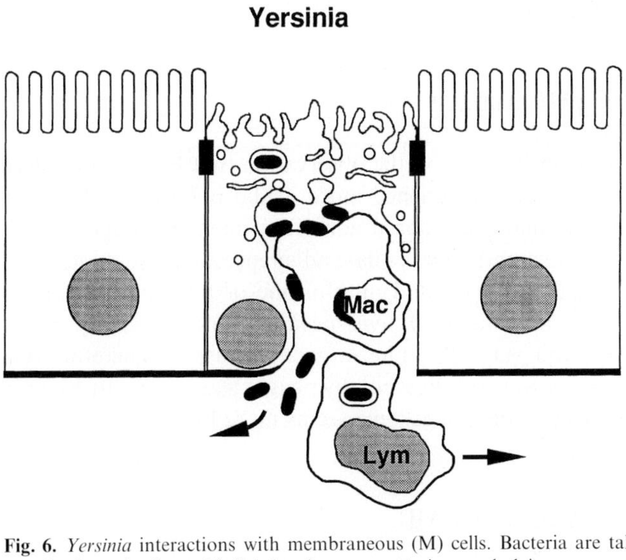

Fig. 6. *Yersinia* interactions with membraneous (M) cells. Bacteria are taken up by M cells, and then transit through the epithelial barrier, encountering underlying macrophages (*Mac*). They prevent phagocytosis and trigger apoptosis in these cells, and can be removed by lymphoid cells (*Lym*). Reproduced with permission from Siebers and Finlay (1996)

cured of the virulence plasmid can still cross the epithelial barrier of the small intestine in the mouse model.

5.1 Crossing the Intestinal Epithelium

The invasive ability of enteropathogenic *Yersinia* spp. correlates with their ability to cause disease. *Y. enterocolitica* and *Y. pseudotuberculosis* are efficiently internalized by cultured cells. Thus far, four factors have been identified to promote association of enteropathogenic *Yersiniae* with cultured mammalian cells. The pH6 antigen (*psaEFABC*) forms fimbrial structures on the bacterial surface mediating adhesion, whereas invasin, Ail (*a*ttachment-*i*nvasion-*l*ocus) and YadA (*Yersinia ad*herence) promote both attachment and entry. Interestingly, the genetic basis for the three invasive phenotypes for *Yersinia* is relatively simple compared with other invasive bacteria. Single genes (*inv*, *ail*, *yadA*) can confer the invasive ability to noninvasive *Escherichia coli*. In fact, all three genes were identified based on this potential (Isberg and Falkow 1985; Miller and Falkow 1988; Yang and Isberg 1993).

At the moment, it remains unclear which of the identified adhesion factors are essential for efficient passage of the intestinal epithelium. Invasin is thought to be the primary bacterial factor promoting penetration of the intestinal epithelium and efficient colonization of the Peyer's patches, but additional factors may be able to substitute for invasin in its absence. Once the epithelial barrier is crossed, the bacteria face several host defense mechanisms. Both YadA and Ail appear to play an overlapping role in resisting the host defense and, therefore, may contribute to survival within the Peyer's patch.

5.2 Adhesion–PsaA or pH6 Antigen

The name of this pilus adhesin is based on the observation that PsaA is produced maximally at pH6. Since bacteria experience more acidic pH values after oral ingestion, it appears possible that the fimbrial adhesions might be induced while passing the gastrointestinal tract and then mediate adherence to intestinal mucosal cells. However, the fact that PsaA is found only in one of the enteropathogens (*Y. pseudotuberculosis*, but not so far in *Y. enterocolitica*) and also in *Y. pestis*, which enters through the skin via a flea bite, argues against it. Therefore, the induction of pH6 antigen in response to ingestion by phagocytes, where the bacteria can experience low pH values in the phagosome seems more likely.

5.3 Invasion – Invasin, YadA, and Ail

The role of *inv*, yadA and *ail* has been studied in detail for *Y. enterocolitica*. Therefore, we will focus on the characteristics of the gene products in this *Yersinia* species, although the contribution of invasin, YadA and Ail to *Yersinia* pathogenesis varies in the two enteropathogenic *Yersinia* spp. Whereas invasins from *Y. enterocolitica* and *Y. pseudotuberculosis* appear to be both structurally and functionally very similar, mutations in *yadA* have significantly less striking effects in *Y. pseudotuberculosis* than seen in *Y. enterocolitica* (PEPE et al. 1995; ROSQVIST et al. 1988).

5.3.1 Invasin

The *inv* gene from *Y, pseudotuberculosis* represents the first *Yersinia* invasion locus cloned (ISBERG and FALKOW 1985; ISBERG et al. 1987). Subsequently, sequences homologous to *inv* have been detected in all *Yersinia* strains (MILLER and FALKOW 1988), although they don't appear to be functional in nonpathogenic strains of *Y. enterocolitica*. Invasin is a 108 kDa protein exposed on the bacterial cell surface. The C-terminal 192 amino acids are necessary and sufficient for binding to and uptake by mammalian host cells. From membrane extracts of HEp-2 cells, beta1 integrins could be identified as the receptors for invasin (ISBERG and LEONG 1990). Although invasin binding to purified alpha5beta1 integrins can be inhibited by RGD-containing peptides, invasin does not contain the RGD sequences characteristic of integrin ligands. However, two aspartate residues, D811 and D911, could be identified to be important for the interaction of invasin with beta1 integrins (LEONG et al. 1995).

In vivo, invasin is required for the early interaction of enteropathogenic *Yersinia spp.* with the Peyer's-patch epithelium, since lack of invasin expression in otherwise wild-type organisms leads to preferred binding of the bacteria to mucus (MARRA and ISBERG 1997).

5.3.2 Yersinia Adherence Protein (YadA)

Although *inv* mutants are drastically reduced in their ability to enter host cells, they retain a low level of invasiveness in tissue-culture assays, indicating that an additional factor, mediation adhesion and invasion, might exist. This low level of invasion is dependent on the presence of a plasmid-encoded invasion factor YadA. YadA can also mediate binding to collagen (SCHULZE-KOOPS et al. 1992), fibronectin and mucus. As with invasin, beta1 integrins have been shown to serve as receptors for YadA, although YadA mediated adhesion may be the indirect result of YadA binding to fibronectin (SCHULZE-KOOPS et al. 1993). YadA also confers complement resistance.

In vivo, YadA has been shown to be required for persistent survival of *Y. enterocolitica* in Peyer's patches (PEPE et al. 1995). *YadA* mutants are avirulent in an animal model system of infection.

5.3.3 Attachment-Invasion-Locus (Ail)

At the same time as *inv*, a second invasion gene, called *ail* (attachment invasion locus) was cloned *from Y. enterocolitica* (Miller and Falkow 1988). When expressed in *E. coli*, Ail mediates high level attachment to many cell types, although invasion levels vary significantly depending on the cell type analyzed. The *ail* gene is only found in pathogenic serotypes of *Y. enterocolitica*, and in *Y. pestis* and *Y. pseudotuberculosis*. However, in *Y. pseudotuberculosis*, Ail does not function as an adhesin/invasin (Yang and Isberg 1993).

In addition, Ail has a significant amount of sequence identity with several other bacterial outer membrane proteins. Like Rck, a protein encoded by the virulence plasmid of *Salmonella typhimurium* that contributes to serum resistance, Ail can confer resistance to the bactericidal effects of complement (Bliska and Falkow 1992). The fact that the *ail* gene product has multiple activities means that its biologically relevant function remains unclear. However, epidemiological analyses suggest that Ail is relevant for pathogenesis, since a close correlation exists between the presence of the *ail* gene and the ability of different *Y. enterocolitica* isolates to cause disease.

5.4 *Yersinia's* Life Outside of Cells

The ability of *Yersinia* species to resist internalization after close association with the surface of host cells is mediated by several plasmid-encoded proteins, called Yops, which are translocated into the cytoplasm of the target cell by type-III secretion. In the cytoplasm, the Yops interfere with the host-cell signal transduction pathways in a way that leads to the neutralization of the phagocytic properties of macrophage. The inhibition of phagocytosis by enteropathogenic *Yersinia* species seems to be crucial for pathogenesis, since invasive *Yersinia* strains are avirulent (Rosqvist et al. 1990).

The process that allows directed secretion of effector molecules from the bacterial pathogen into the target cell of the host requires a TTSS (reviewed in Cornelis and Wolfwatz 1997). Besides the many structural proteins involved in the transport of effectors across the bacterial cell envelope (YscC-G, YscI-L, YscO-U, VirG), additional players are necessary for the coordination of the TTSS, including energy generators (YscN), regulators (LcrD, G, Q, YopN, VirF), chaperones (SycD, E, H) and translocators (YopB, D).

The effectors transported by this complex secretion machinery alter mammalian signaling cascades according to the pathogens' needs. Although most of the cellular targets of the Yops remain to be identified, some of the phenotypes observed in host cells in response to Yop activity and sequence homologies to other proteins suggest possible functions: YopO (YpkA) shares amino acid sequence homology with serine/threonine protein kinases and disrupts the cytoskeleton of epithelial host cells (Galyov et al. 1993); YopJ induces programmed cell death in phagocytes, but not in epithelial cells (Mills et al. 1997; Monack et al. 1997);

YopM prevents blood platelet aggregation (LEUNG and STRALEY 1989); the cytotoxin YopE disrupts the host cell cytoskeleton by causing the collapse of actin stress fibers (ROSQVIST et al. 1991) and shows homology to the *N*-terminus of *Pseudomonas aeruginosa* exoenzyme S, which ADP-ribosylates small GTPases; and YopH is a phosphotyrosine phosphatase (PTPase) that can dephosphorylate proteins containing SH2 and SH3 (src homology) domains in vitro (ZHANG et al. 1992).

Recently, CAS, a 125–135 kDa protein that accumulates at focal adhesions, has been identified as an in vivo target of YopH (BLACK and BLISKA 1997; PERSSON et al. 1997). In addition, pp125FAK is dephosphorylated when host cells are infected with *Yersinia* strains expressing YopH. The dephosphorylation of CAS and pp125FAK may disrupt signaling in the host-cell cytoskeleton. Therefore, YopH and YopE are believed to mediate paralysis of the cytoskeletal dynamics required for the rearrangements necessary for *Yersinia* internalization into target cells.

Yops are essential virulence factors for *Yersinia*, since strains carrying a mutation in one of the Yops described (except YopJ) fail to establish fatal infections in mice. In addition, at least seven more Yops are produced by *Yersinia*. However, their role in pathogenesis is unclear (CORNELIS and WOLFWATZ 1997).

6 Conclusions

It should be apparent from the above that the four enteropathogens discussed use specific and specialized mechanisms to subvert mucosal epithelial-cell (including M-cell) functions to cause infections. Although the specific mechanisms differ among the pathogens, all use a TTSS to deliver bacterial molecules onto or into the epithelial cell, and cause alterations in the epithelial cell cytoskeleton. In at least two cases (*S. typhimurium* and *S. flexneri*), small GTP-binding proteins participate in bacterial invasion. The other two pathogens (EPEC and *Yersinia* species) block uptake into host cells instead and remain as extracellular pathogens. A major challenge in the future is to verify that the processes that occur in tissue-culture models actually occur in vivo, and to determine how these processes contribute to disease. Additionally, the bacterial components involved in these processes are logical targets for therapeutics. Understanding the complex interplay that occurs at the mucosal surface between these pathogens and the host immune system will be key to defining how these pathogens cause disease, and how to exploit this knowledge to develop new vaccines.

Acknowledgements. Work in BBF's lab is supported by a Howard Hughes International Research Scholar Award, the Canadian Bacterial Diseases Network Center of Excellence, the Medical Research Council of Canada, and the National Sciences and Engineering Research Council of Canada. BBF is a MRC Scientist. Work in SF's laboratory is supported by the National Institute of Health and unrestricted gifts from Praxis and Microcide Pharmaceuticals, Inc. BR gratefully acknowledges a postdoctoral fellowship from the DFG (Germany) and a Pilot Study Award from the Digestive Disease Center, Stanford, CA (USA).

References

Adam T, Arpin M, Prevost MC, Gounon P, Sansonetti PJ (1995) Cytoskeletal rearrangements and the functional role of T-plastin during entry of Shigella flexneri into HeLa cells. J Cell Biol 129:367–81

Adam T, Giry M, Boquet P, Sansonetti P (1996) Rho-dependent membrane folding causes Shigella entry into epithelial cells. EMBO J 15:3315–3321

Alpuche-Aranda CM, Racoosin EL, Swanson JA, Miller SI (1994) Salmonella stimulate macrophage macropinocytosis and persist within spacious phagosomes. J Exp Med 179:601–8

Baumler AJ, Tsolis RM, Heffron F (1996) The lpf fimbrial operon mediates adhesion of Salmonella typhimurium to murine Peyer's patches. Proc Natl Acad Sci USA 93:279–83

Baumler AJ, Tsolis RM, Valentine PJ, Ficht TA, Heffron F (1997) Synergistic effect of mutations in invA and lpfC on the ability of Salmonella typhimurium to cause murine typhoid. Infect Immun 65:2254–9

Berendson R, Cheney CP, Schad PA, Boedeker EC (1983) Species-specific binding of purified pili (AF/R1) from the Escherichia coli RDEC-1 to rabbit intestinal mucosa. Gastroenterology 85:837–45

Black DS, Bliska JB (1997) Identification of p130Cas as a substrate of Yersinia YopH (Yop51), a bacterial protein tyrosine phosphatase that translocates into mammalian cells and targets focal adhesions. EMBO J 16:2730–44

Bliska JB, Falkow S (1992) Bacterial resistance to complement killing mediated by the Ail protein of Yersinia enterocolitica. Proc Natl Acad Sci USA 89:3561–5

Carter PB,Collins FM (1974) The route of enteric infection in normal mice. J Exp Med 139:1189–203

Chen L-M, Hobbie S, Galan JE (1996) Requirement of CDC42 for Salmonella-induced cytoskeletal and nuclear responses. Science 274:2115–2118

Clark MA, Jepson MA, Simmons NL, Hirst BH (1994) Preferential interaction of Salmonella typhimurium with mouse Peyer's patch M cells. Res Microbiol 145:543–52

Clerc P, Sansonetti PJ (1987) Entry of Shigella flexneri into HeLa cells: evidence for directed phagocytosis involving actin polymerization and myosin accumulation. Infect Immun 55:2681–8

Collazo CM, Galan JE (1997) The invasion-associated type III system of Salmonella typhimurium directs the translocation of Sip proteins into the host cell. Mol Microbiol 24:747–56

Cornelis GR, Wolfwatz H (1997) The Yersinia Yop virulon – a bacterial system for subvertin eukaryotic cells. Mol Microbiol 23:861–867

Dehio C, Prevost MC, Sansonetti PJ (1995) Invasion of epithelial cells by Shigella flexneri induces tyrosine phosphorylation of cortactin by a pp60c-src-mediated signalling pathway. EMBO J 14:2471–82

Donnenberg MS, Kaper JB, Finlay BB (1997) Interactions between enteropathogenic Escherichia coli and host epithelial cells. Trends Microbiol 5:109–114

Dytoc M, Fedorko L, Sherman PM (1994) Signal transduction in human epithelial cells infected with attaching and effacing Escherichia coli in vitro. Gastroenterology 106:1150–61

Fiederling F, Boury M, Petit C, Milon A (1997) Adhesive factor/rabbit 2, a new fimbrial adhesion and a virulence factor from Escherichia coli O103, a serogroup enteropathogenic for rabbits. Infect Immun 65:847–51

Finlay BB (1994) Molecular and cellular mechanisms of Salmonella pathogenesis. Curr Top Micrbiol Immunol 192:163–85

Finlay BB, Falkow S (1990) Salmonella interactions with polarized human intestinal Caco-2 epithelial cells. J Infect Dis 162:1096–106

Finlay BB, Gumbiner B, Falkow S (1988) Penetration of Salmonella through a polarized Madin-Darby canine kidney epithelial cell monolayer. J Cell Biol 107:221–30

Finlay BB, Ruschkowski S, Dedhar S (1991) Cytoskeletal rearrangements accompanying Salmonella entry into epithelial cells. J Cell Sci 99:283–96

Finlay BB, Rosenshine I, Donnenberg MS, Kaper JB (1992) Cytoskeletal composition of attaching and effacing lesions associated with enteropathogenic Escherichia coli adherence to HeLa cells. Infect Immun 60:2541–3

Foubister V, Rosenshine I, Finlay BB (1994) A diarrheal pathogen, enteropathogenic Escherichia coli (EPEC), triggers a flux of inositol phosphates in infected epithelial cells. J Exp Med 179:993–8

Francis CL, Ryan TA, Jones BD, Smith SJ, Falkow S (1993) Ruffles induced by Salmonella and other stimuli direct macropinocytosis of bacteria. Nature 364:639–42

Galan JE (1996) Molecular genetic bases of Salmonella entry into host cells. Mol Microbiol 20:263–271

Galan JE, Pace J, Hayman MJ (1992) Involvement of the epidermal growth factor receptor in the invasion of cultured mammalian cells by Salmonella typhimurium. Nature 357:588–9

Galyov EE, Hakansson S, Forsberg A, Wolf-Watz H (1993) A secreted protein kinase of Yersinia pseudotuberculosis is an indispensable virulence determinant. Nature 361:730–2

Garcia-del Portillo F, Finlay BB (1994) Salmonella invasion of nonphagocytic cells induces formation of macropinosomes in the host cell. Infect Immun 62:4641–5

Ginocchio CC, Galan JE (1995) Functional conservation among members of the Salmonella typhimurium InvA family of proteins. Infect Immun 63:729–32

Ginocchio CC, Olmsted SB, Wells CL, Galan JE (1994) Contact with epithelial cells induces the formation of surface appendages on Salmonella typhimurium. Cell 76:717–24

Hermant D, Menard R, Arricau N, Parsot C, Popoff MY (1995) Functional conservation of the Salmonella and Shigella effectors of entry into epithelial cells. Mol Microbiol 17:781–789

Hueck CJ, Hantman MJ, Bajaj V, Johnston C, Lee CA, Miller SI (1995) Salmonella typhimurium secreted invasion determinants are homologous to Shigella Ipa proteins. Mol Microbiol 18:479–90

Isberg RR, Falkow S (1985) A single genetic locus encoded by Yersinia pseudotuberculosis permits invasion of cultured animal cells by Escherichia coli K-12. Nature 317:262–4

Isberg RR, Leong JM (1990) Multiple beta 1 chain integrins are receptors for invasin, a protein that promotes bacterial penetration into mammalian cells. Cell 60:861–71

Isberg RR, Voorhis DL, Falkow S (1987) Identification of invasin: a protein that allows enteric bacteria to penetrate cultured mammalian cells. Cell 50:769–78

Jerse AE, Yu J, Tall BD, Kaper JB (1990) A genetic locus of enteropathogenic Escherichia coli necessary for the production of attaching and effacing lesions on tissue culture cells. Proc Natl Acad Sci USA 87:7839–43

Jones BD, Paterson HF, Hall A, Falkow S (1993) Salmonella typhimurium induces membrane ruffling by a growth factor-receptor-independent mechanism. Proc Natl Acad Sci USA 90:10390–4

Jones BD, Ghori N, Falkow S (1994) Salmonella typhimurium initiates murine infection by penetrating and destroying the specialized epithelial M cells of the Peyer's patches. J Exp Med 180:15–23

Kaniga K, Trollinger D, Galan JE (1995a) Identification of two targets of the type III protein secretion system encoded by the inv and spa loci of Salmonella typhimurium that have homology to the Shigella IpaD and IpaA proteins. J Bacteriol 177:7078–85

Kaniga K, Tucker S, Trollinger D, Galan JE (1995b) Homologs of the Shigella IpaB and IpaC invasins are required for Salmonella typhimurium entry into cultured epithelial cells. J Bacteriol 177:3965–71

Kenny B, Devinney R, Stein M, Reinscheid DJ, Frey EA, Finlay BB (1997) Enteropathogenic E. coli (EPEC) transfers its receptor for intimate adherence into mammalian cells. Cell 91:511–520

Kenny B, Finlay BB (1998) Intimin-dependent binding of enteropathogenic Escherichia coli to host cells triggers novel signaling events, including tyrosine phosphorylation of phospholipase C gamma. Infection and Immunity 65:2528–2536

Kerneis S, Bogdanova A, Kraehenbuhl JP, Pringault E (1997) Conversion by Peyer's patch lymphocytes of human enterocytes into M cells that transport bacteria [see comments]. Science 277:949–52

Kohbata S, Yokoyama H, Yabuuchi E (1986) Cytopathogenic effect of Salmonella typhi GIFU 10007 on M cells of murine ileal Peyer's patches in ligated ileal loops: an ultrastructural study. Micrbiol Immunol 30:1225–37

Lee CA (1996) Pathogenicity islands and the evolution of bacterial pathogens. Infectious Agents Dis 5:1–7

Leong JM, Morrissey PE, Marra A, Isberg RR (1995) An aspartate residue of the Yersinia pseudotuberculosis invasin protein that is critical for integrin binding. EMBO J 14:422–31

Leung KY, Straley SC (1989) The yopM gene of Yersinia pestis encodes a released protein having homology with the human platelet surface protein GPIb alpha. J Bacteriol 171:4623–32

Lockman HA, Curtiss R (1992) Isolation and characterization of conditional adherent and non-type 1 fimbriated Salmonella typhimurium mutants. Mol Microbiol 6:933–45

Marra A, Isberg RR (1997) Invasin-dependent and invasin-independent pathways for translocation of Yersinia pseudotuberculosis across the Peyer's patch intestinal epithelium. Infect Immun 65:3412–21

McDaniel TK, Jarvis KG, Donnenberg MS, Kaper JB (1995) A genetic locus of enterocyte effacement conserved among diverse enterobacterial pathogens. Proc Natl Acad Sci USA 92:1664–8

McNeil A, Dunstan SJ, Clark S, Strugnell RA (1995) Salmonella typhimurium displays normal invasion of mice with defective epidermal growth factor receptors. Infect Immun 63:2770–2

Menard R, Dehio C, Sansonetti PJ (1996a) Bacterial entry into epithelial cells: the paradigm of Shigella. Trends Microbiol 4:220–225

Menard R, Prevost MC, Gounon P, Sansonetti P, Dehio C (1996b) The secreted Ipa complex of Shigella flexneri promotes entry into mammalian cells. Proc Natl Acad Sci USA 93:1254–8

Miller VL, Falkow S (1988) Evidence for two genetic loci in Yersinia enterocolitica that can promote invasion of epithelial cells. Infect Immun 56:1242–8

Mills SD, Bolan A, Sory MP, Vandersmissen P, Kerbourch C, Finlay BB, Cornelis GR (1997) Yersinia enterocolitica induces apoptosis in macrophages by a process requiring functional type III secretion and translocation mechanisms and involving YopP, presumably acting as an effector protein. Proc Natl Acad Sci U S A 94:12638–12643

Monack DM, Mecsas J, Ghori N, Falkow S (1997) Yersinia signals macrophages to undergo apoptosis and YopJ is necessary for this cell death. Proc Natl Acad Sci USA 94:10385–90

Mounier J, Vasselon T, Hellio R, Lesourd M, Sansonetti PJ (1992) Shigella flexneri enters human colonic Caco-2 epithelial cells through the basolateral pole. Infect Immun 60:237–48

Pace J, Hayman MJ, Galan JE (1993) Signal transduction and invasion of epithelial cells by S. typhimurium. Cell 72:505–14

Penheiter KL, Mathur N, Giles D, Fahlen T, Jones BD (1997) Non-invasive Salmonella typhimurium mutants are avirulent because of an inability to enter and destroy M cells of ileal Peyer's patches. Mol Microbiol 24:697–709

Pepe JC, Wachtel MR, Wagar E, Miller VL (1995) Pathogenesis of defined invasion mutants of Yersinia enterocolitica in a BALB/c mouse model of infection. Infect Immun 63:4837–48

Perdomo JJ, Gounon P, Sansonetti PJ (1994a) Polymorphonuclear leukocyte transmigration promotes invasion of colonic epithelial monolayer by Shigella flexneri. J Clin Invest 93:633–43

Perdomo OJ, Cavaillon JM, Huerre M, Ohayon H, Gounon P, Sansonetti PJ (1994b) Acute inflammation causes epithelial invasion and mucosal destruction in experimental shigellosis. J Exp Med 180:1307–19

Persson C, Carballeira N, Wolf-Watz H, Fallman M (1997) The PTPase YopH inhibits uptake of Yersinia, tyrosine phosphorylation of p130Cas and FAK, and the associated accumulation of these proteins in peripheral focal adhesions. EMBO J 16:2307–18

Rosenshine I, Donnenberg MS, Kaper JB, Finlay BB (1992) Signal transduction between enteropathogenic Escherichia coli (EPEC) and epithelial cells: EPEC induces tyrosine phosphorylation of host cell proteins to initiate cytoskeletal rearrangement and bacterial uptake. EMBO J 11:3551–60

Rosenshine I, Ruschkowski S, Foubister V, Finlay BB (1994) Salmonella typhimurium invasion of epithelial cells: role of induced host cell tyrosine protein phosphorylation. Infect Immun 62:4969–74

Rosenshine I, Ruschkowski S, Stein M, Reinscheid DJ, Mills SD, Finlay BB (1996) A pathogenic bacterium triggers epithelial signals to form a functional bacterial receptor that mediates actin pseudopod formation. EMBO J 15:2613–2624

Rosqvist R, Forsberg A, Rimpilainen M, Bergman T, Wolf-Watz H (1990) The cytotoxic protein YopE of Yersinia obstructs the primary host defence. Mol Microbiol 4:657–67

Rosqvist R, Forsberg A, Wolf-Watz H (1991) Intracellular targeting of the Yersinia YopE cytotoxin in mammalian cells induces actin microfilament disruption. Infect Immun 59:4562–9

Rosqvist R, Skurnik M, Wolf-Watz H (1988) Increased virulence of Yersinia pseudotuberculosis by two independent mutations. Nature 334:522–4

Ruschkowski S, Rosenshine I, Finlay BB (1992) Salmonella typhimurium induces an inositol phosphate flux in infected epithelial cells. FEMS Microbiol Lett 74:121–6

Schulze-Koops H, Burkhardt H, Heesemann J, Kirsch T, Swoboda B, Bull C, Goodman S, Emmrich F (1993) Outer membrane protein YadA of enteropathogenic yersiniae mediates specific binding to cellular but not plasma fibronectin. Infect Immun 61:2513–9

Schulze-Koops H, Burkhardt H, Heesemann J, von der Mark K, Emmrich F (1992) Plasmid-encoded outer membrane protein YadA mediates specific binding of enteropathogenic yersiniae to various types of collagen. Infect Immun 60:2153–9

Siebers A, Finlay BB (1996) M cells and the pathogenesis of mucosal and systemic infections. Trends Microbiol 4:22–9

Simonet M, Falkow S (1992) Invasin expression in Yersinia pseudotuberculosis. Infect Immun 60:4414–7

Straley SC, Plano GV, Skrzypek E, Haddix PL, Fields KA (1993) Regulation by Ca2+ in the Yersinia low-Ca2+ response. [Review] Mol Microbiol 8:1005–10

Tran Van Nhieu G, Ben-Ze'ev A, Sansonetti PJ (1997) Modulation of bacterial entry into epithelial cells by association between vinculin and the Shigella IpaA invasin. EMBO J 16:2717–29

Watarai M, Funato S, Sasakawa C (1996) Interaction of ipa proteins of Shigella flexneri with alpha(5)beta(1) integrin promotes entry of the bacteria into mammalian cells. J Exp Med 183:991–999

Watarai M, Kamata Y, Kozaki S, Sasakawa C (1997) Rho, a small gtp-binding protein, is essential for shigella invasion of epithelial cells. J Exp Med 185:281–292

Wood MW, Rosqvist R, Mullan PB, Edwards MH, Galyov EE (1996) SopE, a secreted protein of Salmonella dublin, is translocated into the target eukaryotic cell via a sip-dependent mechanism and promotes bacterial entry. Mol Microbiol 22:327–38

Yang Y, Isberg RR (1993) Cellular internalization in the absence of invasin expression is promoted by the Yersinia pseudotuberculosis yadA product. Infect Immun 61:3907–13

Zhang QY, DeRyckere D, Lauer P, Koomey M (1992) Gene conversion in Neisseria gonorrhoeae: evidence for its role in pilus antigenic variation. Proc Natl Acad Sci USA 89:5366–70

Microbial–Host Interactions at Mucosal Sites. Host Response to Pathogenic Bacteria at Mucosal Sites

A. Phalipon and P.J. Sansonetti

Unite de Pathogenie Microbienne Moleculaire, U389, Institut Pasteur, 28 rue du Dr. Roux, 75015 Paris, France

1 Introduction

The mucosal epithelial layer forms the interface between the external and internal environments in the gastrointestinal tract. This area is the site for digestion and absorption of various essential nutrients, yet it must also function as a barrier against various infectious pathogens. There are many non-immunological factors that protect against such agents, including in particular, at the intestinal level, mucus, glycocalyx, peristaltism and innate humoral factors, such as lactoferrin, peroxidase and defensins. In addition to this physiological barrier, an immunological barrier is maintained which encompasses the intestinal epithelial cells, the gut-associated lymphoid tissue (GALT), immunoreactive cells distributed throughout the intestinal tract and the systemic immune system. These components act in concert to mediate two types of immune responses against enteric pathogens. First of all, an innate immune response is induced following bacterial activation of epithelial cells, macrophages and subsets of T cells which, through the secretion of cytokines, mainly induces an inflammatory process. The inflammatory reaction allows recovery from infection. The primary activation also leads, with a delay in time, to the induction of an adaptive immune response characterized by specific humoral and cellular responses.

Survival into the gastrointestinal tract implies, for the enteric pathogens, the expression of a variety of gene products that help to escape the physiological barrier. The environmental conditions act as signals to turn on or off the specific genes needed for adaptation to distinct microenvironments in the host. If it survives these conditions, the luminal pathogen must establish contact with appropriate host cells either to stabilize and colonize the intestinal mucosa or to invade it. At this stage, the immunological barrier has to be subverted by the pathogen for subsequent triggering of its pathogenic process. Depending on the behavior of the pathogen, invasive or non-invasive, obligate or facultative intracellular bacterium, several subversive strategies have been developed to escape host defense mechanisms. The balance resulting from the struggle between the pathogenic bacterium and the host for their own survival finally depends on a complex series of events, including efficient virulence factors expressed by the pathogen and appropriate immune status of the host.

In this chapter, host responses at the intestinal mucosal site will be illustrated by studies on a non-invasive bacterium *Vibrio cholerae* and invasive bacteria, including *Shigella*, *Salmonella* and *Yersinia*. Each of these represents an example of adaptive strategy for bacterial survival, and innate and specific immune responses elicited by the host to remain alive and protected against further infection following primary infection.

2 Host Response to *Vibrio cholerae*, a Non-Invasive Bacterium

Non-invasive bacteria have evolved adapted strategies to colonize the host mucosa and then develop their pathogenic process. *Vibrio cholerae*, the etiologic agent of cholera, is one of the best known examples of a non-invasive bacterium. It is a highly motile gram-negative microorganism that colonizes the surface of the small intestine and produces an enterotoxin, the cholera toxin (CT), which is largely responsible for the symptoms of cholera, i.e., watery diarrhea and vomiting. These symptoms can be so severe and the ensuing dehydration so rapid, that death of the patient can occur within hours of the onset of symptoms. The profuse rice water stools, which are the hallmark of this infection, contain up to 10^8 *V. cholerae* per milliliter, allowing the bacterium to be rapidly disseminated in the environment and to spread to other people.

There are two serogroups, O1 and O139. The former is composed of two biotypes, classical and El Tor. The classical biotype was responsible for essentially all cholera throughout the world until 1961, when it was replaced primarily by the El Tor biotype. In late 1992, the strain O139 appeared and is believed to have grown from the El Tor biotype. Depending on the endemic areas, both O1 and O139 may coexist. For a complete review of cholera pathogenesis, see KAPER et al. (1995).

2.1 Colonization of the Intestinal Mucosa

The first steps of cholera pathogenesis are similar to those of the invasive entero-pathogenic bacteria, i.e., oral ingestion of the microorganisms from contaminated water or food and survival of the microorganism during passage through the gastric acid barrier of the stomach. The following steps diverge since the invasive bacteria enter the mucosal barrier, whereas the non-invasive *V. cholerae* microorganisms colonize the intestinal mucosa, especially the upper small intestine, and remain localized in microcolonies on the intestinal surface, without any significant invasion to other sites of the host. It should be noticed that studies in a rabbit ligated-loop model revealed that *V. cholerae* colonized the epithelial surfaces of both villi and Peyer's patches. Uptake of *V. cholerae* by membraneous (M) cells occurs with subsequent dissemination of free or macrophage-engulfed bacteria into the follicular dome (OWEN et al. 1986). Unlike the invasive pathogens described above, *V. cholerae* uptake by M cells leads to the killing of this non-invasive bacterium. In this case, the Peyer's patches really play their role of immune surveillance of the intestinal mucosa.

The bacterium efficiently delivers CT to host epithelial-cell toxin receptors upon colonization that occurs through interaction of *V. cholerae* fimbriae to the mucosal surface. The O1 and O139 serotypes of *V. cholerae* produce at least two fimbrial types under different conditions. One of these is expressed under the same growth conditions that elicit the highest levels of toxin production and has thus been termed TCP for toxin *c*oregualted *p*ilus (TAYLOR et al. 1987). The amino terminal region of the mature subunit pilin protein, termed TcpA is highly homologous to a group of pilin proteins called type-4 pilins (for a review, see STROM and LORY 1993). Type-4 pili are implicated in the adhesion of several bacterial species that colonize or interact with mucosal surfaces. The requirement of this organelle in the colonization of humans has been established (HERRINGTON et al. 1988). The precise molecular mechanisms by which TCP functions to promote colonization are not known, but appear to involve the ability to both adhere and resist bactericidal mechanisms (CHIANG et al. 1995). Another set of genes termed the *acf* (accessory colonization factor) genes appear to have a role in colonization. They are expressed in a manner similar to the toxin and TCP genes (PETERSON and MEKALANOS 1988). Other soluble or cell-associated hemagglutinins (i.e., MSHA) are produced, but their roles in colonization remain unclear (ATTRIDGE et al. 1996).

2.2 Interaction of *Vibrio cholerae* with the Host Innate Immune System

2.2.1 Epithelial-Cell Interactions

The main virulence determinant of *V. cholerae* is the secreted exotoxin (for a review, see KAPER et al. 1995). CT is composed of one A subunit and five identical B subunits. The B subunits are required for secretion of the toxin out of the bacterial cell, and for its interaction with target cell-surface ganglioside GM1 receptors. The pentameric structure also likely participates in entry of the toxin A subunit into the target cell, after which, the A subunit is proteolytically nicked to become active. Its amino-terminal fragment catalyzes the transfer of the ADP-ribose moiety from nicotinamide adenine dinucleotide (NAD) to the regulatory G protein, which then constitutively activates the mammalian adenylate cyclase leading to increased cyclic adenosine monophosphate (cAMP) levels. This, in turn, activates cAMP-dependent protein kinase A, which phosphorylates proteins involved in intestinal ion transport and produces the characteristicly severe, watery diarrhea associated with cholera.

In addition, there is compelling evidence that prostaglandins and the enteric nervous system (ENS) are involved in the response to CT (KAPER et al. 1995). A model has been suggested in which cAMP levels increased by CT serve not only to activate protein kinase A, but also to regulate transcription of a phospholipase or a phospholipase-activating protein. The activated phospholipase may act on membrane phospholipids to produce arachidonic acid, a precursor of prostaglandins and leukotrienes (PETERSON et al. 1991). In addition to the fact that they induce fluid loss and electolyte secretion (PETERSON and OCHOA 1989; PETERSON et al. 1991), prostaglandins also play a role in the intestinal immune response with a variety of effects on lymphocytes, macrophages and dendritic cells (HWANG 1989).

The ENS, which plays an important role in intestinal secretion and absorption, has been shown to also play a role in secretion due to CT (LUNDGREN 1988). It has been estimated that 60% of the effect of CT on intestinal fluid transport could be attributed to nervous mechanisms (CASSUTO et al. 1981). The overall hypothesis, consistent with several studies, is that CT binds to enterochromaffin cells which release serotonin. Serotonin activates dendrite-like structures located beneath the intestinal epithelium. This leads to the release of vasoactive intestinal peptide (VIP), resulting in electrolyte and fluid secretion. Besides the direct secretory effect, there is also evidence that CT increases intestinal motility and could thereby contribute to diarrhea (MATHIAS et al. 1976).

2.2.2 Immune–Cell Interactions

CT is one of the most potent oral immunogens ever studied (LYCKE and HOLMGREN 1986). Vigorous immune responses are engendered not only against CT delivered orally, but also against unrelated antigens delivered orally with CT (see FREYTAG and CLEMENTS this volume). While the B subunit by itself has been reported to have an adjuvant effect, probably due to its avid binding to GM1, the most potent adjuvant effect of CT is due to the ADP-ribosyltransferase activity of the A subunit. CT has a variety of effects on cells of the immune system, including stimulating interleukin-1 (IL-1) proliferation and enhancing antigen presentation by macrophages (BROMANDER et al. 1991), promoting B-cell isotype differentiation (LYCKE and STROBER 1989) and inhibiting Th1 cells (MUNOZ et al. 1990). It has also been suggested that the adjuvant action of CT is due to increased intestinal permeability in response to CT, perhaps providing increased access of antigens to the gut mucosal immune system (LYCKE et al. 1991).

2.3 Specific Host Response to *Vibrio cholerae* Infection

Several studies have demonstrated the existence of infection-derived immunity to *V. cholerae* (for a review, see KAPER et al. 1995). Protection against disease conferred by an initial clinical infection, with a classical strain lasts for at least 3 years, the longest interval tested within the same biotype (LEVINE et al. 1981). However, a striking difference seems to exist between biotypes (CLEMENS et al. 1991).

Although strong protective immunity is conferred by an infection with *V. cholerae*, the identity of the crucial protective antigens is unclear. The components of the immune response have been recently reviewed (SVENNERHOLM et al. 1994). Both antibacterial immunity and antitoxic immunity exist, with antibacterial immunity being the most important component. However, a synergistic protective effect has been noted when bacterial antigens and toxoids are combined compared with what is observed with either component alone (SVENNERHOLM and HOLMGREN 1976).

The best correlation with protection is a serum vibriocidal antibody response. This assay measures the killing of *V. cholerae* bacteria in the presence of immune sera and complement. The majority of vibriocidal antibodies are directed against lipopolysaccharide (LPS), which has been reported to play a major role in protection in a field trial with a parenteral vaccine consisting of purified LPS (MOSLEY et al. 1970). In contrast, the prevalence and titer of serum antitoxin antibody do not correlate with protection from cholera. It is not believed that vibriocidal antibodies in the serum are the actual mediators of protective immunity, but rather the presence of such antibodies serves as a marker for the presence of secretory immunoglobulin A (IgA) intestinal antibodies, which may be directed against the same antigens.

In recent years, the importance of the intestinal immune system has been recognized. Since *V. cholerae* colonizes the intestinal mucosal surface without invasion of enterocytes, the protective immune response is believed to reside at the mucosal surface, without a major contribution from serum antibodies. Secretory IgA have been reported to protect against infection in mice (Winner et al. 1991; Apter et al. 1993; Lee et al. 1994). Interestingly, in these studies, anti-LPS secretory IgA are much more effective than anti-CT IgA in prevention of *V. cholerae*-induced diarrheal disease. Therefore, two types of vaccines are developed, both of which are delivered orally: killed whole-cell toxoid or live attenuated bacterial strains. They replace the parenteral vaccines, previously developed, that had modest and short-lived efficacy (Levine and Pierce 1992).

3 Host Response to *Shigella*, an Invasive, Facultative Intracellular Bacterium that Escapes from the Phagocytic Vacuole

Shigellosis is a severe form of bloody diarrhea which reflects the capacity of the causative microorganism, *Shigella*, a gram-negative enterobacteria, to invade the colonic and rectal mucosa of humans. *S. flexneri* and *S. sonnei* are responsible for the endemic form of the disease, whereas *S. dysenteriae 1* accounts for devastating epidemics. The molecular and cellular bases of *Shigella* invasion of the intestinal mucosa have been intensely studied over the last years (Parsot and Sansonetti 1996). A complex series of events leads to translocation through the epithelial lining, invasion of epithelial cells, and elicitation of an intense inflammatory reaction which eventually causes tissue destruction, thus provoking the dysenteric symptoms. Usually, the bacterial infection remains mucosally localized. Systemic dissemination occurs essentially in malnourished children infected by *S. dysenteriae 1*.

3.1 Passage of the Intestinal Barrier

It has been shown that in confluent monolayers of human colonic epithelial cell lines, *S. flexneri* is unable to penetrate via the apical pole of the cells, thus indicating that the machinery required for bacterial internalization is only present on the basolateral side of the cells (Mounier et al. 1992). The validity of this observation seems to be confirmed in vivo (Wassef et al. 1988; Sansonetti et al. 1991; Sansonetti et al. 1996; Perdomo et al. 1994a) indicating that colonic or rectal epithelial cells are not the first target for *Shigella*, and raising the question of the route of access of bacteria to the subepithelial zone in the intestinal mucosa.

To cross the epithelium and reach the basolateral side of the cells that is permissive for entry, *Shigella* uses the M cells (Wassef et al. 1988; Sansonetti et al. 1991; Sansonetti et al. 1996; Perdomo et al. 1994a) that are present in the follicular-associated epithelium (FAE) which covers the lymphoid nodules associated with the mucosa (see Chap. 2). Those cells are not killed by the invasive microorganism which is rapidly translocated to the subepithelial tissue of the follicular dome. In this area, which initially corresponds to the pocket formed by M cells, a majority of bacteria appears extracellular, even if physically closely associated with the mononuclear cells that populate these zones (Sansonetti et al. 1996). These observations indicate that a significant part of *Shigella* life inside tissues is extracellular.

Once the mucosa is crossed, *Shigella* may invade the intestinal epithelial cells, as it has been reported following studies using epithelial cell lines. Invasion of these cells by the bacterium can be summarized as follows (for a review, Parsot and Sansonetti 1996). *Shigella* enters the cell, essentially through the basolateral pole, via a process of bacteria-directed phagocytosis that involves bacterial virulence factors and subsequent major cell cytoskeleton rearrangements. A few minutes after entry, invasive *Shigella* lyses the membrane-bound phagocytic vacuole and escapes to the cytoplasm. Access to the cytoplasmic compartment allows immediate and rapid intracellular growth of the pathogen, and interaction with the host-cell cytoskeleton, which permits intracellular movement and cell-to-cell spread of this otherwise non-motile microorganism. These steps are largely described by Raupach et al. this volume.

3.2 Interaction of *Shigella* with the Host Innate Immune System

3.2.1 Macrophages and Polymorphonuclear Cells

To avoid the natural host defense mechanisms occurring within the mucosa-associated lymphoid follicles that are used as a port of entry by the bacteria, *Shigella* has evolved an interesting strategy. In this area, bacteria are engulfed by resident macrophages, possibly dendritic cells and locally recruited monocytes. The infected mononuclear cells are killed by the invasive microorganisms (Perdomo et al. 1994a; Sansonetti et al. 1996; Zychlinsky et al. 1996). These early events are followed by a wave of inflammation which is characteristic of shigellosis. The structure of the lymphoid follicle and associated FAE is destroyed within 8 h in the rabbit ileal-loop model and extends to the surrounding villous tissues.

A significant part of this inflammatory process seems to be caused by apoptotic death of infected macrophages, a process that is observed in the rectal mucosa of patients with shigellosis (Islam et al. 1997). It has been shown in vitro that macrophages infected by invasive shigellae undergo programmed cell death within 2–3 h of infection (Zychlinsky et al. 1992). If macrophages are pretreated with LPS, within 1 h, large quantities of mature IL-1β are released (Zychlinsky et al. 1994a), reflecting direct interaction between the IpaB invasin which is the molecule causing apoptosis (Zychilnsky et al. 1994b) and the IL-1 cleavage enzyme, the cysteine protease ICE or caspase 1 (Chen et al. 1996). This interaction is likely to explain the dual response: cell death–IL-1β release. The consequence of early IL-1 release is initiation of the inflammatory process. In the rabbit ligated-loop model of infection, concurrent administration of the IL-1 receptor antagonist (IL-1ra) with the development of infection almost abolishes the histopathological symptoms of experimental shigellosis (Sansonetti et al. 1995). In addition, the number of bacteria present in the tissues is significantly diminished, suggesting that early inflammation is an essential factor in disrupting the epithelial permeability, and facilitating invasion of the mucosa. In support of this view, in vitro modeling, using polarized colonic epithelial cells of human origin grown on filters, has shown that bacteria deposited on the apical side of these cells are unable to invade, as already mentioned. If human polymorphonuclear (PMN) leukocytes are concurrently added on the basal side, efficient invasion is observed, confirming that PMN leukocytes may play a major role in promoting infection of the epithelium (Perdomo et al. 1994b). In addition, the pretreatment of animals with a monoclonal antibody directed against CD18, and β2 subunit of the MAC1 integrin of PMN leukocytes, abrogates all histopathological symptoms of tissue destruction, due to neutralization of the inflammatory response, and also consequently diminishes entry of bacteria within the mucosa (Perdomo et al. 1994a).

These experiments raised the interesting concept according to which inflammation may account both for disrupting the permeability of the epithelium, thus facilitating bacterial invasion, and for causing tissue destruction which, more than bacterial invasion alone, is characteristic of the symptoms observed in shigellosis. Interestingly, this inflammatory process mediated by IL-1β release, through ICE interaction seems to also play a central role in the recovery of primary infection by the infected host. Recently, it has been shown that ICE-knock out mice had a delayed inflammatory response, and were then unable to recover from infection. In contrast, their wild-type counterparts recovered from infection with the development of an inflammatory reaction very early following infection (Sansonetti et al. unpublished observations). The efficient killing of *Shigella* by PMN has been reported (Renesto et al. 1996; Mandic-Mulec et al. 1997), suggesting that these inflammatory cells may not only contribute initially to the severe tissue damage characteristic of shigellosis, but also ultimately participate in clearance and resolution of infection. This, therefore, illustrates the dual role that innate immune responses may play throughout the course of infection.

3.2.2 Others Cells

It has recently been reported that, in response to bacterial invasion, intestinal epithelial cells can provide signals that are important for the initiation and amplification of an acute mucosal inflammatory response (for a review, Hedges et al. 1995). For instance, infection of monolayers of human colonic epithelial cell lines with *S. dysenteriae 1* results in the coordinate expression and upregulation of a specific array of pro-inflammatory cytokines, IL-8, MCP-1, GM-CSF and tumor necrosis factor alpha (TNFα). Interestingly, an identical array of cytokines and

IL-6 are also expressed by freshly isolated human colon epithelial cells (Jung et al. 1995). The contribution of the intestinal epithelial cells in mucosal inflammation, elicited following bacterial invasion, remains to be established in vivo.

In addition to the major role of macrophage-released IL-1β in promoting inflammation and subsequent invasion, extensive cytokine production and secretion has been reported in the rectal mucosa of infected patients (Raqib et al. 1995a; Raqib et al. 1995b). The presence of TNFα, TNFβ, IL-4, IL-6, IL-8, IL-10, transforming growth factor beta (TGFβ) and IL-1 receptor antagonist, in addition to IL-1 α- and β-producing cells has been reported in biopsies of infected patients during the acute phase, with a predominance of IL-1β-, TGFβ-, IL-4- and IL-10-producing cells. Cytokine levels correlated with increased numbers of cytokine-producing cells and with the severity of disease. In contrast, interferon gamma (IFN-γ) levels, depressed at the disease onset, progressively increased during the convalescent stage.

During this convalescent stage of the disease, gradual accumulation of mRNA for all the above cited cytokines was observed (Raqib et al. 1996). The frequency of cytokine mRNA-expressing cells varied in the range of 3 to 100-fold higher than that of the corresponding protein-synthesizing cells. A selective downregulation of the receptors for IFN-γ, TNF (type-I receptor), IL-1, IL-4 and TGFβ (type-I receptor) at the onset of the disease, with a gradual reappearance during the convalescent stage has also been reported (Raqib et al. 1995c). Although these data do not address the respective roles of these cytokines in the pathogenesis of shigellosis, they indicate that severe disease activity is accompanied by high concentrations of pro-inflammatory cytokines in stools and late-occurring IFN-γ production. They suggest that suppressed IFN-γ production may play a role in delaying the resolution of infection. Conversely, depressed production may be a result of infection itself and may be related to the development of immunity to *Shigella* infection that appears to be mainly of the TH2 type.

3.3 Specific Host Response to *Shigella* Infection

The type of immunity elicited by the host following *Shigella* infection is obviously related to the bacterial behavior which exhibits extracellular and intracellular stages. Both humoral and cellular immune responses are, in fact, elicited in natural and experimental *Shigella* infections. Rare data are available on cell-mediated immunity (Zwillich et al. 1989; Islam et al. 1995; Islam et al. 1996). For instance, the potential role of cell-mediated immunity in lysing *Shigella*-infected cells remains to be investigated. In contrast, the targets and the effectors of the humoral response, as well as their role in protection, have been studied extensively. Secretory IgA antibodies produced in local secretions and serum IgG are both directed against LPS and some virulence plasmid-encoded proteins, among which the Ipa antigens are the most consistently recognized (for review, see Phalipon and Sansonetti 1995a). Following natural or experimental infection, hosts become resistant to subsequent infection with homologous serotypes (Mel et al. 1968;

DuPont et al. 1972), pointing to LPS as the primary target antigen for protective immunity. The role of anti-Ipa antibodies is much more difficult to assess. Although high titers of secretory IgAs directed against Ipa proteins have been correlated with a decrease in the duration of the illness (Oberhelmann et al. 1991), the demonstration of a direct effect of anti-Ipa immunity in protecting hosts against infection has not yet been established.

Following mucosal infection, local and systemic responses are usually elicited. Local immunity developed in response to a primary infection with a given mucosal pathogen is believed to play the major role in protecting the host against a secondary infection with the same pathogen. For instance, with *Shigella* infection, it is assumed that mucosal, rather than systemic, immunity plays a major role since *Shigella* infection mainly remains localized to the colonic mucosa. Anti-LPS and anti-Ipa secretory IgA antibodies, detected in colostrum and breast-milk samples of mothers living in *Shigella*-endemic regions, are believed to account for the observed beneficial effect of breast feeding on the incidence and severity of shigellosis in these countries (Clemens et al. 1986), thus emphasizing the protective role of local immunity.

The first experimental demonstration of the major role of the anti-LPS mucosal immunity has recently been established (Phalipon et al. 1995b). Nevertheless, protection may occur following parenteral vaccination with a LPS-protein conjugated vaccine (Cohen et al. 1996). Such particulate vaccine is expected, essentially, to raise a systemic humoral response following parenteral administration. Although such data are inconsistent with the "old" observations showing that killed whole-cell vaccines administered parenterally were not protective, they re-open the possibility that systemic humoral immunity could be somewhat protective. Recently, the homing potentials of mucosally and parenterally induced antibody-secreting cells (ASC) were compared by examining the homing receptor expression of circulating specific ASC in the blood of volunteers vaccinated orally or parenterally with the same antigen (Kantele et al. 1997). Results indicate that all the circulating ASC, after oral vaccination, are committed to migrate to the mucosal compartment. However, ASC induced by parenteral vaccination are mostly directed to the systemic compartment, yet a part of them also has mucosal homing properties. This might explain, in part, the role of systemic immunity in protection of mucosal sites, especially when infection remains mucosally localized.

4 Host Response to *Salmonella*, an Invasive, Facultative Intracellular Bacterium that Does Not Escape from the Phagocytic Vacuole

Salmonella species cause various diseases in humans, depending on the serovar. *Salmonella typhi* is the agent of typhoid fever, a life-threatening septicemic infection, whereas *Salmonella typhimurium* induces a self-limiting gastroenteritis with only occasional bacteremia. Interestingly, a septicemic disease develops in mice infected with *S. typhimurium*. As there is no animal model that reflects typhoid fever accurately, most investigators have, instead, focused on the septicemia produced by *S. typhimurium* in mice. The microorganisms that reach the intestinal tract cross the intestinal epithelium to reach the lamina propria, where they replicate or proceed to deeper tissues, presumably carried within non-activated macrophages (for review, see Galan 1996). They subsequently drain through the lymphatics to the thoracic duct into the blood and, ultimately, infect the liver and spleen (Miller et al. 1995).

4.1 Passage of the Intestinal Barrier

As for shigellae, M cells have often been suggested as the primary invasion site for *Salmonella* spp. (Clark et al. 1994; Jones et al. 1994). This hypothesis has been largely supported by experiments carried out in the rabbit ligated ileal-loop model, but the correlation of these findings with natural infection is less clear.

In contrast to *Shigella* infection, M cells invaded by *S. typhimurium* die. There is an apparent interaction with adjacent enterocytes during the early phase of infection. A mutant of *S. typhimurium* that does not invade tissue-culture cells also does not invade or disrupt M cells (Jones et al. 1994). This is similar to what has been observed with a non-invasive mutant of *Shigella*, that does not enter the dome of the Peyer's patches as efficiently as a wild-type strain (Sansonetti et al. 1996). Many organisms have been reported to preferentially infect or to associate with M cells (Siebers and Finlay 1996), but, M-cell uptake is not as active as that seen for *Salmonella* species in all cases. Whether microorganisms target a specific M-cell receptor by using specific adhesins or, instead, interact predominantly with M cells because of their underlying biological properties, remains to be established (see Neutra, this volume). In *S. typhimurium*, the *lpf* fimbrial operon is involved in targeting the pathogen to murine Peyer's patches and may mediate specific adherence to M cells (Baumler et al. 1996).

It seems that *Salmonella* can also breach the intestinal barrier through the absorptive columnar epithelial cells of the small intestine (Takeuchi 1967). Unlike *Shigella*, *Salmonella* is able to enter epithelial cells through the apical pole (Finlay et al. 1988). Once inside, again, in contrast to *Shigella*, *Salmonella* remains inside the phagocytic vacuole (for details of bacteria–epithelial cell interactions, see the review by Raupach et al. this volume).

To summarize, it is likely that the relative importance of the different intestinal epithelial cells in *Salmonella* invasion of the host may be largely dependent on the species of the infected hosts as well as on the *Salmonella* serotype.

4.2 Interaction of *Salmonella* with the Host Innate Immune System

4.2.1 Macrophages

Following passage through the epithelium of the Peyer's patches, virulent *Salmonella* strains enter the environment of the follicular dome, which is populated with host lymphocytes and macrophages. To move deeper into tissues, these bacteria must avoid killing mechanisms of professional phagocytes following internalization. *S. typhimurium* entry into macrophages is associated with membrane ruffling, macropinocytosis, and the formation of "spacious phagosomes" where the organism resides (Alpuche-Aranda et al. 1992).

There is a direct correlation between the ability to form spacious phagosomes and the susceptibility of the host to infection with various *Salmonella* serotypes (Alpuche-Aranda et al. 1992). In vitro studies have shown that *Salmonella* develop two types of interaction with macrophages: on the one hand, it is able to survive within the phagocytic vacuole and, on the other hand, as observed for *Shigella*, it triggers macrophage apoptosis in at least a significant proportion of these cells (Monack et al. 1996). Recently, it has been reported that *Salmonella* resides intracellularly inside macrophages in the liver of mice infected intravenously with low infectious doses and, at the same time, triggers cell death of phagocytes (Richter-Dahlfors et al. 1997). The existence of different populations of *S. typhimurium* in macrophages has previously been proposed (Gahring et al. 1990). This may reflect the existence of alternative entry pathways into phagocytes, with more than a single pathway operating simultaneously in the same cell. This may also reflect a difference in the bacteria–macrophage interaction in the time course of infection. It seems that survival within the phagocytic vacuole occurs at the early stages of invasion of the intestinal mucosa by *Salmonella*, whereas bacterium-induced macrophage apoptosis occurs at the later stages of the infectious process. In contrast, *Shigella* rapidly escapes the phagocytic vacuole upon internalization, and triggers macrophage apoptosis, especially at the early stage of invasion of the mucosa-associated lymphoid follicles.

Functional conservation of the *Salmonella* and *Shigella* effectors of entry into epithelial cells has recently been reported (Hermant et al. 1995; Kaniga et al. 1995). For instance, *Shigella* IpaB invasin and *Salmonella* SipB invasin are highly related to each other (65% identity and 81% similarity), especially in the central region of the proteins. However, the *N* and *C* termini of these proteins are less conserved. We may speculate that those regions may confer SipB and IpaB specificity of function. This might explain how such similar proteins could mediate such different interactions with macrophages.

In the case of *Salmonella* survival within the phagocytic vacuole, once the vacuole is formed, its composition changes over time, suggesting that *Salmonella* directs the components of the vacuole for its own needs (for review, Finley and Falkow 1997). Concerning vacuole-acidification, results are still controversial, but it may be necessary for survival and replication (Rathman et al. 1996). Several bacterial factors enhance *S. typhimurium* survival within macrophages, including the PhoP/PhoQ system which activates at least five bacterial products and represses others (MILLER 1991). One of the phenotypes that PhoP/PhoQ regulates is the capacity to survive bactericidal cationic peptides, involved in killing intracellular bacteria (PARRA-LOPEZ et al. 1994). The *phoP* locus also appears to inhibit antigen processing and presentation of bacterial antigens expressed by intracellular bacteria (WICK et al. 1995), which may enhance virulence. Finally, the response of *Salmonella* to an intracellular acidic environment is associated with the rapid synthesis of several bacterial gene products. The bacterial genes involved in the survival of *Salmonella* within phagocytic cells have been recently reviewed (JONES and FALKOW 1996).

An interesting feature of *Salmonella* infection in mice is that different strains of mice show different levels of susceptibility to infection. While resistance to disease, whether in animal or in humans, is multigenic, there has been considerable emphasis on the *Ity* gene (PLANT and GLYNN 1979) which has a major effect on susceptibility to *Salmonella* infection. Mice are either *Salmonella*-sensitive, Ity^{s}, or they carry the dominant Ity^{r} allele. Ity^{s} mice succumb to overwhelming sepsis following the parenteral injection of very few microorganisms. The major effect of Ity^{r} is an almost complete inhibition of bacterial growth in mice (BENJAMIN et al. 1990). In vitro studies suggest that the *Ity* effect is due to differences in the rate of intracellular killing of *Salmonella* by resident phagocytic cells (VAN DISSEL et al. 1986). Some experiments suggest that *Ity* encodes a macrophage-specific membrane transport function (VIDAL et al. 1993).

4.2.2 Other Cells

Both IFN-γ and TNFα appear to play important roles in stimulating early host defenses against the virulent bacteria. Nonphagocytic tissue-culture cells display an increased resistance to invasion when treated with these cytokines (DEGRE et al. 1989), and IFN-γ increases the fusion of phagosomes containing bacteria with lysosomes (ISHIBASHI and ARAI 1990). The primary role of IFN-γ appears to be the inhibition of bacterial growth rather than induction of host-mediated killing mechanisms (MUOTIALA and MAKELA 1990; NAUCIEL and ESPINASSE 1992). The lymphoid cells within intestinal Peyer's patches produce large amounts of IFN-γ in response to stimulation with *S. typhimurium* (RAMARATHINAM et al. 1991). The levels of TNFα also increase in response to infection with *S. typhimurium* and inhibition of the TNFα response increases the susceptibility of the mouse to infection (NAKANO et al. 1990).

Interestingly, the host immune response to virulent *Salmonella* appears to induce a form of immunosuppression (LEE et al. 1985) mediated by phagocytic cells. *Salmonella*-infected macrophages actually produce large quantities of nitric oxide that poison lymphocytes (AL-RAMADI et al. 1992). A reduction in numbers of functional lymphocytes appears to be the direct cause of the suppression of the mouse immune response observed in systemic murine typhoid (EISENSTEIN et al. 1994).

As shown for *Shigella*, colonic epithelial cells produce pro-inflammatory cytokines, including IL-8, MCP-1, GM-CSF, and TNFα, in response to *Salmonella* invasion (JUNG et al. 1995). Like *Shigella*, addition of *S. typhimurium* to the apical surface of polarized epithelial cells also causes recruitment of PMNs from the basolateral surface across an in vivo polarized epithelial barrier (MC CORMICK et al. 1993). PMN transmigration requires active protein synthesis by both bacteria and epithelial cells, and induces IL-8 production although IL-8 and *N*-methionyl-1-leucyl-1-phenyl-alanine are not necessary for PMN migration. This may provide an additional pathway for invading bacteria to penetrate across the epithelial barrier. Interestingly, some reports substantiate the concept that, unlike *Shigella*, survival of *Salmonella* within PMN's can contribute to the pathogenicity of the organism at later stages of infection. Different microbial defense mechanisms protecting *Salmonella* from oxidative and non-oxidative killing by PMNs have been identified (DUNLAP et al. 1992).

Another type of cell that is present at the mucosal site is the γδ T cell. Dominant γδ T-cell response to infections with various microbial pathogens suggests that at least a significant fraction of γδ T cells represent a first line of defense against infections (for a review, HAAS et al. 1993). It has recently been reported that *Salmonella*-infected macrophages produce IL-15 (NISHIMURA et al. 1996), a cytokine with stimulatory activities for NK cells and B cells (CARSON et al. 1994; ARMITAGE et al. 1995). Interestingly, IL-15 also has a stimulatory activity for γδ T cells (NISHIMURA et al. 1996). It has been speculated that a subset of naive γδ T cells begins to express IL-2 receptors after stimulation with relevant antigen, and expands rapidly in response to IL-15 released from infected macrophages, resulting in early appearance of γδ T cells during infection with intracellular bacteria. This may provide the first line of defense against infection with various pathogens well before antigen-specific αβ T cells expand clonally through autocrine pathway by IL-2 production. The γδ T cells stimulated by IL-15 may play a role covering the gap between the phagocytic system and the highly evolved type of immune response mediated by αβ T cells in host defense against microbial infections.

4.3 Specific Host Response to *Salmonella* Infection

The majority of individuals that survive typhoid fever generally acquire immunity to re-infection. Because of the complex nature of the pathogenesis of *S. typhi* clinical infection, human protection is probably played by secretory intestinal antibody (in preventing mucosal invasion) (SARASOMBATH et al. 1987; CHAU et al. 1981), circulating antibodies (against bacteremic organisms), and cell-mediated immunity (to eliminate intracellular bacilli) (LEVINE and HORNICK 1981). Studies in a mouse model of infection have shown that the secretory IgA-mediated local response plays a major role in protection (MICHETTI et al. 1992; MICHETTI et al. 1994). The *Ity*^{s} mouse strain has been used to study *Salmonella* immunity, since both humoral and cell-mediated immunity are required for protection, while innately resistant mice control low doses of virulent *Salmonella* with a moderate, nonspecific immune response (COLLINS 1974, HORMAECHE et al. 1990; MASTROENI et al. 1993).

Transfer of immunity experiments have demonstrated that CD4+ cells, CD8+ cells and serum are all required to protect naive mice from challenge with virulent *S. typhimurium* strain (Mastroeni et al. 1993). The humoral response to *S. typhimurium* in mice parallels the antibody response observed in human typhoid fever, which is directed against LPS and a number of other undefined antigens (Brown and Hormaeche 1989; O'Callaghan et al. 1990). Similar to the development of *Shigella* vaccines, two types of vaccines have been developed against *Salmonella*: one is based on the development of live attenuated strains, administered orally, whereas the other is based on purified Vi antigen, a protective antigen administered parenterally (for review, see Levine 1994). Both actually induce protection in humans. We may expect that the live attenuated strain, by stimulating the local immunity efficiently, protects against the mucosal infection, whereas the Vi-based vaccine protects against the systemic phase of infection.

5 Host Response to Enteropathogenic *Yersinia*, an Invasive, Extracellular Bacterium

The enteropathogenic *Yersinia* are two gram-negative bacterial species related to the causative agent of the bubonic plague, *Y. pestis. Y. Enterocolitica,* and essentially cause a variety of intestinal diseases. *Y. pseudotuberculosis* causes mild or unapparent bacterial infections, but triggers, following translocation of the intestinal barrier, a number of autoimmune disorders. These include mesenteric lymphadenitis and reactive arthritis, particularly in individuals harboring the HLA-B27 histocompatibility allele (for review, see Isberg 1996). The primary site of infection of these orally transmitted pathogens is the lymphoid follicles of the small intestine. For a long time, *Yersinia* was considered to be an intracellular pathogen, but recent findings have shown that the pathogen proliferates in the extracellular fluid during infection and prevents its uptake process by professional phagocytes, a mechanism termed antiphagocytosis (Forsberg et al. 1994). Therefore, its "intracellular" life seems to be restricted to its passage of the epithelial barrier.

5.1 Passage of the Intestinal Barrier

The most extensively studied animal models for this enteric infection are the rabbit and mouse infection assays, in which the disease proceeds by translocation across the epithelium of the ileum or the colon to the submucosal region (Carter 1975). Invasin-dependent and invasin-independent pathways for translocation of *Y. pseudotuberculosis* across the Peyer's patches have recently been reported (Marra and Isberg 1997). It appears that the bacterial adhesive factors control the site of bacterial interaction within the intestinal environment. *Y. enterocolitica* and

Y. pseudotuberculosis are found within M cells in the FAE of the Peyer's patches (Autenrieth and Firshing 1996; Grutzkau et al. 1990; Hanski et al. 1989).

Fine interactions of the pathogen with the M cells that translocate through it are described by Raupach et al. this volume. Internalization appears rather patchy, with some intestinal cells showing large numbers of bacteria and adjoining cells showing none, indicating that the mammalian cell receptors for the bacteria may be unevenly distributed throughout the intestine. Bacteria can be found in the Peyer's patches shortly thereafter, with microorganisms both within and outside cells. In animal models, the bacteria are drained into mesenteric lymph nodes, and a large number can be found in these locations within 24 h after initial infection (Carter 1975). In human infections, the disease often limits itself to infection of mesenteric lymph nodes (O'Loughlin et al. 1990). After oral infection, susceptible animals are unable to control replication of *Yersinia* at this site, and bacterial growth continues after localization into the liver and spleen. The bacteria are exclusively located extracellularly and remain extracellular for the duration of the disease. Subsequently, this process leads to pyogenic lesions and production of IL-1 in the Peyer's patches (Autenrietg and Firsching 1996a; Beuscher et al. 1992).

Unlike *Salmonella* which enters the epithelial cells via the apical side, *Yersinia*, like *Shigella*, enters the epithelial cells via the basolateral pole, which is reached by the bacterium only after its translocation through the epithelium via M cells. It has been shown that the Inv protein, responsible for the entry of the bacteria within epithelial cell lines (for more details, see Raupach et al. this volume), is also involved in vivo at the early stages of the pathogenic process, i.e., the interaction with the Peyer's patches (Pepe and Miller 1993). In contrast to what has been observed in vitro, it appears that, in vivo, *Yersinia* remains extracellular, but firmly affixed to the host cell surface. Therefore, the Inv protein seems to function in vivo much more like an adhesin than an invasin. Its attachment to β1 integrins is the key step in the contact-dependent secretion of plasmid virulence factors and the subsequent translocation of several bacterial proteins into the host cell cytoplasm that leads to pathogenesis.

5.2 Interaction of *Yersinia* with the Host Innate Immune System

5.2.1 Antiphagocytic Activity of *Yersinia*

To avoid killing by macrophages and, therefore, allow its extracellular survival, *Yersinia* exerts an antiphagocytic activity that is likely to protect it from phagocytosis (Forsberg et al. 1994). This is a major virulence mechanism that is shared by the three *Yersinia* species, which possess a common conserved virulence plasmid that encodes about 11 secreted proteins called Yops (*Y*ersinia *o*uter membrane *p*rotein*s*). Among these proteins, two are directly involved in antiphagocytosis, YopE and YopH (Forsberg et al. 1994). These Yops affect a general phagocytic mechanism, including uptake involving β1-integrin interactions as well as uptake mediated via Fc receptors and complement receptors (Fallamn et al. 1995; Ruckdeschel et al. 1996).

Interestingly, this blocking effect of *Yersinia* is not restricted to professional phagocytes; other normally non-phagocytic cells are affected in a similar manner by the pathogen. YopE disrupts the F-actin network of the target cell, but the molecular mechanisms underlying this effect are not known. Given the homology that YopE shares with the membrane translocation domain of exoenzyme S of *Pseudomonas aeruginosa*, it is possible that YopE targets small GTP-binding proteins, since exoenzyme S has been shown to modify these signaling molecules (Coburn et al. 1989; Kulich et al. 1994). The other effector involved in antiphagocytosis, YopH, is homologous to eukaryotic tyrosine phosphatases (Guan and Dixon 1990) and has the highest activity of all tyrosine phosphatases known today (Zhang et al. 1992). High expression of YopH alone is sufficient to block internalization of the bacteria (Persson et al. 1997). It is likely that YopH acts by dephosphorylating phosphotyrosine proteins involved in the initial signal transduction events triggering the uptake of bacteria.

The molecular targets for YopH were recently identified as p130^{Cas} (Crk-associated substrate) and focal adhesion kinase (FAK) in non phagocytic cells (Persson et al. 1997), and with p130^{Cas} in macrophages (N. Carballeira and M. Fällman, unpublished results). Fällman et al. (1997) have recently proposed a model of blockage of bacterial uptake including the three components, YopH, p130^{Cas} and FAK. To summarize, it seems that *Yersinia* has chosen a key protein involved in a general phagocytic mechanism important for the uptake by professional, as well as non-professional phagocytes, as a suitable target to obstruct phagocytosis. Additionally, another bacterial product, YpkA, shares homology with host serine/threonine kinases and presumably disrupts host signaling pathways in phagocytic cells (Galyov et al. 1993).

Besides this antiphagocytic activity, *Yersinia*, like *Shigella* and *Salmonella*, also induces macrophage apoptosis (Ruckdeschel et al. 1997a; Mills et al. 1997; Monack et al. 1997). The actual function of this cell programmed death-induced process remains to be evaluated in vivo.

It should be noticed that to increase its chance of extracellular survival, in addition to its antiphagocytic activity, *Yersinia* species resists to complement killing. This resistance is mediated by the Ail protein that also mediates high levels of adherence to epithelial cells (Bliska and Falkow 1992).

5.2.2 Others Cells

Studies focusing on defense mechanisms against parenteral *Yersinia* infection have implicated both T cells and macrophages, including TNFα and IFN-γ, as essential components in protecting the host against primary infection (Authenrieth and Heesemann 1992; Authenrieth et al. 1993, 1994). Recently, mechanisms of the local host defense in Peyer's patches have also been investigated (Autenrieth et al. 1996b). Treatment with anti-IFN-γ, and particularly with anti-TNFα antibodies caused a dramatic increase of bacterial growth in infected organs followed by morphological tissue alterations, including necrosis of *Yersinia*-infected Peyer's patches. These results suggest that both cytokines play a central role in local defense mechanisms in Peyer's patches and mesenteric lymph nodes, possibly by activation of macrophages. As a matter of fact, a recent report has provided evidence that TNFα production in Peyer's patches may be suppressed by virulence factors expressed by *Yersinia*, especially YopB (Beuscher et al. 1995; Burdack et al. 1997). Macrophage TNFα production is prevented by inhibition of ERK1/2, p38 and JNK (C-Jun NH_2-terminal) kinase activities (Ruckdeschel et al. 1997b). Although suppression of this single cytokine response is probably not sufficient to facilitate survival of the infecting organisms, suppression of TNFα production by YopB significantly contributes to the evasion of the bacteria from antibacterial host defense.

IL-12 is also a key cytokine in the protective response to *Yersinia* infection, especially in spleen and liver infections (Bohn and Autenrieth 1996). IL-12 mediates *Yersinia*-triggered IFN-γ production by both NK cells and CD4+ T cells. However, IL-12 plays only a minor role in defense mechanisms against the bacteria in Peyer's patches (Bohn and Autenrieth 1996).

Interestingly, T-cell activation is triggered by Inv, the *Yersinia* invasin protein that renders the bacteria able to invade the mammalian through interaction with β1 integrin. Among the different members of the very late antigen (VLA) integrin family of adhesion molecules expressed by CD4+ T cells, VLA-4 is the T cell invasin receptor (Ennis et al. 1993). The observation that an outer membrane protein of a bacterium that is associated with reactive arthritis and other autoimmune spondyloarthropathies can act as a T-cell co-stimulus may have implications for the etiology of these diseases.

T-cell activation may also occur via superantigens. *Yersinia pseudotuberculosis* isolated from patients manifesting acute and systemic symptoms produces Yersinia-pseudotuberculosis-derived mitogen (YPM) that exhibits superantigenic properties (i.e., abilities to bind directly to MHC class-II molecules and selectively stimulate T-cell populations bearing particular Vβ elements in T-cell receptors) (Uchiyama et al. 1993). We may speculate that this is one more strategy developed by the pathogen to overcome local host defense systems.

As previously mentioned for *Shigella* and *Salmonella*, invasion of colonic epithelial cells by *Yersinia* results in the secretion of IL-8, MCP-1, GM-CSF and TNFα (Jung et al. 1995). In addition, it has been reported that virulent *Y. enterocolitica* induced a significantly lower level of IL-8 secretion than a corresponding non-virulent strain. The reduced secretion of IL-8 is due to the presence of Yop proteins, especially YopB and YopD which are required for the suppressive effect (Schulte et al. 1996).

5.3 Specific Host Response to *Yersinia* Infection

To interrupt the transmission cycle, numerous efforts have been made to protect rodents and humans against the plague bacillus by immunization. In 1963, Lawton et al. (1963) reported on protective immunity in mice after immunization with the V antigen, a 37-kDa secreted protein encoded by the virulence plasmid of pathogenic bacteria. WAKE and SUTOH (1983) described long-term immunity to the plague after immunization with live *Y. pseudotuberculosis* and ALONSO et al. (1980) demonstrated cross reactivity to *Y. pestis* after oral immunization of mice with *Y. pseudotuberculosis*. However, this cross protection was only partial due to the fact that the V antigen differs among species (ROGGENKAMP et al. 1997).

Whereas the V antigen is shared among the three *Yersinia* species, the protein YadA, an adhesin, is produced exclusively by the two enteropathogenic species (SKURNICK and WOLF-WATZ 1989; VOGEL et al. 1993) and has been shown to be a protective antigen. YadA is a surface-exposed protein forming a capsule-like coat around the pathogen. Thus, protective antibodies directed against this protein most likely function by opsonizing extracellular bacteria. Moreover, anti-YadA serum may inhibit the cell adherence of the bacteria.

A large number of studies have demonstrated the importance of T cells in the immune response to *Yersinia*. A natural epitope derived from YopH is presented by MHC class I, despite the fact that these organisms are found mostly extracellularly and, when seen in an intracellular compartment, remain entirely within the vacuole. Such presentation suggests that *Yersinia* bound to host cells have the ability to translocate into the cytoplasm of those cell proteins that contribute to their virulence and, in the process, subject the infected cell to recognition by cytotoxic lymphocytes (CTL) (STARNBACH and BEVAN 1994).

6 Conclusion

Diarrheal diseases of bacterial origin remain a major cause of morbidity and mortality worldwide. In spite of significant improvements, insufficient public health education and hygiene facilities, as well as uncontrolled use of antibiotics, contribute to persistence of the threat of and establishment of conducive conditions for the emergence of new pathogens (i.e., *V. cholerae* of the O139 serotype) and of antibiotic resistant strains (i.e., *S. typhi*). Vaccination remains the most efficient way to overcome this problem. Over the last 10 years, rising hope relies on the slow, but sustained, development and improvement of vaccines. This has particularly been the case for typhoid fever and cholera. Efforts have been put into basic research for a better understanding of not only the physiopathological processes of, but also the host response to, infection. Such data are required for the rational development of new vaccine strategies. For example, the relative role of systemic and local immunity in protecting a host against enteric infection must be established in order to develop efficient vaccines.

This chapter has reviewed the basis of innate and specific immunity against specific enteric pathogens that have been chosen as examples of particular behaviors at the intestinal level, i.e., *V. Cholerae,* which remains luminal; *Shigella,* which invades the mucosa, but remains essentially local; *Yersinia,* which invades, but rarely spreads beyond local lymph nodes; and *Salmonella,* which, from the intestinal zones of invasion, can spread systemically. These various examples underline the diverse strategies utilized by these pathogens and emphasize the difficulty in elaborating a single strategy for immunization.

In recent years, studies have focused on the molecular mechanisms of interaction of these bacteria with epithelial and immune cells in in vitro assays. A better understanding of bacteria–cell cross talk has allowed the identification of essential virulence genes and products and also the characterization of cellular components involved in the pathogenicity. However, the understanding of the in vivo situation remains much more difficult to assess. Two major investigating strategies are currently emerging. The first strategy involves in vitro modeling of particular areas or properties of the epithelial barrier. In these systems, epithelial cells are combined with immune cells in order to address specific issues, such as differentiation of epithelial cells into M-like cells in the presence of lymphocytes with restoration of capacity for bacterial translocation (Kerneis et al. 1997). Such systems have also been used to study destabilization of the epithelial lining integrity by basal PMN cells in the presence of apical bacteria, thus facilitating epithelial invasion (Perdomo et al. 1994b).

The second strategy that has already been very useful in examining in vivo events involves the development of adapted animal models of infection. This encompasses the increasing use of knock-out and transgenic animals, as well as xenotransference of human embryonic intestine into immunocompromized mice (Savidge et al. 1995). These different systems establish a bridge between pure models of cell infection and more or less sophisticated models of infectious diseases, in which the actual basis of non-specific and specific immune protection can be studied. Ultimately, these data need to be compared with those obtained from patients and volunteers in vaccine trials. Therefore, studies of the host response to enteric pathogens at the mucosal level are likely to become an increasing area of interest until truly efficient vaccines become readily available.

References

Alonso JM, Vilmer D, Mazigh D, Mollaret HH (1980) Mechanisms of acquired resistance to plague in mice infected by Yersinia enterocolitica O:3 Curr Microbiol 4:117–122

Alpuche-Aranda CM, Swanson JA, Loomis WP, Miller SI (1992) Salmonella typhimurium activates virulence gene transcription within acidified macrophage phagosomes. Proc Natl Acad Sci USA 89:10079–10083

Al-Ramadi BK, Meissler JJJ, Huang D, Eisenstein TK (1992) Immunosuppression induced by nitric oxide and its inhibition by interleukin-4. Eur J Immunol 22:2249–2254

Apter FM, Michetti P, Winner L III, Mack J, Mekalanos JJ, Neutra MR (1993) Analysis of the roles of antipolysaccharide and anti-cholera toxin immunoglobulin A (IgA) antibodies in protection against Vibrio cholerae and cholera toxin by use of monoclonal IgA antibodies in vivo. Infect Immun 61:5279–5285

Armitage RJ, Macduff BM, Eisenman R, Paxton R, Grabstein KH (1995) IL-15 has a stimulatory activity for the induction of B cell proliferation and differentiation. J Immunol 154:483–489

Attridge SR, Manning PA, Holmgren J, Jonson G (1996) Relative significance of mannose-sensitive hemagglutinin and toxin-coregulated pili in colonization of infant mice by Vibrio cholerae El Tor. Infect Immun 64:3369–3373

Autenrieth IB, Heesemann J (1992) In vivo neutralization of tumor necrosis factor alpha and interferon-gamma abrogates resistance to Yersinia enterocolitica in mice. Med Microbiol Immunol 181:333–338

Autenrieth IB, Vogel U, Preger S, Heymer B, Heesemann J (1993) Experimental Yersinia enterocolitica infection in euthymic and T-cell deficient athymic mice nude C57BL/6 mice: comparison of time course, histomorphology, and immune response. Infect Immun 61:2585–2595

Autenrieth IB, Beer M, Bohn E, Kaufmann SHE, Heesemann J (1994) Immune responses to Yersinia enterocolitica in susceptible BALB/c and resistant C57BL/6 mice: an essential role for gamma interferon. Infect Immun 62:2590–2599

Autenrieth IB, Firsching R (1996a) Penetration of M cells and destruction of Peyer's patches by Yersinia enterocolitica: an ultrastructural and histological study. J Med Microbiol 44:285–294

Autenrieth IB, Kempf V, Sprinz T, Preger S, Schnell A (1996b) Defence mechanisms in Peyer's patches and mesenteric lymph nodes against Yersinia enterocolitica involve integrins and cytokines. Infect Immun 64:1357–1368

Baumler AJ, Tsolis RM, Heffron F (1996) The lpf fimbrial operon mediates adhesion of Salmonella typhimurium to murine Peyer's patches. Proc Natl Acad Sci USA 93:279–283

Benjamin WJ, Hall P, Roberts SJ, Briles DE (1990) The primary effect on the Ity locus is on the rate of growth of Salmonella typhimurium that are relatively protected from killing. J Immunol 144:3143–3151

Beuscher HU, Rausch UP, Otterness IG, Röllinghoff M (1992) Transition from IL-1 beta to IL-1 alpha production during maturation of inflammatory macrophages in vivo. J Exp Med 175:1793–1797

Beuscher HU, Rödel F, Forsberg A, Röllinghoff M (1995) Bacterial evasion of host immune response defence: Yersinia enterocolitica encodes a suppressor for tumor necrosis factor alpha expression. Infect Immun 63:1270–1277

Bliska JB, Falkow S (1992) Bacterial resistance to complement killing mediated by the Ail protein of Yersinia enterocolitica. Proc Natl Acad Sci USA 89:3561–3565

Bohn E, Autenrieth IB (1996) IL-12 is essential for resistance against Yersinia enterocolitica by triggering IFN-γ production in NK cells and CD4+ T cells. J Immunol 156:1458–1468

Bromander AJ, Holmgren J, Lycke N (1991) Cholera toxin stimulates IL-1 production and enhances antigen presentation by macrophages in vitro. J Immunol 146:2908–2914

Brown A, Hormaeche CE (1989) The antibody response to salmonellae in mice and humans studied by immunoblots and ELISA. Microbiol Pathog 6:445–454

Burdack S, Schmidt A, Knieschies E, Rollinghoff M, Beuscher HU (1997) Tumor necrosis factor-alpha expression induced by anti-YopB antibodies coincides with protection against Yersinia enterocolitica infection in mice Med Microbiol Immunol 185:223–229

Carson WE, Giri JG, Lindemann ML, Linett ML, Ahdieh M, Paxton R, Anderson D, Eisenmann J, Grabstein K, Caligiuri MA (1994) Interleukin (IL) 15 is a novel cytokine that activates human natural killer cells via components of the IL-2 receptor. J Exp Med 180:1395–1399

Carter PB (1975) Pathogenicity of Yersinia enterocolitica for mice. Infect Immun 11:164–170

Cassuto J, Jodal M, Tuttle R, Lundgren O (1981) On the role of intramural nerves in the pathogenesis of cholera toxin-induced intestinal secretion. Scand J Gastroenterol 16:377–384

Chau PY, Tsang RSW, Lam SK, La Brooy JT, Rowley D (1981) Antibody response to the lipopolysaccharide and protein antigens of S. typhi during typhoid infection. Clin Exp Immunol 46:515–520

Chen Y, Smith MR, Thirumalai K, Zychlinsky A (1996) A bacterial invasin induces macrophage apoptosis by binding directly to ICE. EMBO J 15:3853–3860

Chiang S, Taylor R, Koomey M, Mekalanos J (1995) Single amino acid substitutions in the N-terminus of Vibrio cholerae TcpA affect colonization, autoagglutination, and serum resistance. Mol Microbiol 17:1133–1142

Clark MA, Jepson MA, Simmons NL, Hirst BH (1994) Preferential interaction of Salmonella typhimurium with mouse Peyer's patch M cells. Res Microbiol 145:543–552

Clemens JD, Stanton B, Stoll B, Shahid NS, Banu H, Alunddin Chowdhury AKM (1986) Breast feeding as a determinant of severity in shigellosis. Am J Epidemiol 123:710–720

Clemens JD, Van Loon F, Sack DA, Rao MR, Kay BA, Khan MR, Yunus M, Harris JR, Svennerholm AM, Holmgren J (1991) Biotype as determinant of immunising effect of cholera. Lancet 337:883–884

Coburn J, Wyatt RT, Iglewski BH, Gill M (1989) Several binding proteins, including p21c-H-ras are preferred substrates of Pseudomonas aeruginosa exoenzyme S. J Biol Chem 264:9004–9008

Cohen D, Ashkenazi S, Green S, Gdalevich M, Robin G, Slepan R, Yovzori M, Orr N, Block C, Ashkenazi J, Shemer J, Taylor D, Hale T, Sadoff J, Pavliakova D, Schneerson R, Robbins JB (1997) Double-blind vaccine-controlled randomized efficacy trial of an investigational Shigella sonnei conjugate vaccine in young adults. Lancet 349:155–158

Collins FM (1974) Vaccines and cell-mediated immunity. Bacteriol Rev 38:371–402

Degre M, Bukholm G, Czamlecki CW (1989) In vitro treatment of Hep-2 cells with human tumor necrosis factor-alpha and human interferons reduces invasineness of Salmonella typhimurium. J Biol Regul Homeost Agents 3:1–7

Dunlap NE, Benjamin WH, Berry AK, Elridge JH, Briles DE (1992) A "safe-site" for Salmonella typhimurium is within splenic polymorphonuclear cells. Microb Pathog 13:181–190

DuPont HL, Hornick RB, Snyder MJ, Libonati JP, Formal SB, Gangarosa EJ (1972) Immunity in shigellosis. II. Protection induced by oral live vaccine or primary infection. J Infect Dis 125:12–16

Eisenstein TK, Huang D, Meissler JJJ, Al-Ramadi B (1994) Macrophage nitric oxide mediates immunosuppression in infectious inflammation. Immunobiology 191:493–502

Ennis E, Isberg RR, Shimizu Y (1993) Very late antigen 4-dependent adhesion and costimulation of resting human T cells by the bacterial beta 1 integrin ligand invasin J Exp Med 177:207–212

Fällman M, Andersson K, Hakansson S, Magnusson KE, Stendahl O, Wolf-Watz H (1995) Yersinia pseudotuberculosis inhibits Fc receptor-mediated phagocytosis in J774 cells. Infect Immun 63:3117–3124

Fällmann M, Persson C, Wolf-Watz H (1997) Yersinia proteins that target host cell signalling pathways. J Clin Invest 99:1153–1157

Finlay BB, Gumbiner B, Falkow W (1988) Penetration of Salmonella through a polarized Madin-Darby canine kidney epithelial cell monolayer. J Cell Biol 107:221–230

Finlay BB, Falkow S (1997) Common themes in microbial pathogenicity revisited. Microbiol Mol Biol Rev 61:136–169

Forsberg A, Rosqvist R, Wolf-Watz H (1994) Regulation and polarized tranfer of the Yersinia outer proteins (Yops) involved in antiphagocytosis. Trends Microbiol 2:14–19

Gahring LC, Heffron F, Finlay BB, Falkow S (1990) Invasion and replication of Salmonella typhimurium in animal cells. Infect Immun 58:443–448

Galan JE (1996) Molecular and cellular bases of Salmonella entry into host cells. Curr Top Microbiol Immunol 209:43–60

Galyov EE, Hakansson S, Forsberg A, Wolf-Watz H (1993) A secreted protein kinase of Yersinia pseudotuberculosis is an indispensable virulence determinant. Nature 361:730–732

Grutzkau A, Hanski C, Hahn H, Riecken EO (1990) Involvement of M cells in the bacterial invasion of Peyer's patches: a common mechanism shared by Yersinia enterocolitica and the enteroinvasive bacteria. Gut 31:1011–1015

Guan K, Dixon JE (1990) Protein tyrosine phosphatase act as an essential virulence determinant in Yersinia. Science 24:554–556

Hanski C, Kutschka U, Schmoranzer HP, Naumann M, Stallmach A, Hahn H, Menge H, Riecken EO (1989) Immunohistochemical and electron microscopic study of interaction of Yersinia enterocolitica serotype O8 with intestinal mucosa during experimental enteritis. Infect Immun 57:673–678

Hass W, Pereira P, Tonegawa S (1993) g/d cells. Annu Rev Immunol 11:637–656

Hedges SR, Agace WA, Svanborg C (1995) Epithelial cytokine responses and mucosal cytokine networks Trends in Microbiol 7:266–270

Hermant D, Menard R, Arricau N, Parsot C, Popoff MY (1995) Functional conservation of the Salmonella and Shigella effectors of entry into epithelial cells. Mol Microbiol 17:781–789

Herrington DR, Hall R, Losonsky G, Mekalanos J, Taylor R, Levine MM (1988) Toxin, toxin-coregulated pili, and the toxR regulon are essential for Vibrio cholerae pathogenesis in humans J Exp Med 168:1487–1492

Hormaeche CE, Mastoeni P, Arena A, Uddin J, Joysey HS (1990) T cells do not mediate the early suppression of a Salmonella infection in the RES. Immunol 70:247–250

Hwang D (1989) Essential fatty acids and immune response FASEB J 3:2052–2055

Isberg RR (1996) Uptake of enteropathogenic Yersinia by mammalian cells. Curr Top Microbiol Immunol 209:1–24

Ishibashi Y, Arai T (1990) Specific inhibition of phagosome-lysosome fusion in murine macrophages mediated by Salmonella typhimurium infection. FEMS Micriobiol Immunol 2:35–43

Islam D, Bardhan PK, Lindberg AA, Christensson B (1995) Shigella infection induces cellular activation of T and B cells, and distinct species-related changes in peripheral blood lymphocyte subsets during the course of the disease. Infect Immun 63:2941–2949

Islam D, Wretlind B, Lindberg AA, Christensson B (1996) Changes in the peripheral blood T-cell receptor V beta repertoire in vivo and in vitro during shigellosis. Infect Immun 64:1391–1399

Islam D, Veress B, Bardhan PK, Lindberg AA, Christensson B (1997) In situ characterization of inflammatory response in the rectal mucosae of patients with shigellosis. Infect Immun 65:739–749

Jones BD, Ghori N, Falkow S, (1994) Salmonella typhimurium initiates murine infection by penetrating and destroying the specialized epithelial M cells of the Peyer's patches. J Exp Med 180:15–23

Jones BJ, Falkow S (1996) Salmonellosis: host immune responses and bacterial virulence determinants. Annu Rev Immunol 14:533–561

Jung HC, Eckmann L, Yang SK, Panja A, Fierer J, Morzycka-Wroblewska E, Kagnoff MF (1995) A distinct array of proinflammatory cytokines is expressed in human colon epithelial cells in response to bacterial invasion. J Clin Invest 95:55–65

Kaniga K, Tucker S, Trollinger D, Galan JE (1995) Homologs of the Shigella IpaB and IpaC invasins are required for S. typhimurium entry into cultured epithelial cells. J Bacteriol 177:3965–3971

Kaper JB, Morris JM, Levine MM (1995) Cholera. Clin Microbiol Review 8:48–86

Kantele A, Kantele JM, Savilahti E, Westerholm M, Arvilommi H, Lazarovits A, Butcher EC, Makela PH (1997) Homing potentials of circulating lymphocytes in humans depend on the site of activation: oral, but not parenteral typhoid vaccination induces circulating antibody-secreting cells that all bear homing receptors directing them to the gut. J Immunol 158:574–579

Kerneis S, Bogdanova A, Kraehenbuhl JP, Pringault E (1997) Conversion by Peyer's patch lymphocytes of human enterocytes into M cells that transport bacteria. Science 277:949–952

Kulich SM, Yahr TL, Mende-Mueller LM, Barbieri JT, Frank DW (1994) Cloning the structural gene for the 49-kDa form of exoenzyme S (exoS) from Pseudomonas aeruginosa strain 388. J Biol Chem 269:10431–10437

Lawton WD, Erdman RL, Surgella ML (1963) Biosynthesis and purification of V and W antigen in Pasteurella pestis. J Immunol 91:179–184

Lee JC, Gibson CW, Eisenstein TK (1985) Macrophage-mediated mitogenic suppression induced in mice of the C3H lineage by a vaccine strain of Salmonella typhimurium. Cell Immunol 91:75–91

Lee CK, Weltzin R, Soman G, Georgakopoulos KM, Houle DM, Monath TM (1994) Oral administration of polymeric immunoglobulin A prevents colonization with Vibrio cholerae in neonatal mice. Infect Immun 62:887–891

Levine MM (1994) Typhoid fever vaccines. In: SA Plotkin, EA Mortimer (eds) Vaccines. Saunders, Philadelphia, pp 597–633

Levine MM, Hornick RB (1981a) Immunology of enteric pathogens, Salmonella, Shigella, and Escherichia coli. In: Nahmias AJ, O'Reilly RJ (eds) Immunology of human infection. Plenum, New York, pp 249–290

Levine MM, Black RE, Clements ML, Cisneros L, Nalin DR, Young CR (1981b) Duration of infection-derived immunity to cholera. J Infect Dis 143:818–820

Levine MM, Pierce NF (1992) Immunity and vaccine development. In: D. Barua, WB Greenough, III (eds) Cholera. Plenum, New York, pp 285–327

Lundgren O (1988) G factors controlling absorption and secretion in the small intestine. In: W Donachie, E Griffiths, J Stephen (eds) Bacterial infection of respiratory and gastrointestinal mucosae. IRL Press, Oxford, pp 97–112

Lycke N, Holmgren J (1986) Strong adjuvant properties of cholera toxin on gut mucosal immune responses to orally presented antigens. Immunology 59:301–308

Lycke N, Karlsson U, Sjölander A, Magnusson KE (1991) The adjuvant action of cholera toxin is associated with an increased intestinal permeability for luminal antigens. Scand J Immunol 33:691–698

Lycke N, Strober W (1989) Cholrea toxin promotes B cell isotype differentiation. J Immunol 142:3781–3787

Mc Cormick BA, Colgan SP, Delp-Archer C, Miller SI, Madara JL (1993) Salmonella typhimurium attachment to human intestinal epithelial monolayers: transcellular signalling to subepithelial neutrophils. J Cell Biol 12:895–907

Mandic-Mulec I, Weiss J, Zychlinsky A (1997) Shigella flexneri is trapped in polymorphonuclear leukocyte vacuoles and efficiently killed Infect Immun 65:110–115

Marra A, Isberg RR (1997) Invasin-dependent and invasin-independent pathways for translocation of Y. pseudotuberculosis across the Peyer's patch intestinal epithelium. Infect Immun 65:3412–3421

Mastroeni P, Villarreal-Ramos B, Hormaeche CE (1993) Adoptive transfer of immunity to oral challenge with virulent Salmonellae in innately susceptible BALB/c mice requires both immune serum and T cells. Infect Immun 61:3981–3984

Mathias JR, Carlson GM, DiMarino AJ, Bertiger G, Morton HE, Cohen S (1976) Intestinal myoelectric activity in response to live Vibrio cholerae and cholera enterotoxin. J Clin Invest 58:91–96

Mel DM, Arsic BL, Nikolic BD, Radovanovic ML (1968) Studies on vaccination against bacillary dysentery. IV. Oral immunization with live monotypic and combined vaccines. Bull World Health Organ 39:375–380

Michetti P, Mahan MJ, Slauch JM, Mekalanos JJ, Neutra MR (1992) Monoclonal secretory IgA protects mice against oral challenge with invasive Salmonella typhimurium. Infect Immun 60:1786–1792

Michetti P, Porta N, Mahan MJ, Slauch JM, Mekalanos JJ, Blum AL, Kraehenbuhl JP, Neutra MR (1994) Monoclonal immunoglobulin A prevents adherence and invasion of polarized epithelial cell monolayers by Salmonella typhimurium. Gastroenterology 107:915–923

Miller SI (1991) PhoP/PhoQ: macrophage-specific modulators of Salmonella virulence? Mol Microbiol 5:2073–2078

Miller SI, Hohmann E, Pegues D (1995) Salmonella (including S. typhimurium). In: G Mandell, J Benett, R Dolin (eds) Principles and practice of infectious diseases. Churchill Livingstone, New York, pp 2013–2033

Mills SD, Boland A, Sory MP, Van der Smissen P, Kerbourch C, Finley BB, Cornelis GR (1997) Yersinia enterocolitica induces apoptosis in macrophages by a process requiring functional type-III secretion and translocation mechanisms and involving YopP, presumably acting as an effector protein Proc Natl Acad Sci USA 94:12638–12643

Monack DM, Raupach B, Hromocky AE, Falkow S (1996) Salmonella typhimurium invasion induces apoptosis in infected macrophages. Proc Natl Acad Sci USA 93:9833–9838

Monack DM, Mecsas J, Ghori N, Falkow S (1997) Yersinia signals macrophages to undergo apoptosis and YopJ is necessary for this cell death Proc Natl Acad Sci USA 94:10385–10390

Mosley WH, Woodward WE, Azia KMS, et al (1970) The 1968–1969 cholera vaccine field trial in rural East Pakistan. Effectiveness of monovalent Ogawa and Inaba vaccines and a purified Inaba antigen, with comparative results of serological and animal protection tests. J Infect Dis 121[suppl]:S1-S9

Mounier J, Vasselon T, Hellio R, Lesourd M, Sansonetti PJ (1992) Shigella flexneri enters human colonic Caco-2 epithelial cells through the basolateral pole. Infect Immun 60:237–248

Muotiala A, Makela PH (1990) The role of IFN-gamma in murine Salmonella typhimurium infection. Microb Pathog 8:135–141

Munoz E, Zubiaga AM, Merrow M, Sauter NP, Huber BT (1990) Cholera toxin discriminates between T helper 1 and T helper 2 cells in T cell receptor mediated activation: role of cAMP in T cell proliferation. J Exp Med 172:95–103

Nishimura H, Hiromatsu K, Kobayashi N, Grabstein KH, Paxton R, Sugamura K, Bluestone JA, Yoshikai Y (1996) IL-15 is a novel growth factor for murine gd T cells induced by Salmonella infection.

Nakano Y, Onozuka K, Terada Y, Shinomiya H, Nakano M (1990) Protective effect of recombinant tumor necrosis factor-alpha in murine salmonellosis. J Immunol 144:1935–1941

Nauciel C, Espinasse MF (1992) Role of gamma interferon and tumor necrosis factor alpha in resistance to Salmonella typhumurium infection. Infect Immun 60:450–454

Oberhelman RA, Kopecko DJ, Salazar-Lindo E, Gotuzzo E, Buysse JM, Venkatesan MM, Yi A, Fernandez-Prada C, Guzman M, Leon-Barua R, Sack RB (1991) Prospective study of systemic and mucosal immune responses in dysenteric patients to specific Shigella invasion plasmid antigens and lipopolysaccharide. Infect Immun 59:2341–2350

O'Callaghan D, Maskell D, Tite J, Dougan G (1990) Immune responses in BALB/c mice following immunization with aromatic compound or purine-dependent Salmonella typhimurium strains. Immunol 69:184–189

O'Loughlin EV, Gall DG, Pai CH (1990) Yersinia enterocolitica: mechanisms of microbial pathogenesis and pathophysiology of diarrhoea. J Gastroenterol Hepatol 5:173–179

Owen RL, Pierce NF, Apple RT, Cray WC Jr (1986) M-cell transport of Vibrio cholerae from the intestinal lumen into Peyer's patches: a mechanism for antigen sampling and for microbial transepithelial transmigration J Infect Dis 153:1108–1118

Parra-Lopez C, Lin R, Aspedon A, Groisman EA (1994) A Salmonella protein that is required for resistance to antimicrobial peptides and transport of potassium. EMBO J 13:3964–3972

Parsot C, Sansonetti PJ (1996) Invasion and the pathogenesis of Shigella infections. Curr Top Microbiol Immunol 209:25–43

Plant J, Glynn AA (1979) Locating Salmonella resistance gene on mouse chromosome 1. Clin Exp Immunol 37:1–6

Pepe JC, Miller VL (1993) Yersinia enterocolitica invasin: a primary role in the initiation of infection. Proc Natl Acad Sci USA 90:6473–6477

Perdomo JJ, Cavaillon JM, Huerre M, Ohayon H, Gounon P, Sansonetti PJ (1994a) Acute inflammation causes epithelial invasion and mucosal destruction in experimental shigellosis. J Exp Med180:1307–1319

Perdomo OJJ, Gounon P, Sansonetti PJ (1994b) Polymorphonuclear leukocyte transmigration promotes invasion of colonic epithelial monolayer by Shigella flexneri. J Clin Invest 93:633–643

Persson CN, Carballeira N, Wolf-Watz H, Fällman M (1997) The PTPase YopH inhibits uptake of Yersinia, tyrosine phosphorylation of p130Cas and FAK, and the associated accumulation of these proteins in peripheral focal adhesions. EMBO J 16:2307–2318

Peterson K, Mekalanos J (1988) Characterization of the Vibrio cholerae ToxR regulon: identification of novel genes involved in intestinal colonization. Infect Immun 56:2822–2829

Peterson JW, Ochoa LG (1989) Role of prostaglandins and cAMP in the secretory effects of cholera toxin. Science 245:857–860

Peterson JW, Reitmeyer JC, Jackson CA, Ansari GAS (1991) Protein synthesis is required for cholera toxin-induced stimulation of arachidonic acid metabolism. Biochim Biophys Acta 1092:79–84

Phalipon A, Sansonetti PJ (1995a) Live attenuated Shigella flexneri mutants as vaccine candidates against shigellosis and vectors for antigen delivery. Biologicals 23:125–134

Phalipon A, Kaufmann M, Michetti P, Cavaillon JM, Huerre M, Sansonetti PJ, Kraehenbuhl JP (1995b) Monoclonal IgA antibody directed against serotype-specific epitope of Shigella flexneri lipopolysaccharide protects against murine experimental shigellosis. J Exp Med 182:769–778

Ramarathinam L, Shaban RA, Niesel DW, Klimpel GR (1991) Interferon gamma production by gut-associated lymphoid tissue and spleen following oral Salmonella typhimurium challenge. Microb Pathog 11:347–356

Raqib R, Wretlind B, Andersson J, Lindberg AA (1995a) Cytokine secretion in acute shigellosis is correlated to disease activity and directed more to stool than to plasma. J Infect Dis 171:376–384

Raqib R, Lindberg AA, Wretlind B, Bardahn PK, Andersson U, Andersson J (1995b) Persistance of local cytokine production in shigellosis in acute and convalescent stages. Infect Immun 63:289–296

Raqib R, Lindberg A, Björk L, Bardhan PK, Wretlind B, Andersson U, Andersson J (1995c) Down-regulation of gamma-interferon, tumor necrosis factor, interleukin 1 (IL-1), IL-3, IL-4, and transforming growth factor-beta type I receptors at the local site during the acute phase of Shigella infection. Infect Immun 63:3079–3087

Raqib R, Ljungdahl A, Lindberg AA, Wretling B, Andersson U, Andersson J (1996) Dissociation between cytokine mRNA expression and protein production in shigellosis. Eur J Immunol 26:1130–1138

Rathman M, Sjaastad MD, Falkow S (1996) Acidification of phagosomes containing Salmonella typhimurium in murine macrophages. Infect Immun 64:2765–2773

Renesto P, Mounier J, Sansonetti PJ (1996) Induction of adherence and degranulation of polymorphonuclear leukocytes: a new expression of the invasive phenotype of Shigella flexneri. Infect Immun 64:719–723

Richter-Dahlfors A, Buchan AMJ, Finlay BB (1997) Murine salmonellosis studied by confocal microscopy: Salmonella typhimurium resides intracellularly inside macrophages and exerts a cytotoxic effect on phagocytes in vivo. J Exp Med 186:569–580

Roggenkamp A, Geiger AM, Leitritz L, Kessler A, Heesemann J (1997) Passive immunity to infection with Yersinia spp mediated by anti-recombinant V antigen is dependent on polymorphism of V antigen Infect Immun 65:446–451

Ruckdeschel K, Roggenkamp A, Schubert S, Heesemann J (1996) Differential contribution of Yersinia enterocolitica virulence factors to evasion of microbicidal action of neutrophils. Infect Immun 64:724–733

Ruckdeschel K, Roggenkamp A, Lafont V, Mangeat P, Heesemann J, Rouot B (1997a) Interaction of Yersinia enterocolitica with macrophages leads to macrophage cell death through apoptosis. Infect Immun 65:4813–4821

Ruckdeschel K, Machold J, Roggenkamp A, Schubert S, Pierre J, Zumbihl R, Liautard JP, Heesemann J, Rouot B (1997b) Yersinia enterocolitica promotes deactivation of macrophage mitogen-activated protein kinases extracellular signal-regulated kinase-1/2, p38, and c-Jun NH2-terminal kinase. Correlation with its inhibitory effect on tumor necrosis factor alpha production J Biol Chem 272:15920–15927

Sansonetti PJ, Arondel J, Fontaine A, d'Hauteville H, Bernardini ML (1991) OmpB (osmo-regulation) and icsA (cell to cell spread) mutants of S. flexneri: vaccine candidates and probes to study the pathogenesis of shigellosis. Vaccine 9:416–421

Sansonetti PJ, Arondel J, Cavaillon JM, Huerre M (1995) Role of IL-1 in the pathogenesis of experimental shigellosis. J Clin Invest 96:884–892

Sansonetti PJ, Arondel J, Cantey R, Prévost MC, Huerre M (1996) Infection of rabbit peyer's patches by Shigella flexneri: effect of adhesive or invasive bacterial phenotypes on follicle-associated epithelium. Infect Immun 64:2752–2764

Sarasombath S, Banchuin N, Sukosol T, Rungpitarangsi B, Manasatit S (1987) Systemic and intestinal immunities after natural typhoid infection. J Clin Microbiol 25:1088–1093

Savidge T, Morey AL, Ferguson DJ, Fleming KA, Shmakov AN, Phillips AD (1995) Human intestinal development in a severe-combined immunodeficient xenograft model. Differentiation 58:361–371

Schulte R, Wattiau P, Hartland EL, Robins-Browne RM, Cornelis GR (1996) Differential secretion of IL-8 by human epithelial cell lines upon entry of virulent or nonvirulent Yersinia enterocolitica. Infect Immun 64:2106–2113

Siebers A, Finlay BB (1996) M cells and the pathogenesis of mucosal and systemic infections. Trends Microbiol 4:22–29

Skurnick M, Wolf-Watz H (1989) Analysis of the yop A gene encoding the Yop1 virulence determinants of Yersinia spp. Mol Microbiol 3:517–529

Starnbach MN, Bevan MJ (1994) Cells infected with Yersinia present an epitope to class I MHC-restricted CTL. J Immunol 153:1603–1612

Strom M, Lory S (1993) Structure-function and biogenesis of the type IV pili. Ann Rev Microbiol 47:567–596

Svennerholm AM, Holmgren J (1976) Synergistic protective effect in rabbits of immunization with Vibrio cholerae lipopolysaccharide and toxin/toxoid. Infect Immun 13:735–740

Svennerholm AM, Jonson G, Holmgren J (1994) Immunity to Vibrio cholerae infection. In: IK Wachsmuth, PA Blake, O Olsvik (eds) Vibrio cholerae and cholera: molecular to global perspectives. ASM Press, Washington, pp 257–271

Takeuchi A (1967) Electron microscopic studies of experimental Salmonella infection. 1. Penetration into the intestinal epithelium by Salmonella typhimurium. Am J Pathol 50:109–136

Taylor RK, Miller VL, Furlong D, Mekalanos J (1987) Use of phoA gene fusion to identify a pilus colonization factor coordinately regulated with cholera toxin Proc Natl Acad Sci USA 84:2833–2837

Uchiyama T, Miyoshi-Akiyama T, Kato H, Fujimaki W, Imanishi K, Yan XJ (1993) Superantigenic properties of a novel mitogenic substance produced by Yersinia pseudotuberculosis isolated from patients manifesting acute and systemic symptoms. J Immunol 151:4407–4413

Van Dissel JT, Stikkelbroeck IJ, Sluiter W, Leijh PCJ, Van Furth R (1986) Differences in initial rate of intracellular killing of Salmonella typhimurium by granulocytes of Salmonella-suceptible C57BL/10 mice and Salmonella-resistant CBA mice. J Immunol 136:1074–1080

Vidal SM, Malo D, Vogan K, Skamene E, Gros P (1993) Natural resistance to infection with intracellular parasites: isolation of a candidate for BCG. Cell 73:469–485

Vogel U, Autenrieth IB, Berner R, Heesemann J (1993) Role of plasmid-encoded antigens of Yersinia enterocolitica in humoral immunity against secondary Y. enterocolitica infection in mice. Microb Pathog 15:23–36

Wake A, Sutoh Y (1983) Mechanisms of long and short term immunity to plague. Immunol 34:1045–1052

Wassef, JS, Keren DK, Mailloux JM (1989) Role of M cell in initial uptake and in ulcer formation in the rabbit intestinal loop model of shigellosis. Infect Immun 57:858–863

Wick MJ, Hardling CV, Twesten NJ, Normack SJ, Pfeifer JD (1995) The phoP locus influences processing and presentation of Salmonella typhimurium antigens by activated macrophages. Mol Microbiol 16:465–476

Winner L III, Mack J, Weltzin R, Mekalanos JJ, Kraehenbuhl JP, Neutra MR (1991) New model for analysis of mucosal immunity: intestinal secretion of specific monoclonal immunoglobulin A from hybridoma tumors protects agaisnt Vibrio cholerae infection. Infect Immun 59:977–982

Zhang ZY, Clemens JC, Schubert HL, Stuckey JA, Fischer MW, Hume DM, Saper MA, Dixon JE (1992) Expression, purification and physiochemical characterization of a recombinant Yersinia protein tyrosine phosphatase. J Biol Chem 267:23759–23766

Zwillich SH, Duby AD, Lipsky PE (1989) T-lymphocyte clones responsive to Shigella flexneri. J Clin Microbiol 27:417–421

Zychlinsky A, Prevost MC, Sansonetti PJ (1992) Shigella flexneri induces apoptosis in infected macrophages. Nature 358:167–169

Zychlinsky A, Fitting C, Cavaillon JM, Sansonetti PJ (1994a) Interleukin 1 is released by murine macrophages during apoptosis induced by Shigella flexneri. J Clin Invest 94:1328–1332

Zychlinsky A, Kenny B, Ménard R, Prévost MC, Holland IB, Sansonetti PJ (1994b) IpaB mediates macrophage apoptosis induced by Shigella flexneri. Mol Microbiol 11:619–627

Zychlinsky A, Thirumachi K, Arondel J, Cantey JR, Aliprantis AO, Sansonetti PJ (1996) In vivo apoptosis in Shigella flexneri infection. Infect Immun 64:5357–5365

Host Responses to Respiratory Virus Infection and Immunization

J.E. Crowe, Jr.

1 Host Cell Targets of Infection

1.1 Replication of Respiratory Viruses In Vivo

The major viral causes of lower respiratory tract (LRT) disease in humans are respiratory syncytial virus (RSV), parainfluenza viruses (PIV), and the influenza viruses. Despite significant advances in our understanding of the basic virology and

---

Division of Pediatric Infectious Diseases, Department of Pediatrics, Vanderbilt University Medical School, D-7235 Medical Center North, 1161 21st Avenue South, Nashville, TN 37232–2581, USA

molecular biology of these viruses, relatively little is known about the life cycle and spread of these respiratory viruses in vivo. Infection is transmitted primarily by fomites or large-particle aerosols; therefore the initiation site of natural infection is presumed to be the nasopharynx (NP). It is probable that these viruses predominately infect respiratory epithelial cells.

However formal demonstration of epithelial localization of infection in situ in immunocompetent humans is limited. Upper respiratory tract (URT) epithelial cells, obtained by superficial scraping of turbinate tissue, support replication of RSV (Becker et al. 1993). Superficial nasal mucosal epithelial cells, obtained by means of a scraping technique or nasal wash, are used in commercial methods of rapid identification of RSV or influenza-A virus by antigen detection. Epithelial cells of adenoid origin grown in tissue culture remained highly differentiated, with sub-populations of cells retaining active ciliary motility and others demonstrating specialized secretory functions (Endo et al. 1996). These mixed epithelial-cell cultures were permissive for the growth of influenza-A virus. The susceptibility to RSV infection of primary adult human nasal epithelium, primary adult human bronchial epithelium and a human bronchial epithelial cell line (BEAS-2B) was similar; however, bronchial cells released less infectious RSV than did nasal cells (Becker et al. 1992). In fetal human tracheal organ culture, RSV replication occurred in a population of ciliated epithelial cells, whereas other cells in the epithelial layer were spared (Henderson et al. 1978).

The mechanism for spread of infection to the LRT is unknown. The rapid time course of spread in vivo suggests that aspiration of infected secretions results in direct inoculation of the LRT. Autopsy cases of fatal RSV bronchiolitis have demonstrated the presence of relatively little virus antigen (Gardner et al. 1970), and pathologic changes in fatal cases can be patchy in distribution (Aherne et al. 1970). Bronchoalveolar lavage and tissue specimens from bone marrow transplant patients, autopsies evaluated by direct immunofluorescence or immunoperoxidase stains revealed epithelial cells and macrophages that stain positive for RSV, and electron microscopy revealed occasional epithelial cells with cytoplasmic inclusions composed of filamentous virions (Neilson and Yunis 1990; Parham et al. 1993; Panuska et al. 1992). There is limited autopsy data from PIV-infected (Downham et al. 1975; Delage et al. 1979; Jarvis et al. 1979) or influenza-infected humans (Louria et al. 1959; Hers and Mulder 1961; McQuillin et al. 1970; Yousuf et al. 1997). In summary, the population of cells that is the primary target of respiratory virus infections in humans is likely to be epithelial in origin, but has not been fully defined in normal hosts.

1.2 Virus Receptors on Host Cells

The hemagglutinin protein (HA) of influenza virus mediates both virus attachment to a glycoconjugate terminating in sialic acid and fusion with an intracellular membrane at low pH. Recent studies have also shown that annexin V, a non-glycosylated calcium-dependent, phospholipid binding protein, contributes to sialic acid-independent binding of influenza-A and B virus strains to cultured cell lines

(Otto et al. 1994; Huang et al. 1996). The PIV hemagglutinin-neuraminidase (HN) glycoprotein mediates attachment to sialic-acid containing host-cell receptors (Markwell and Paulson 1980). The RSV surface protein that mediates attachment is the RSV G (glycosylated) glycoprotein (Levine et al. 1987); its cellular receptor is unknown. The viability of a mutant RSV strain, lacking both the G protein and the short hydrophobic (SH) membrane protein, indicates that the attachment protein is not strictly required for replication in cell culture or in rodents (Karron et al. 1997).

2 The Host Response Contributes to Pathology

2.1 Inflammatory Mediators Induced in Virus-Infected Epithelial Cells

The magnitude and character of the inflammatory response in the airway appear to be highly regulated by epithelial cell-derived cytokines and chemokines that attract and activate specific subsets of leukocytes associated with airway inflammation. Many investigators have examined cytokine secretion patterns and levels in vitro during infection of established cell lines, such as the BEAS-2B line (a human bronchial epithelium from a normal subject transformed with an adenovirus 12-SV40 virus hybrid) (Ke et al. 1988) or the A549 lung alveolar cell type-II carcinoma line (Giard et al. 1973). Cytokine secretion appears to be highly regulated by nuclear factor kappa B (NF-κB) transcription factors, dimers of structurally related proteins that are retained in the cytoplasm by association with the inhibitory κB proteins. Upon various cellular stimulations, usually related to stress or pathogens, the inhibitors are degraded and the NF-κB factors translocate to the nucleus where they bind to κB DNA elements to induce transcription of a large number of genes, especially those associated with immune responses (Baldwin 1996).

Nuclear factor-interleukin 6 (NF-IL-6) is a factor that can regulate expression of cytokine and adhesion molecule genes without increased transcription. RSV infection induces increased expression of a number of cytokines, including IL-6 and IL-8, that are transcriptionally regulated by NF-κB and NF-IL-6, in cell lines including A549 cells (Mastronare et al. 1996; Mastronare et al. 1995; Jaaluin et al. 1996; Arnold et al. 1994). This effect could be decreased in the presence of an antioxidant (Mastronare et al. 1995) or the viral replication inhibitor ribavirin (Fieler et al. 1996).

Increased transcriptional activation of the IL-8 gene in RSV-infected A549 cells can be specifically associated with nuclear translocation of the NF-κB transactivating subunit RelA (Garoalo et al. 1996b). Infection with an H3N2 influenza-A virus also induced IL-6 and IL-8 in the NCI-H292 bronchial epithelial cell line (Matsukura et al. 1996). In A549 cells, RSV stimulated messenger RNA (mRNA) accumulation and protein production of IL-11 (a potent inducer of airway bronchospasm) in a time- and dose-dependent fashion (Elias et al. 1994; Einarsson et al. 1995), while RSV and PIV3 stimulated stromal cells to secrete IL-11

(Einarsson et al. 1996). Epithelial cell-derived cytokine levels are elevated in the nasal secretions of children during acute upper respiratory virus infections (Noah et al. 1995).

The chemokine RANTES (regulated upon activation, normal T cell expressed and secreted) is a chemo-attractant for eosinophils, monocytes, T cells and basophils. Enhanced recruitment of eosinophils to the airways by RANTES and other soluble factors has been hypothesized to be a common factor in the wheezing of RSV-induced bronchiolitis and allergen-induced reactive airways disease. RSV infection induces RANTES expression in a variety of respiratory epithelial cell types including primary cultures of human nasal and adenoid-derived epithelial cells, nasal epithelial cell explants, normal human primary bronchial epithelial cell cultures and the BEAS-2B cell line (Saito et al. 1997; Becker et al. 1997). RANTES is also found at increased levels in the nasal wash secretions of RSV-infected children (Becker et al. 1997). Infection with an H3N2 influenza-A virus induces RANTES in the bronchial epithelial cell line NCI-H292 (Matsukura et al. 1996). The inducible translational control of important transcription factors and expression of inflammatory factors during virus infection suggests that epithelial cells rapidly and specifically respond to infection in a regulated fashion.

2.2 Increased Expression of Surface Markers on Virus-Infected Epithelial Cells

RSV infection induces increased surface expression of intercellular adhesion molecule-1 (ICAM-1, CD54) and CD18 in A549 cells (Patel et al. 1995; Arnold and Konig 1996; Stark et al. 1996) or human nasal epithelial cells (Matsuzaki et al. 1996), thus promoting adhesion of neutrophils and eosinophils (Stark et al. 1996). Human PIV2 also increases the ability of neutrophils to adhere to airway epithelial monolayer cultures of both primary human tracheal epithelial cells and A549 or BEAS-2B cells in a virus dose-dependent fashion (Tosi et al. 1992; Stark et al. 1992). RSV-infected A549 cells or normal human bronchial epithelial cells exhibited an increased expression of class-I major histocompatibility complex (MHC), as determined by flow cytometry or immunoprecipitation of class-I MHC from metabolically radiolabeled cells (Garoalo et al. 1996a). Cells derived from the adenoids of children with URT infections and otitis media with effusion, exhibited the presence of MHC class-II positive ciliated epithelial cells (van Nieuwkerk et al. 1990).

2.3 Eosinophils, Mast Cells and Their Products

Eosinophils that are recruited and retained in tissues by the soluble and surface molecules discussed in Sect. 2.1 and Sect. 2.2 contribute to pathogenesis of respiratory virus infections. Eosinophil cationic protein (ECP), a cytotoxic protein contained in the granules of eosinophils, is higher in samples of NP secretions from children with RSV bronchiolitis than in infected children without bronchiolitis (Garoalo et al. 1992; Siurs et al. 1994; Inra et al. 1995). NP ECP levels correlate better with disease severity in RSV bronchiolitis than do peripheral blood eosinophil counts (Garoalo et al. 1992). Immunization of mice with native RSV G protein or a vaccinia virus (VV) recombinant expressing RSV G protein primes for an eosinophilic inflammatory reaction in the lungs on subsequent RSV challenges. This suggests a unique property of the G glycoprotein, the mechanism of which is not understood (Alwan and Openshaw 1993; Alwan et al. 1993 Hancock et al. 1996).

RSV can bind directly to human peripheral blood eosinophils in culture (Kimpen et al. 1992). Electron microscopy of eosinophils incubated with live RSV has demonstrated virions in phagocytic vacuoles at the periphery of the cells and eosinophil activation characterized by degranulation (Kimpen et al. 1996). Binding of RSV activates release of inflammatory mediators, as demonstrated by an increase in superoxide production or priming for enhanced leukotriene (LT) C4 release from eosinophils or a macrophage-like cell line (Kimpen et al. 1992; Ananaba and Anderson 1991). Histamine and eicosanoid products, including LTs and prostaglandins from mast cells, eosinophils, and macrophages, cause bronchoconstriction following inhalation.

RSV-infected children with bronchiolitis have increased levels of LT and histamine in their respiratory tract that peak at 3–8 days after onset of illness (Welliver et al. 1981; Volovitz et al. 1988). Increased airway levels of histamine and other mast-cell products are associated with more severe RSV disease in children. RSV F- or G-specific immunoglobulin E (IgE) antibodies, that can cause mast-cell degranulation, are detected in the nasal wash fluids or on shed epithelial cells of children following RSV infection (Welliver et al. 1980a; Welliver et al. 1985; Welliver et al. 1989; Russi et al. 1993). RSV-specific IgE antibodies appear to be secreted locally at the mucosa since they are more concentrated in nasal-wash specimens than in simultaneously collected sera (Welliver et al. 1985). In summary, increased local levels of eosinophil mast-cell products and IgE during virus infection suggests a role for these factors in the pathogenesis of virus-associated reactive airways disease.

2.4 Immunopathogenesis Associated with T Lymphocytes

The immune response associated with clearance of the virus during acute infection (discussed below in Sect. 3.2) causes pathologic changes that can be detected by histology and clinical disease expression. Both CD4+ and CD8+ T lymphocytes contribute to illness in RSV-infected mice (Cannon et al. 1988; Graham et al. 1991b; Alwan et al. 1994). Immunization of RSV seronegative infants with a formalin-inactivated RSV (FI-RSV) preparation resulted in enhanced severe disease on subsequent naturally-acquired infection (Chin et al. 1969; Fulginiti et al. 1969; Kim et al. 1969; Kapikian et al. 1969). Prior immunization with FI-RSV induced increased lymphocyte-transformation (LTF) activity in peripheral blood

(Kim et al. 1976), but induced an antibody response characterized by a low ratio of virus neutralizing titer to binding titer (Murphy et al. 1986b).

Subsequent rodent studies have demonstrated that FI-RSV-induced enhanced illness in mice occurs in the context of a predominant T-helper (Th)2 pattern of cytokine expression associated with lung CD4+ lymphocytes which differs from the predominant Th1 cytokine pattern induced by live RSV infection (Connors et al. 1992b; Graham et al. 1993a). Enhanced histopathology or illness in mice induced by FI-RSV can be abrogated by IL-4 and IL-10 depletion (Tang and Graham 1994; Connors et al. 1994).

3 Immune Effectors and Their Role in Resolution or Prevention of Infection

The relative importance of specific immune effectors differs between prevention and resolution of infection, and between the URT and LRT. Therefore, careful distinctions of these characteristics will be made as mechanisms of immunity are considered in these settings.

3.1 Humoral Immunity

3.1.1 Antibodies

3.1.1.1 Antibodies May Contribute to the Resolution of Acute Primary Infection

RSV- or influenza-virus-infected rodents that are depleted of B cells by anti-IgM antibodies exhibit delayed virus clearance and increased morbidity (Iwasaki and Nozima 1977; Kris et al. 1985; Graham et al. 1991a). The viral LD_{50} (dose that causes mortality in 50% of those subjects treated) of pathogenic influenza viruses is decreased 10- to 100-fold in B cell-deficient (Ig-/- μ-chain knockout) mice (Graham and Braciale 1996; Gerhard et al. 1997).

Antibodies alone can cure influenza virus infection of T cell-deficient (Virelizier 1975; Scherle et al. 1992) or severe combined immunodeficient (SCID) mice (Palladino et al. 1995). However, some passive-antibody transfer studies in influenza-infected nude or T cell-depleted mice resulted in temporary or no reduction of disease (Schulman et al. 1977; Kris et al. 1988). IgA and IgG RSV antibodies are present in the nasal secretions of infants early during infection (McIntosh et al. 1978; Hornsleth et al. 1984), and the decrease in virus shedding in infants shedding RSV was temporally associated with the detection of RSV-specific IgA in nasal washes (McIntosh et al. 1978). However, intravenous transfer of standard or RSV-immunoglobulin to hospitalized RSV-infected infants was clinically ineffective (Hemming et al. 1987; Rodriguez et al. 1997). This was probably because these patients presented to the hospital relatively late in infection, when pathologic changes were already present and virus replication was waning.

3.1.1.2 Antibodies Prevent Reinfection

Laboratory studies and clinical trials provide strong evidence supporting the dominant role of antibodies in protection against reinfection. The relative sparing of infants from virus-associated LRT disease, in the first weeks or months of life, correlates with the level of passively-acquired maternal antibodies to RSV (Parrott et al. 1973; Lamprecht et al. 1976; Glezen et al. 1981; Ogilvie et al. 1981), PIV3 (Glezen et al. 1984) or influenza viruses (Puck et al. 1980). Passive-transfer studies in rodents demonstrate the protective efficacy in the LRT of polyclonal immune sera and neutralizing RSV monoclonal antibodies (mAbs) (Walsh et al. 1984; Taylor et al. 1984; Prince et al. 1985a; Prince et al. 1987).

The results of prophylactic clinical trials in humans with RSV human immune globulin suggest similar results (Groothuis et al. 1993; Groothuis et al. 1995). Delivery of adequate quantities of antibodies is critical to effective in vivo neutralization in both rodents and humans (Prince et al. 1985b; Crowe et al. 1994; Groothuis et al. 1993). In passive-transfer studies, serum RSV-neutralizing-antibody titer correlates well with the level of protection against LRT virus replication. In contrast, serum neutralizing RSV or influenza antibodies usually have little effect on virus replication in the URT (Ramphal et al. 1979; Walsh et al. 1984; Taylor et al. 1984; Prince et al. 1985b), unless non-physiological levels of antibodies are administered (Sami et al. 1995).

Neutralizing antibodies induced by active immunization or wild-type (*wt*) virus infection protect rodents against reinfection. β2 microglobulin-deficient mice were protected by immunization with VV recombinants, expressing influenza virus HA and neuraminidase (NA) proteins when CD4+ T cells were depleted at the time of challenge. This suggests that preformed antibodies protect in this setting (Epstein et al. 1993). Mice immunized with VV recombinants expressing the RSV F or G glycoproteins resist RSV challenge after depletion of both CD4+ and CD8+ T cells prior to challenge, again suggesting the sufficiency of virus glycoprotein induced antibodies in protection (Connors et al. 1992a).

The specific mechanisms by which antibodies neutralize respiratory viruses were reviewed recently (Crowe 1996). In humans, antibodies appear in both the serum and respiratory secretions, within days of primary respiratory virus infection (McIntosh et al. 1978; Welliver et al. 1980b). Post-infection serum antibody levels to respiratory viruses often wane several months after primary infection if infection occurs at a very young age (Frank et al. 1979; Welliver et al. 1980b; Frank and Taber 1983). However, serum antibodies are generally long-lived after a second or third infection (Wagner et al. 1989). Both young age and passively-acquired antibodies inhibit the primary antibody response to virus infection (Parrott et al. 1973; McIntosh et al. 1978; Watt et al. 1986; Murphy et al. 1986a; Gruber et al. 1997).

Antibody-mediated suppression of antibody response to immunization is particularly evident when the route of immunization is parenteral (Belshe et al. 1982; Murphy et al. 1988; Murphy et al. 1991). Administration of vaccines by the intranasal route (Murphy et al. 1989) or in increased doses can overcome the suppressive effects of antibody in some cases (Murphy et al. 1991). A large number of clinical trials of live attenuated RSV, PIV3 and influenza-virus strains have identified promising vaccine candidates that are safe and immunogenic, even in young infants (recently reviewed in Crowe 1998a). High levels of maternal antibodies may suppress secretory IgA responses in the NP secretions of infants following primary infection (Yamazaki et al. 1994), but to a lesser degree than systemic IgG responses (Tsutsumi et al. 1995). Infection of seronegative chimpanzees with live attenuated RSV vaccines, following passive transfer of human RSV Ig, resulted in decreased primary serum antibody response compared with immunization of animals not treated with Ig, but enhanced the antibody response to a second infection (Crowe et al. 1995). The mechanism for this enhanced immunogenicity is unknown.

3.1.2 The Unique Role of Secretory Immunoglobulin A (sIgA)

Secretory IgA (sIgA) is uniquely suited for antiviral activity at the respiratory mucosa. Mucosal antibody secretion occurs locally and polymeric IgA is taken up specifically by the poly-Ig receptor (pIgR) at the base of the epithelial cell, transcytosed in the apical direction, and secreted onto the mucosa. This transport mechanism enhances the antiviral activity of sIgA in two ways. First, IgA appears to effect intracellular virus neutralization by complexing with virions or viral components in the intracellular environment during IgA transcytosis. In polarized cell monolayers expressing pIgR, Sendai-, or influenza-virus infection initiated at the apical cell surface is decreased by virus-specific IgA (but not IgG), presented at the basal cell surface, and IgA and viral proteins co-localized as detected by immunofluorescence (Mazanec et al. 1992a; Mazanec et al. 1995). Second, sIgA is actively transported onto the mucosa and into the respiratory lumen, while IgG antibodies reach the mucosa by passive transudation, which is a much less efficient mechanism of transfer. Comparisons of the relative efficacy of IgA or IgG antibodies against Sendai or influenza viruses for mucosal protection suggest that these isotypes are equivalent if topically delivered to the mucosa in similar concentrations (Mazanec et al. 1992b; Renegar and Small 1991b).

In vivo data suggest that sIgA plays a principal role in protection against reinfection. Nasal immunity to reinfection in mice with influenza virus was abrogated by the intranasal administration of anti-IgA, but not anti-IgG or anti-IgM antiserum (Renegar and Small 1991a). This suggests that under physiological conditions sIgA is the dominant mediator of nasal immunity to influenza reinfection in the mouse. Nasal wash IgA responses in humans occur soon after intranasal infection or immunization with RSV, PIV or influenza virus (McIntosh et al. 1978; Welliver et al. 1980b; Yanagihara and McIntosh 1980; Hornsleth et al. 1984; Johnson et al. 1985), and exhibit memory responses upon reinfection (Wright et al. 1983). The IgA response of infants may differ from that of adults, since most of the mucosal IgA secreted by infants following first infection with PIV1, PIV2, or RSV is non-neutralizing (McIntosh et al. 1978, Yanagihara and McIntosh 1980).

The relatively transient persistence of high levels of specific sIgA in secretions may account for the observation that paramyxoviruses readily re-infect the NP of individuals of all ages throughout life. A correlation of virus-specific nasal wash IgA levels with protection against infection has been observed in some human studies (Johnson et al. 1986; Friedewald et al. 1968), but not in others (Hall et al. 1991). Technical difficulties in achieving reproducible quantitative recovery of Igs from the NP secretions render interpretation of these studies difficult.

3.1.3 Complement

The complement system is an important component of innate immunity to respiratory viruses. Cells infected with influenza virus or RSV activate complement by both the classical and alternative complement pathways (Smith et al. 1981; Edwards et al. 1986; Reading et al. 1995). Titers of serum samples, obtained sequentially from infants and young children with RSV infection at all phases of illness determined by complement-enhanced RSV neutralizing antibody assays, are greater than those obtained in the absence of complement (Kaul et al. 1981). Some RSV- or PIV-specific mAbs that are non-neutralizing in vitro and non-protective in animals lacking complement, can neutralize virus in vitro and protect in vivo when complement is present (Vasantha et al. 1988; Corbeil et al. 1996).

Influenza-virus clearance is inhibited in de-complemented or C5-deficient mice (Hicks et al. 1978). Antibody-mediated complement fixation is not strictly required for the antiviral effect of all respiratory virus antibodies, since antigen-binding fragments of antibodies [F(ab′)$_2$, Fab fragments] or even a linear peptide derived from antibody heavy-chain complementarity determining region 3 of some RSV antibodies are neutralizing in vitro and therapeutic in RSV-infected mice (Prince et al. 1990; Crowe et al. 1994; Crowe 1998; Bourgeois et al. 1998).

3.2 Cellular Immunity

3.2.1 Antiviral T Lymphocytes in Rodent Models

3.2.1.1 CD8+ T Lymphocytes

The kinetics of activated cytotoxic T lymphocytes (CTLs) following virus clearance suggest that they do not play a major role in prevention of reinfection by respiratory viruses. In the mouse, primary anti-RSV CTL activity peaks early following infection, but becomes undetectable several weeks after infection (Connors et al. 1991). Influenza-virus memory T-cell responses are long-lived, but do not appear to provide protection against a second infection (Liang et al. 1994; Bachmann et al. 1997). In contrast, CTLs play a central role in the resolution of established infections by respiratory viruses. RSV, PIV, or influenza-virus replication in immunosuppressed rodents is prolonged (Anderson et al. 1980; Johnson et al. 1982; Wong et al. 1985; Bender et al. 1992).

Influenza-virus pneumonia in mice can be cured by a CD8+ T-cell response (Allan et al. 1990; Eichelberger et al. 1991; Bender et al. 1992; Tripp et al. 1995). Passive transfer of virus-specific CTL lines or clones into RSV, influenza- or Sendai virus-infected immunodeficient mice results in restriction of virus replication (Yap et al. 1978; Wells et al. 1981; Lukacher et al. 1984; Kast et al. 1986; Cannon et al. 1987; Cannon et al. 1988; Mackenzie et al. 1989; Munoz et al. 1991). Clearance of respiratory virus infections by CD8+ T-cell effectors appears to require direct contact-dependent recognition of virus-infected target cells (Mackenzie et al. 1989; Hou and Doherty 1995), followed by lysis of infected cells (Kagi et al. 1995).

Recent studies have defined two major effector mechanisms of CTL; (1) exocytosis from CTL onto target cells of the contents of cytoplasmic granules containing the pore-forming protein perforin and a family of serine proteases termed granzymes, and (2) engagement of a tumor necrosis factor receptor (TNFR)-like molecule on the target cell (Fas [CD95] or TNFR) by FasL or TNF of the CTL (Smyth and Trapani 1998). Experiments with perforin gene knockout mice demonstrated that perforin is not required for resolution of influenza virus infection (Kagi et al. 1995; Kagi and Hengartner 1996). In this setting, CTL effectors may mediate lysis via Fas (CD95) ligation. PIV3 can inhibit cell-mediated cytotoxicity by selective inhibition of granzyme B mRNA (Sieg et al. 1995), an immune evasion strategy that could prolong replication in vivo.

3.2.1.2 CD4+ T Lymphocytes

Mice can recover from influenza infection in the absence of MHC class I-restricted CD8+ T cells (Eichelberger et al. 1991; Scherle et al. 1992; Bender et al. 1992; Allan et al. 1993). Sendai virus-specific CD4+ CTL are detected in acutely infected mice lacking class I-restricted CD8+ T cells (Hou et al. 1992). The viral clearance that such cells mediate is delayed. Cytolytic CD4+ T cells or Th1-type CD4+ T-cell clones can mediate influenza-virus clearance of infected mice (Lightman et al. 1987; Taylor et al. 1990; Graham et al. 1994).

Adoptive transfer experiments using bone marrow chimeras that express MHC class-II glycoproteins in the lymphoid compartment, but not pulmonary epithelial cells, indicate that the CD4+ effect is contact-independent, in contrast to the contact-dependent effect of CD8+ cells (Topham et al. 1996). Such contact-independent antiviral effects are probably mediated by Th activity for virus-specific Ig production that is delivered in lymphoid tissue (Palladino et al. 1995; Mozdzanowska et al. 1997). B cell-deficient (Ig-/-μMT) mice depleted of CD8+ T cells by mAb treatment, develop vigorous influenza-specific CD4+ T-cell responses with long-term persistence of increased Th precursor frequency, but limited ability to clear subsequent influenza virus infection in the absence of B cells (TOPHAM and DOHERTY 1998). Influenza-virus clearance following adoptive transfer of CD4+ Th1 cells into *nu/nu* mice, but not SCID mice that do not make antibodies, also suggests that the contribution of CD4+ T cells to virus clearance is mediated principally through help for production of antiviral antibodies (SCHERLE et al. 1992). When CD8+ cells are present, CD4+ cells also contribute to viral clearance by promoting the clonal expansion of PIV1- or influenza-specific CD8+ CTL in the lymph nodes or spleen (MO et al. 1997b; SARAWAR et al. 1993; TRIPP et al. 1995).

3.2.2 Antiviral T Lymphocytes in Humans

It is likely that cell-mediated immunity plays a primary role in the clearance of established respiratory virus infection in humans. The best evidence to support this concept is the fact that patients with defects in cellular immunity, such as recipients of bone-marrow transplantation or patients with cancer who are undergoing immunosuppressive chemotherapy, suffer more prolonged virus shedding or more frequent and more severe illnesses with RSV (FISHAUT et al. 1980; HALL et al. 1986; ENGLUND et al. 1988; HARRINGTON et al. 1992; WHIMBEY et al. 1995; WHIMBEY et al. 1996), influenza virus (WHIMBEY et al. 1994; YOUSUF et al. 1997), or PIV3 (WHIMBEY et al. 1993; LEWIS et al. 1996). However, the role of cell-mediated immune effectors in preventing disease in humans who have been re-infected or who encounter virus challenge following immunization is less clear.

3.2.2.1 Lymphocyte Transformation Activity in Humans

Increased lymphocyte transformation activity has been detected in infants following bronchiolitis associated with RSV infection (SCOTT et al. 1978; WELLIVER et al. 1979; CRANAGE and GARDNER 1980; BERTOTTO et al. 1980; SCOTT et al. 1984). This follows either RSV F glycoprotein subunit immunization of seropositive children (WELLIVER et al. 1994) or FI-RSV immunization and subsequent infection characterized by enhanced disease (KIM et al. 1976). A higher percentage of infants less than 6 months of age with RSV bronchiolitis exhibit increased LTF activity than rises in specific serum IgG or IgA (SCOTT et al. 1978). It is not known if the cellular activity measured by this technique protects human subjects or contributes to the pathogenesis of the disease during infection. Patients with RSV or PIV3 bronchiolitis have higher levels of LTF activity than patients without wheezing (WELLIVER et al. 1986), and recipients of the FI-RSV vaccine, many of whom developed enhanced disease, had increased LTF activity (KIM et al. 1976).

Much work has been done with rodents to examine the cellular and molecular basis for enhanced disease during infection following FI-RSV immunization. The bulk of these studies suggest the induction of an altered immune response that correlates with aberrant induction of Th2 cytokines. There is no cytokine data from those humans who received the FI-RSV vaccine. LTF activity is usually short-lived (LAZAR et al. 1980); however, increased influenza-specific LTF activity could be detected in those individuals whose last exposure to H1N1 virus was 20 years prior (Lazar and Wright 1980).

3.2.2.2 Cytotoxic T Lymphocytes in Humans

Virus-specific CTL activity, detected by chromium release of infected target cells, can be demonstrated in the peripheral blood of infants or adults following RSV infection (Bangham et al. 1986; Isaacs et al. 1987; Chiba et al. 1989), naturally-acquired *wt* influenza virus infection (McMichael et al. 1983a; Hildreth et al. 1983; McMichael et al. 1986), experimental *wt* influenza virus challenge (McMichael et al. 1983b), or immunization with live or inactivated influenza-virus vaccines (Ennis et al. 1981b; McMichael et al. 1981; Fries et al. 1993). The role of human CTLs in protection against disease upon reinfection, however, is unclear. Adults previously infected with influenza virus appear fully susceptible to pandemic strains that presumably retain CTL determinants in the conserved internal virus proteins present in previously circulating strains. This epidemiological observation suggests that CTL activity induced by *wt* virus infection does not protect, to a clinically significant degree, against disease upon reinfection in humans.

3.2.3 Natural Killer Cells and Antibody-Dependent Cellular Cytotoxicity

Bronchoalveolar lavage (BAL) populations recovered from mice with secondary influenza pneumonia contain γ-δ TCR-NK1.1+ NK cell populations that have a high level of NK activity (Eichelberger and Doherty 1994). Depletion of NK cells in mice prior to intranasal influenza-virus challenge results in a 10- to 100-fold increase in viral replication, indicating a modest role in modulation of infection by these cells (Stein-Streilein et al. 1988). The fact that decline of NK cell activity against influenza virus in aged mice, which are more susceptible to lethal infection, is partially restored by vitamin E supplementation also suggests a role for NK cells in recovery from influenza virus infection (Hayek et al. 1997). NK cell activity increases transiently after influenza infection (Ennis et al. 1981a; Lewis et al. 1986). NK cell antiviral activity can be enhanced by antibody-mediated mechanisms. RSV antibody-dependent cellular cytotoxicity (ADCC) activity, mediated by NK cells, has been detected in monkey and human peripheral blood, using antibodies in serum or human NP secretions (Scott et al. 1977; Meguro et al. 1979; Cranage et al. 1981; Kaul et al. 1982; Koff et al. 1983; Okabe et al. 1983). Serum antibodies that enhance ADCC appear to be able to rise and fall in human subjects independently of virus neutralizing activity in the serum (Meguro et al. 1979).

3.2.4 Monocytes/Macrophages

Macrophages are abundant in the alveolar spaces of the lung. In vitro and in vivo data suggest that the resident pulmonary alveolar-macrophage population actively suppresses the antigen-presenting function of lung-dendritic cells (DCs) in situ

(Holt et al. 1993), utilizing multiple mechanisms including transforming growth factor beta (TGF-β) secretion (Lipscomb et al. 1993). Low numbers of DCs and high numbers of macrophages were noted in the human alveolar compartment, suggesting that the alveolar compartment may be less efficient in mounting an immune response (van Haarst et al. 1994). Human alveolar macrophages recovered during RSV infection can yield RSV by infectious center assays (Midulla et al. 1993). However, replication of RSV in alveolar macrophages is restricted in both mice (Franke et al. 1994) and humans (Panuska et al. 1990; Becker et al. 1992; Cirino et al. 1993). Inoculation of human alveolar macrophages with live or inactivated RSV induced increased TNF, IL-6, and IL-8 expression in one study (Becker et al. 1991), and IL-10 in another (Panuska et al. 1995).

3.2.5 Cytokines Derived from Cellular Immune Effectors

Type-I cytokines, including interferon gamma (IFN) γ, IFNα and tumor necrosis factor (TNF)-α, exhibit potent antiviral effects in vivo (Ramsay et al. 1993). Interferon is induced during influenza infection, and influenza virus is sensitive to interferon (Hill et al. 1972; Murphy et al. 1973; Richman et al. 1976). Neutralization of IFN-γ in CD8+ T-cell deficient (β2 microglobulin knockout) mice delayed recovery from influenza virus infection (Sarawar et al. 1994). Influenza infection of IFN-γ-deficient (-/-) mice, however, demonstrated that this cytokine is not required for an effective humoral or cellular response (Graham et al. 1993b).

Sendai-virus elimination from the respiratory tract in IFN-γ-deficient mice occurred with normal kinetics (Mo et al. 1997b). TNF-α combined with IFN-β aborted RSV replication in A549 cells (Merolla et al. 1995). Induction of several type-II cytokines correlates with immunopathogenesis in RSV animal models, without contributing to virus clearance (discussed in Sect. 2.4). Homozygous disruption of the IL-4 gene did not appear to alter the severity of infection or viral clearance in Sendai-virus infected 129/J mice; however, decreased CTL precursors were noted (Mo et al. 1997a). IL-5-deficient mice with influenza virus infection exhibit no significant effect of this deficiency on antibody response or antiviral killer cell immunity (Kopf et al. 1996). In contrast, IL-6 deficiency does cause major effects on the development of systemic and mucosal antiviral antibody responses to influenza virus, predominately affecting IgA and IgG responses (Ramsay et al. 1994). Recombinant VV encoded IL-6 restored the mucosal immune defect in IL-6-deficient mice, confirming a major role for this factor in IgA responses.

3.2.6 Dendritic Cells

Dendritic cells (DCs) are widely distributed in the lung where they are distinguished by their morphology and class-II MHC antigen expression. DCs serve as potent pulmonary antigen-presenting cells; however, little is currently known about how these cells respond to specific respiratory viruses at the mucosa. Recruitment of a wave of DCs into the respiratory tract mucosa appears to be a feature of the acute cellular response to local challenge with viral protein antigens or PIV infection

(McWilliam et al. 1996; McWilliam et al. 1997). Children with adenoidal hypertrophy exhibit a marked increase in the number of intra-epithelial γ-δ TCR+ lymphocytes and dendritic HLA-DR+, S-100+ dendritic cells in the lymphoid tissue underlying the epithelium (Bani et al. 1994). The adenoids of children who have otitis media with effusion exhibit accumulations of DCs in extra-follicular areas (van Nieuwkerk et al. 1990, 1992).

Acknowledgements. The author gratefully acknowledges support from the Memorial Foundation, Goodlettsville, the NIAID/NIH Mucosal Pathogens Research Unit (NIH-NIAID N01-AI-65298), and the NICHD/NIH (R29 HD36311–01). Dr. Crowe is a Pfizer Faculty Scholar.

References

Aherne W, Bird T, Court SD, Gardner PS, McQuillin J (1970) Pathological changes in virus infections of the lower respiratory tract in children. J Clin Pathol 23:7–18

Allan W, Tabi Z, Cleary A, Doherty PC (1990) Cellular events in the lymph node and lung of mice with influenza: consequences of depleting CD4+ T cells. J Immunol 144:3980

Allan W, Carding SR, Eichelberger M, Doherty PC (1993) hsp65 mRNA+ macrophages and gamma delta T cells in influenza virus-infected mice depleted of the CD4+ and CD8+ lymphocyte subsets. Microb Pathog 14:75–84

Alwan WH, Openshaw PJ (1993) Distinct patterns of T- and B-cell immunity to respiratory syncytial virus induced by individual viral proteins. Vaccine 11:431–437

Alwan WH, Record FM, Openshaw PJ (1993) Phenotypic and functional characterization of T cell lines specific for individual respiratory syncytial virus proteins. J Immunol 150:5211–5218

Alwan WH, Kozlowska WJ, Openshaw PJ (1994) Distinct types of lung disease caused by functional subsets of antiviral T cells. J Exp Med 179:81–89

Ananaba GA, Anderson LJ (1991) Antibody enhancement of respiratory syncytial virus stimulation of leukotriene production by a macrophage-like cell line. J Virol 65:5052–5060

Anderson MJ, Pattison JR, Cureton RJ, Argent S, Heath RB (1980) The role of host responses in the recovery of mice from Sendai virus infection. J Gen Virol 46:373–379

Arnold R, Werner F, Humbert B, Werchau H, Konig W (1994) Effect of respiratory syncytial virus-antibody complexes on cytokine (IL-8, IL-6, TNF-α) release and respiratory burst in human granulocytes. Immunology 82:184–191

Arnold R, Konig W (1996) ICAM-1 expression and low-molecular-weight G-protein activation of human bronchial epithelial cells (A549) infected with RSV. J Leukoc Biol 60:766–771

Bachmann MF, Kundig TM, Hengartner H, Zinkernagel RM (1997) Protection against immunopathological consequences of a viral infection by activated but not resting cytotoxic T cells: T cell memory without "memory T cells"? Proc Natl Acad Sci USA 94:640–645

Baldwin AS Jr (1996) The NF-κB and IκB proteins: new discoveries and insights. Annu Rev Immunol 14:649–683

Bangham CR, Openshaw PJ, Ball LA, King AM, Wertz GW, Askonas BA (1986) Human and murine cytotoxic T cells specific to respiratory syncytial virus recognize the viral nucleoprotein (N), but not the major glycoprotein (G), expressed by vaccinia virus recombinants. J Immunol 137:3973–3977

Bani D, Gallo O, Fini-Storchi O (1994) Intraepithelial lymphocyte subpopulations and dendritic accessory cells in normal and hypertrophic adenoids. Laryngoscope 104:869–873

Becker S, Quay J, Soukup J (1991) Cytokine (tumor necrosis factor, IL-6, and IL-8) production by respiratory syncytial virus-infected human alveolar macrophages. J Immunol 147:4307–4312

Becker S, Soukup J, Yankaskas JR (1992) Respiratory syncytial virus infection of human primary nasal and bronchial epithelial cell cultures and bronchoalveolar macrophages. Am J Respir Cell Mol Biol 6:369–374

Becker S, Koren HS, Henke DC (1993) Interleukin-8 expression in normal nasal epithelium and its modulation by infection with respiratory syncytial virus and cytokines tumor necrosis factor, interleukin-1, and interleukin-6. Am J Respir Cell Mol Biol 8:20–27

Becker S, Reed W, Henderson FW, Noah TL (1997) RSV infection of human airway epithelial cells causes production of the β-chemokine RANTES. Am J Physiol 272:L512-L520

Belshe RB, Van Voris LP, Mufson MA (1982) Parenteral administration of live respiratory syncytial virus vaccine: results of a field trial. J Infect Dis 145:311–319

Bender BS, Croghan T, Zhang L, Small PA, Jr. (1992) Transgenic mice lacking class I major histocompatibility complex-restricted T cells have delayed viral clearance and increased mortality after influenza virus challenge. J Exp Med 175:1143–1145

Bertotto A, Stagni G, Caprino D, Sonaglia F, Velardi A (1980) Cell-mediated immunity in RSV bronchiolitis. J Pediatr 97:334–335

Bourgeois C, Bour JB, Aho LS, Pothier P (1998) Prophylactic administration of a complementarity-determining region derived from a neutralizing monoclonal antibody is effective against respiratory syncytial virus infection in BALB/c mice. J Virol 72:807–810

Cannon MJ, Stott EJ, Taylor G, Askonas BA (1987) Clearance of persistent respiratory syncytial virus infections in immunodeficient mice following transfer of primed T cells. Immunology 62:133–138

Cannon MJ, Openshaw PJ, Askonas BA (1988) Cytotoxic T cells clear virus but augment lung pathology in mice infected with respiratory syncytial virus. J Exp Med 168:1163–1168

Chiba Y, Higashidate Y, Suga K, Honjo K, Tsutsumi H, Ogra PL (1989) Development of cell-mediated cytotoxic immunity to respiratory syncytial virus in human infants following naturally acquired infection. J Med Virol 28:133–139

Chin J, Magoffin RL, Shearer LA, Schieble JH, Lennette EH (1969) Field evaluation of a respiratory syncytial virus vaccine and a trivalent parainfluenza virus vaccine in a pediatric population. Am J Epidemiol 89:449–463

Cirino NM, Panuska JR, Villani A, Taraf H, Rebert NA, Merolla R, Tsivitse P, Gilbert IA (1993) Restricted replication of respiratory syncytial virus in human alveolar macrophages. J Gen Virol 74:1527–1537

Connors M, Collins PL, Firestone CY, Murphy BR (1991) Respiratory syncytial virus (RSV) F, G, M2 (22 K), and N proteins each induce resistance to RSV challenge, but resistance induced by M2 and N proteins is relatively short-lived. J Virol 65:1634–1637

Connors M, Kulkarni AB, Collins PL, Firestone CY, Holmes KL, Morse HC, Murphy BR (1992a) Resistance to respiratory syncytial virus (RSV) challenge induced by infection with a vaccinia virus recombinant expressing the RSV M2 protein (Vac-M2) is mediated by CD8+ T cells, while that induced by Vac-F or Vac-G recombinants is mediated by antibodies. J Virol 66:1277–1281

Connors M, Kulkarni AB, Firestone CY, Holmes KL, Morse HC, Sotnikov AV, Murphy BR (1992b) Pulmonary histopathology induced by respiratory syncytial virus (RSV) challenge of formalin-inactivated RSV-immunized BALB/c mice is abrogated by depletion of CD4+ T cells. J Virol 66:7444–7451

Connors M, Giese NA, Kulkarni AB, Firestone CY, Morse HC, Murphy BR (1994) Enhanced pulmonary histopathology induced by respiratory syncytial virus (RSV) challenge of formalin-inactivated RSV-immunized BALB/c mice is abrogated by depletion of interleukin-4 (IL-4) and IL-10. J Virol 68:5321–5325

Corbeil S, Seguin C, Trudel M (1996) Involvement of the complement system in the protection of mice from challenge with respiratory syncytial virus Long strain following passive immunization with monoclonal antibody 18A2B2. Vaccine 14:521–525

Cranage MP, Gardner PS (1980) Systemic cell-mediated and antibody responses in infants with respiratory syncytial virus infections. J Med Virol 5:161–170

Cranage MP, Gardner PS, McIntosh K (1981) In vitro cell-dependent lysis of respiratory syncytial virus-infected cells mediated by antibody from local respiratory secretions. Clin Exp Immunol 43:28–35

Crowe JE Jr (1996) The role of antibodies in respiratory viral immunity. Semin Virol 7:273–283

Crowe JE Jr (1998a) Immune responses of infants to infection with respiratory viruses and live attenuated respiratory virus candidate vaccines. Vaccine (in press)

Crowe JE Jr, Murphy BR, Chanock RM, Williamson RA, Barbas CF, Burton DR (1994) Recombinant human respiratory syncytial virus (RSV) monoclonal antibody Fab is effective therapeutically when introduced directly into the lungs of RSV-infected mice. Proc Natl Acad Sci USA 91:1386–1390

Crowe JE Jr, Bui PT, Siber GR, Elkins WR, Chanock RM, Murphy BR (1995) Cold-passaged, temperature-sensitive mutants of human respiratory syncytial virus (RSV) are highly attenuated, immunogenic, and protective in seronegative chimpanzees, even when RSV antibodies are infused shortly before immunization. Vaccine 13:847–855

Crowe JE Jr, Gilmour PS, Murphy BR, Chanock RM, Duan L, Pomerantz RJ, Pilkington GR (1998) Isolation of a second recombinant human respiratory syncytial virus (RSV) monoclonal antibody fragment (Fab RSVF2–5) that exhibits therapeutic efficacy in vivo. J Infect Dis 177:1073–1076

Delage G, Brochu P, Pelletier M, Jasmin G, Lapointe N (1979) Giant-cell pneumonia caused by parainfluenza virus. J Pediatr 94:426–429

Downham MA, Gardner PS, McQuillin J, Ferris JA (1975) Role of respiratory viruses in childhood mortality. BMJ 1:235–239

Edwards KM, Snyder PN, Wright PF (1986) Complement activation by respiratory syncytial virus-infected cells. Arch Virol 88:49–56

Eichelberger M, Allan W, McMickle A, Doherty PC (1991) Clearance of influenza virus respiratory infection in mice lacking class I major histocompatibility complex-restricted CD8+ T cells. J Exp Med 174:875–870

Eichelberger M, Doherty PC (1994) Gamma delta T cells from influenza-infected mice develop a natural killer cell phenotype following culture. Cell Immunol 159:94–102

Einarsson O, Geba GP, Panuska JR, Zhu Z, Landry M, Elias JA (1995) Asthma-associated viruses specifically induce lung stromal cells to produce interleukin-11, a mediator of airways hyperreactivity. Chest 107:132S-133S

Einarsson O, Geba GP, Zhu Z, Landry M, Elias JA (1996) Interleukin-11: stimulation in vivo and in vitro by respiratory viruses and induction of airways hyper-responsiveness. J Clin Invest 97:915–924

Elias JA, Zheng T, Einarsson O, Landry M, Trow T, Rebert N, Panuska J (1994) Epithelial interleukin-11. Regulation by cytokines, respiratory syncytial virus, and retinoic acid. J Biol Chem 269:22261–22268

Endo Y, Carroll KN, Ikizler MR, Wright PF (1996) Growth of influenza A virus in primary, differentiated epithelial cells derived from adenoids. J Virol 70:2055–2058

Englund JA, Sullivan CJ, Jordan MC, Dehner LP, Vercellotti GM, Balfour HH, Jr. (1988) Respiratory syncytial virus infection in immunocompromised adults. Ann Intern Med 109:203–208

Ennis FA, Meager A, Beare AS, Qi YHU, Riley D, Schwarz G, Schild GC, Rook AH (1981a) Interferon induction and increased natural killer-cell activity in influenza infections in man. Lancet 2:891–893

Ennis FA, Rook AH, Qi YH, Schild GC, Riley D, Pratt R, Potter CW (1981b) HLA restricted virus-specific cytotoxic T-lymphocyte responses to live and inactivated influenza vaccines. Lancet 2:887–891

Epstein SL, Misplon JA, Lawson CM, Subbarao EK, Connors M, Murphy BR (1993) Beta 2-microglobulin-deficient mice can be protected against influenza A infection by vaccination with vaccinia-influenza recombinants expressing hemagglutinin and neuraminidase. J Immunol 150:5484–5493

Fiedler MA, Wernke-Dollries K, Stark JM (1996) Inhibition of viral replication reverses respiratory syncytial virus-induced NF-κB activation and interleukin-8 gene expression in A549 cells. J Virol 70:9079–9082

Fishaut M, Tubergen D, McIntosh K (1980) Cellular response to respiratory viruses with particular reference to children with disorders of cell-mediated immunity. J Pediatr 96:179–186

Frank AL, Taber LH, Glezen WP, et al. (1979) Reinfection with influenza A (H3N2) virus in young children and their families. J Infect Dis 140:829–836

Frank AL, Taber LH (1983) Variation in frequency of natural reinfection with influenza A viruses. J Med Virol 12:17–23

Franke G, Freihorst J, Steinmuller C, Verhagen W, Hockertz S, Lohmann-Matthes ML (1994) Interaction of alveolar macrophages and respiratory syncytial virus. J Immunol Methods 174:173–184

Friedewald WT, Forsyth BR, Smith CB, Gharpure MA, Chanock RM (1968) Low-temperature-grown RS virus in adult volunteers. JAMA 203:690–694

Fries LF, Dillon SB, Hildreth JE, Karron RA, Funkhouser AW, Friedman CJ, Jones CS, Culleton VG, Clements ML (1993) Safety and immunogenicity of a recombinant protein influenza A vaccine in adult human volunteers and protective efficacy against wild type H1N1 virus challenge. J Infect Dis 167:593–601

Fulginiti VA, Eller JJ, Sieber OF, Joyner JW, Minamitani M, Meiklejohn G (1969) Respiratory virus immunization. I. A field trial of two inactivated respiratory virus vaccines; an aqueous trivalent parainfluenza virus vaccine and an alum-precipitated respiratory syncytial virus vaccine. Am J Epidemiol 89:435–448

Gardner PS, McQuillin J, Court SD (1970) Speculation on pathogenesis in death from respiratory syncytial virus infection. BMJ 1:327–330

Garofalo R, Kimpen JL, Welliver RC, Ogra PL (1992) Eosinophil degranulation in the respiratory tract during naturally acquired respiratory syncytial virus infection. J Pediatr 120:28–32

Garofalo R, Mei F, Espejo R, Ye G, Haeberle H, Baron S, Ogra PL, Reyes VE (1996a) Respiratory syncytial virus infection of human respiratory epithelial cells up-regulates class I MHC expression through the induction of IFN-β and IL-1α. J Immunol 157:2506–2513
Garofalo R, Sabry M, Jamaluddin M, Yu RK, Casola A, Ogra PL, Brasier AR (1996b) Transcriptional activation of the interleukin-8 gene by respiratory syncytial virus infection in alveolar epithelial cells: nuclear translocation of the RelA transcription factor as a mechanism producing airway mucosal inflammation. J Virol 70:8773–8781
Gerhard W, Mozdzanowska K, Furchner M, Washk G, Maiese K (1997) Role of the B-cell response in recovery of mice from primary influenza virus infection. Immunol Rev 159:95–103
Giard DJ, Aaronson SA, Todaro GJ, Arnstein P, Kersey JH, Dosik H, Parks WP (1973) In vitro cultivation of human tumors: establishment of cell lines derived from a series of solid tumors. J Natl Cancer Inst 51:1417–1423
Glezen WP, Paredes A, Allison JE, Taber LH, Frank AL (1981) Risk of respiratory syncytial virus infection for infants from low-income families in relationship to age, sex, ethnic group, and maternal antibody level. J Pediatr 98:708–715
Glezen WP, Frank AL, Taber LH, Kasel JA (1984) Parainfluenza virus type 3: seasonality and risk of infection and reinfection in young children. J Infect Dis 150:851–857
Graham BS, Bunton LA, Rowland J, Wright PF, Karzon DT (1991a) Respiratory syncytial virus infection in anti-mu-treated mice. J Virol 65:4936–4942
Graham BS, Bunton LA, Wright PF, Karzon DT (1991b) Role of T lymphocyte subsets in the pathogenesis of primary infection and rechallenge with respiratory syncytial virus in mice. J Clin Invest 88:1026–1033
Graham BS, Henderson GS, Tang YW, Lu X, Neuzil KM, Colley DG (1993a) Priming immunization determines T helper cytokine mRNA expression patterns in lungs of mice challenged with respiratory syncytial virus. J Immunol 151:2032–2040
Graham MB, Dalton DK, Giltinan D, Braciale VL, Stewart TA, Braciale TJ (1993b) Response to influenza infection in mice with a targeted disruption in the interferon gamma gene. J Exp Med 178:1725–1732
Graham MB, Braciale VL, Braciale TJ (1994) Influenza virus-specific CD4+ T helper type 2 T lymphocytes do not promote recovery from experimental virus infection. J Exp Med 180:1273–1282
Graham MB, Braciale TJ (1996) Influenza virus clearance in B lymphocyte deficient mice. In: Brown LE, Hampson AW, Webster RG (eds) Options for control of influenza III. Elsevier, Amsterdam, pp 166–169
Groothuis JR, Simoes EA, Levin MJ, Hall CB, Long CE, Rodriguez WJ, Arrobio J, Meissner HC, Fulton DR, Welliver RC (1993) Prophylactic administration of respiratory syncytial virus immune globulin to high-risk infants and young children. The Respiratory Syncytial Virus Immune Globulin Study Group. N Engl J Med 329:1524–1530
Groothuis JR, Simoes EA, Hemming VG (1995) Respiratory syncytial virus (RSV) infection in preterm infants and the protective effects of RSV immune globulin (RSVIG). Respiratory Syncytial Virus Immune Globulin Study Group. Pediatrics 95:463–467
Gruber WC, Darden PM, Still JG, Lohr J, Reed G, Wright PF, Wyeth-Ayerst ca Influenza Vaccine Study Group (1997) Evaluation of bivalent live attenuated influenza A vaccines in children 2 months to 3 years of age: safety, immunogenicity and dose-response. Vaccine 15:1379–1384
Hall CB, Powell KR, MacDonald NE, Gala CL, Menegus ME, Suffin SC, Cohen HJ (1986) Respiratory syncytial viral infection in children with compromised immune function. N Engl J Med 315:77–81
Hall CB, Walsh EE, Long CE, Schnabel KC (1991) Immunity to and frequency of reinfection with respiratory syncytial virus. J Infect Dis 163:693–698
Hancock GE, Speelman DJ, Heers K, Bortell E, Smith J, Cosco C (1996) Generation of atypical pulmonary inflammatory responses in BALB/c mice after immunization with the native attachment (G) glycoprotein of respiratory syncytial virus. J Virol 70:7783–7791
Harrington RD, Hooton TM, Hackman RC, Storch GA, Osborne B, Gleaves CA, Benson A, Meyers JD (1992) An outbreak of respiratory syncytial virus in a bone marrow transplant center. J Infect Dis 165:987–993
Hayek MG, Taylor SF, Bender BS, Han SN, Meydani M, Smith DE, Eghtesada S, Meydani SN (1997) Vitamin E supplementation decreases lung virus titers in mice infected with influenza. J Infect Dis 176:273–276
Hemming VG, Rodriguez W, Kim HW, Brandt CD, Parrott RH, Burch B, Prince GA, Baron PA, Fink RJ, Reaman G (1987) Intravenous immunoglobulin treatment of respiratory syncytial virus infections in infants and young children. Antimicrob Agents Chemother 31:1882–1886

Henderson FW, Hu SC, Collier AM (1978) Pathogenesis of respiratory syncytial virus infection in ferret and fetal human tracheas in organ culture. Am Rev Respir Dis 118:29–37

Hers JF, Mulder J (1961) Broad aspects of the pathology and pathogenesis of human influenza. Am Rev Respir Dis 83 [Suppl]:54–67

Hicks JT, Ennis FA, Kim E, Verbonitz M (1978) The importance of an intact complement pathway in recovery from a primary viral infection: influenza in de complemented and in C5-deficient mice. J Immunol 121:1437–1445

Hildreth JE, Gotch FM, McMichael AJ (1983) Plasma membranes from influenza virus-infected cells induce in vitro human secondary virus-specific cytotoxic T lymphocyte responses. Mol Biol Med 1:225–233

Hill DA, Baron S, Perkins JC, Worthington M, Van Kirk JE, Mills J, Kapikian AZ, Chanock RM (1972) Evaluation of an interferon inducer in viral respiratory disease. JAMA 219:1179–1184

Holt PG, Oliver J, Bilyk N, McMenamin C, McMenamin PG, Kraal G, Thepen T (1993) Downregulation of the antigen presenting cell function(s) of pulmonary dendritic cells in vivo by resident alveolar macrophages. J Exp Med 177:397–407

Hornsleth A, Friis B, Grauballe PC, Krasilnikof PA (1984) Detection by ELISA of IgA and IgM antibodies in secretion and IgM antibodies in serum in primary lower respiratory syncytial virus infection. J Med Virol 13:149–161

Hou S, Doherty PC, Zijlstra M, Jaenisch R, Katz JM (1992) Delayed clearance of Sendai virus in mice lacking class I MHC-restricted CD8+ T cells. J Immunol 149:1319–1325

Hou S, Doherty PC (1995) Clearance of Sendai virus by CD8+ T cells requires direct targeting to virus-infected epithelium. Eur J Immunol 25:111–116

Huang RT, Lichtenberg B, Rick O (1996) Involvement of annexin V in the entry of influenza viruses and role of phospholipids in infection. FEBS Lett 392:59–62

Ingram JM, Rakes GP, Hoover GE, Platts-Mills TA, Heymann PW (1995) Eosinophil cationic protein in serum and nasal washes from wheezing infants and children. J Pediatr 127:558–564

Isaacs D, Bangham CR, McMichael AJ (1987) Cell-mediated cytotoxic response to respiratory syncytial virus in infants with bronchiolitis. Lancet 2:769–771

Iwasaki T, Nozima T (1977) Defense mechanisms against primary influenza virus infection in mice. I. The roles of interferon and neutralizing antibodies and thymus dependence of interferon and antibody production. J Immunol 118:256–263

Jamaluddin M, Garofalo R, Ogra PL, Brasier AR (1996) Inducible translational regulation of the NF-IL6 transcription factor by respiratory syncytial virus infection in pulmonary epithelial cells. J Virol 70:1554–1563

Jarvis WR, Middleton PJ, Gelfand EW (1979) Parainfluenza pneumonia in severe combined immunodeficiency disease. J Pediatr 94:423–425

Johnson PR, Feldman S, Thompson JM, Mahoney JD, Wright PF (1986) Immunity to influenza A virus infection in young children: a comparison of natural infection, live cold-adapted vaccine, and inactivated vaccine. J Infect Dis 154:121–127

Johnson PR, Jr., Feldman S, Thompson JM, Mahoney JD, Wright PF (1985) Comparison of long-term systemic and secretory antibody responses in children given live, attenuated, or inactivated influenza A vaccine. J Med Virol 17:325–335

Johnson RA, Prince GA, Suffin SC, Horswood RL, Chanock RM (1982) Respiratory syncytial virus infection in cyclophosphamide-treated cotton rats. Infect Immun 37:369–373

Kagi D, Seiler P, Pavlovic J, Ledermann B, Burki K, Zinkernagel RM, Hengartner H (1995) The roles of perforin- and Fas-dependent cytotoxicity in protection against cytopathic and noncytopathic viruses. Eur J Immunol 25:3256–3262

Kagi D, Hengartner H (1996) Different roles for cytotoxic T cells in the control of infections with cytopathic versus non cytopathic viruses. Curr Opin Immunol 8:472–477

Kapikian AZ, Mitchell RH, Chanock RM, Shvedoff RA, Stewart CE (1969) An epidemiological study of altered clinical reactivity to respiratory syncytial (RS) virus infection in children previously vaccinated with an inactivated RS virus vaccine. Am J Epidemiol 89:405–421

Karron RA, Buonagurio DA, Georgiu AF, Whitehead SS, Adamus JE, Clements-Mann ML, Harris DO, Randolph VB, Udem SA, Murphy BR, Sidhu MS (1997) Respiratory syncytial virus (RSV) SH and G proteins are not essential for viral replication in vitro: clinical evaluation and molecular characterization of a cold-passaged, attenuated RSV subgroup B mutant. Proc Natl Acad Sci USA 94:13961–13966

Kast WM, Bronkhorst AM, de Waal LP, Melief CJ (1986) Cooperation between cytotoxic and helper T lymphocytes in protection against lethal Sendai virus infection. Protection by T cells is MHC-restricted and MHC-regulated; a model for MHC-disease associations. J Exp Med 164:723–738

Kaul TN, Welliver RC, Ogra PL (1981) Comparison of fluorescent-antibody, neutralizing-antibody, and complement-enhanced neutralizing-antibody assays for detection of serum antibody to respiratory syncytial virus. J Clin Microbiol 13:957–962

Kaul TN, Welliver RC, Ogra PL (1982) Development of antibody-dependent cell-mediated cytotoxicity in the respiratory tract after natural infection with respiratory syncytial virus. Infect Immun 37:492–498

Ke Y, Reddel RR, Gerwin BI, Miyashita M, McMenamin M, Lechner JF, Harris CC (1988) Human bronchial epithelial cells with integrated SV40 virus T antigen genes retain the ability to undergo squamous differentiation. Differentiation 38:60–66

Kim HW, Canchola JG, Brandt CD, Pyles G, Chanock RM, Jensen K, Parrott RH (1969) Respiratory syncytial virus disease in infants despite prior administration of antigenic inactivated vaccine. Am J Epidemiol 89:422–434

Kim HW, Leikin SL, Arrobio J, Brandt CD, Chanock RM, Parrott RH (1976) Cell-mediated immunity to respiratory syncytial virus induced by inactivated vaccine or by infection. Pediatr Res 10:75–78

Kimpen JL, Garofalo R, Welliver RC, Ogra PL (1992) Activation of human eosinophils in vitro by respiratory syncytial virus. Pediatr Res 32:160–164

Kimpen JL, Garofalo R, Welliver RC, Fujihara K, Ogra PL (1996) An ultrastructural study of the interaction of human eosinophils with respiratory syncytial virus. Pediatr Allergy Immunol 7:48–53

Koff WC, Caplan FR, Case S, Halstead SB (1983) Cell-mediated immune response to respiratory syncytial virus infection in owl monkeys. Clin Exp Immunol 53:272–280

Kopf M, Brombacher F, Hodgkin PD, Ramsay AJ, Milbourne EA, Dai WJ, Ovington KS, Behm CA, Kohler G, Young IG, Matthaei KI (1996) IL-5-deficient mice have a developmental defect in CD5+ B-1 cells and lack eosinophilia but have normal antibody and cytotoxic T cell responses. Immunity 4:15–24

Kris RM, Asofsky R, Evans CB, Small PA, Jr. (1985) Protection and recovery in influenza virus-infected mice immunosuppressed with anti-IgM. J Immunol 134:1230–1235

Kris RM, Yetter RA, Cogliano R, Ramphal R, Small PA (1988) Passive serum antibody causes temporary recovery from influenza virus infection of the nose, trachea, and lung of nude mice. Immunology 63:349–353

Lamprecht CL, Krause HE, Mufson MA (1976) Role of maternal antibody in pneumonia and bronchiolitis due to respiratory syncytial virus. J Infect Dis 134:211–217

Lazar A, Okabe N, Wright PF (1980) Humoral and cellular immune responses of seronegative children vaccinated with a cold-adapted influenza A/HK/123/77 (H1N1) recombinant virus. Infect Immun 27:862–866

Lazar A, Wright PF (1980) Cell-mediated immune response of human lymphocytes to influenza A/USSR (H1N1) virus infection. Infect Immun 27:867–871

Levine S, Klaiber-Franco R, Paradiso PR (1987) Demonstration that glycoprotein G is the attachment protein of respiratory syncytial virus. J Gen Virol 68:2521–2524

Lewis DE, Gilbert BE, Knight V (1986) Influenza virus infection induces functional alterations in peripheral blood lymphocytes. J Immunol 137:3777–3781

Lewis VA, Champlin R, Englund J, Couch R, Goodrich JM, Rolston K, Przepiorka D, Mirza NQ, Yousuf HM, Luna M, Bodey GP, Whimbey E (1996) Respiratory disease due to parainfluenza virus in adult bone marrow transplant recipients. Clin Infect Dis 23:1033–1037

Liang S, Mozdzanowska K, Palladino G, Gerhard W (1994) Heterosubtypic immunity to influenza type A virus in mice. Effector mechanisms and their longevity. J Immunol 152:1653–1661

Lightman S, Cobbold S, Waldmann H, Askonas BA (1987) Do L3T4+ T cells act as effector cells in protection against influenza virus infection? Immunology 62:139–144

Lipscomb MF, Pollard AM, Yates JL (1993) A role for TGF-β in the suppression by murine bronchoalveolar cells of lung dendritic cell initiated immune responses. Reg Immunol 5:151–157

Louria DB, Blumenfeld HL, Ellis JT, Kilbourne ED, Rogers DE (1959) Studies on influenza in the pandemic of 1957–58. II. Pulmonary complications of influenza. J Clin Invest 38:213–265

Lukacher AE, Braciale VL, Braciale TJ (1984) In vivo effector function of influenza virus-specific cytotoxic T lymphocyte clones is highly specific. J Exp Med 160:814–826

Mackenzie CD, Taylor PM, Askonas BA (1989) Rapid recovery of lung histology correlates with clearance of influenza virus by specific CD8+ cytotoxic T cells. Immunology 67:375–381

Markwell MA, Paulson JC (1980) Sendai virus utilizes specific sialyloligosaccharides as host cell receptor determinants. Proc Natl Acad Sci USA 77:5693–5697

Mastronarde JG, Monick MM, Hunninghake GW (1995) Oxidant tone regulates IL-8 production in epithelium infected with respiratory syncytial virus Amer J Resp Cell Mol Biol 13:237–244 [published erratum 13:629]

Mastronarde JG, He B, Monick MM, Mukaida N, Matsushima K, Hunninghake GW (1996) Induction of interleukin (IL)-8 gene expression by respiratory syncytial virus involves activation of nuclear factor (NF)-kappa B and NF-IL-6. J Infect Dis 174:262–267

Matsukura S, Kokubu F, Noda H, Tokunaga H, Adachi M (1996) Expression of IL-6, IL-8, and RANTES on human bronchial epithelial cells, NCI-H292, induced by influenza virus A. J Allergy Clin Immunol 98:1080–1087

Matsuzaki Z, Okamoto Y, Sarashina N, Ito E, Togawa K, Saito I (1996) Induction of intercellular adhesion molecule-1 in human nasal epithelial cells during respiratory syncytial virus infection. Immunology 88:565–568

Mazanec MB, Kaetzel CS, Lamm ME, Fletcher D, Nedrud JG (1992a) Intracellular neutralization of virus by immunoglobulin A antibodies. Proc Natl Acad Sci USA 89:6901–6905

Mazanec MB, Lamm ME, Lyn D, Portner A, Nedrud JG (1992b) Comparison of IgA versus IgG monoclonal antibodies for passive immunization of the murine respiratory tract. Virus Res 23:1–12

Mazanec MB, Coudret CL, Fletcher DR (1995) Intracellular neutralization of influenza virus by immunoglobulin A anti-hemagglutinin monoclonal antibodies. J Virol 69:1339–1343

McIntosh K, Masters HB, Orr I, Chao RK, Barkin RM (1978) The immunologic response to infection with respiratory syncytial virus in infants. J Infect Dis 138:24–32

McMichael AJ, Gotch F, Cullen P, Askonas B, Webster RG (1981) The human cytotoxic T cell response to influenza A vaccination. Clin Exp Immunol 43:276–284

McMichael AJ, Gotch FM, Dongworth DW, Clark A, Potter CW (1983a) Declining T-cell immunity to influenza 1977–82. Lancet 2:762–764

McMichael AJ, Gotch FM, Noble GR, Beare PA (1983b) Cytotoxic T-cell immunity to influenza. N Engl J Med 309:13–17

McMichael AJ, Michie CA, Gotch FM, Smith GL, Moss B (1986) Recognition of influenza A virus nucleoprotein by human cytotoxic T lymphocytes. J Gen Virol 67:719–726

McQuillin J, Gardner PS, McGuckin R (1970) Rapid diagnosis of influenza by immunofluorescent techniques. Lancet 2:690–695

McWilliam AS, Napoli S, Marsh AM, Pemper FL, Nelson DJ, Pimm CL, Stumbles PA, Wells TN, Holt PG (1996) Dendritic cells are recruited into the airway epithelium during the inflammatory response to a broad spectrum of stimuli. J Exp Med 184:2429–2432

McWilliam AS, Marsh AM, Holt PG (1997) Inflammatory infiltration of the upper airway epithelium during Sendai virus infection: involvement of epithelial dendritic cells. J Virol 71:226–236

Meguro H, Kervina M, Wright PF (1979) Antibody-dependent cell-mediated cytotoxicity against cells infected with respiratory syncytial virus: characterization of in vitro and in vivo properties. J Immunol 122:2521–2526

Merolla R, Rebert NA, Tsiviste PT, Hoffmann SP, Panuska JR (1995) Respiratory syncytial virus replication in human lung epithelial cells: inhibition by tumor necrosis factor alpha and interferon β. Am J Respir Crit Care Med 152:1358–1366

Midulla F, Villani A, Panuska JR, Dab I, Kolls JK, Merolla R, Ronchetti R (1993) Respiratory syncytial virus lung infection in infants: immunoregulatory role of infected alveolar macrophages. J Infect Dis 168:1515–1519

Mo XY, Sangster MY, Tripp RA, Doherty PC (1997a) Modification of the Sendai virus-specific antibody and CD8+ T-cell responses in mice homozygous for disruption of the interleukin-4 gene. J Virol 71:2518–2521

Mo XY, Tripp RA, Sangster MY, Doherty PC (1997b) The cytotoxic T-lymphocyte response to Sendai virus is unimpaired in the absence of gamma interferon. J Virol 71:1906–1910

Mozdzanowska K, Furchner M, Washko G, Mozdzanowski J, Gerhard W (1997) A pulmonary influenza virus infection in SCID mice can be cured by treatment with hemagglutinin-specific antibodies that display very low virus-neutralizing activity in vitro. J Virol 71:4347–4355

Munoz JL, McCarthy CA, Clark ME, Hall CB (1991) Respiratory syncytial virus infection in C57BL/6 mice: clearance of virus from the lungs with virus-specific cytotoxic T cells. J Virol 65:4494–4497

Murphy BR, Baron S, Chalhub EG, Uhlendorf CP, Chanock RM (1973) Temperature-sensitive mutants of influenza virus. IV. Induction of interferon in the nasopharynx by wild type and a temperature-sensitive recombinant virus. J Infect Dis 128:488–493

Murphy BR, Alling DW, Snyder MH, Walsh EE, Prince GA, Chanock RM, Hemming VG, Rodriguez WJ, Kim HW, Graham BS (1986a) Effect of age and preexisting antibody on serum antibody response of infants and children to the F and G glycoproteins during respiratory syncytial virus infection. J Clin Microbiol 24:894–898

Murphy BR, Prince GA, Walsh EE, Kim HW, Parrott RH, Hemming VG, Rodriguez WJ, Chanock RM (1986b) Dissociation between serum neutralizing and glycoprotein antibody responses of infants and children who received inactivated respiratory syncytial virus vaccine. J Clin Microbiol 24:197–202

Murphy BR, Olmsted RA, Collins PL, Chanock RM, Prince GA (1988) Passive transfer of respiratory syncytial virus (RSV) antiserum suppresses the immune response to the RSV fusion (F) and large (G) glycoproteins expressed by recombinant vaccinia viruses. J Virol 62:3907–3910

Murphy BR, Collins PL, Lawrence L, Zubak J, Chanock RM, Prince GA (1989) Immunosuppression of the antibody response to respiratory syncytial virus (RSV) by pre-existing serum antibodies: partial prevention by topical infection of the respiratory tract with vaccinia virus-RSV recombinants. J Gen Virol 70:2185–2190

Murphy BR, Prince GA, Collins PL, Hildreth SW, Paradiso PR (1991) Effect of passive antibody on the immune response of cotton rats to purified F and G glycoproteins of respiratory syncytial virus (RSV). Vaccine 9:185–189

Neilson KA, Yunis EJ (1990) Demonstration of respiratory syncytial virus in an autopsy series. Pediatr Pathol 10:491–502

Noah TL, Henderson FW, Wortman IA, Devlin RB, Handy J, Koren HS, Becker S (1995) Nasal cytokine production in viral acute upper respiratory infection of childhood. J Infect Dis 171:584–592

Ogilvie MM, Vathenen AS, Radford M, Codd J, Key S (1981) Maternal antibody and respiratory syncytial virus infection in infancy. J Med Virol 7:263–271

Okabe N, Hashimoto G, Abo T, Wright PF, Karzon DT (1983) Characterization of the human peripheral blood effector cells mediating antibody-dependent cell-mediated cytotoxicity against respiratory syncytial virus. Clin Immunol Immunopathol 27:200–209

Otto M, Gunther A, Fan H, Rick O, Huang RT (1994) Identification of annexin 33 kDa in cultured cells as a binding protein of influenza viruses. FEBS Lett 356:125–129

Palladino G, Mozdzanowska K, Washko G, Gerhard W (1995) Virus-neutralizing antibodies of immunoglobulin G (IgG) but not of IgM or IgA isotypes can cure influenza virus pneumonia in SCID mice. J Virol 69:2075–2081

Panuska JR, Cirino NM, Midulla F, Despot JE, McFadden ER, Jr., Huang YT (1990) Productive infection of isolated human alveolar macrophages by respiratory syncytial virus. J Clin Invest 86:113–119

Panuska JR, Hertz MI, Taraf H, Villani A, Cirino NM (1992) Respiratory syncytial virus infection of alveolar macrophages in adult transplant patients. Am Rev Respir Dis 145:934–939

Panuska JR, Merolla R, Rebert NA, Hoffmann SP, Tsivitse P, Cirino NM, Silverman RH, Rankin JA (1995) Respiratory syncytial virus induces interleukin-10 by human alveolar macrophages. Suppression of early cytokine production and implications for incomplete immunity. J Clin Invest 96:2445–2453

Parham DM, Bozeman P, Killian C, Murti G, Brenner M, Hanif I (1993) Cytologic diagnosis of respiratory syncytial virus infection in a bronchoalveolar lavage specimen from a bone marrow transplant recipient. Am J Clin Pathol 99:588–592

Parrott RH, Kim HW, Arrobio JO, Hodes DS, Murphy BR, Brandt CD, Camargo E, Chanock RM (1973) Epidemiology of respiratory syncytial virus infection in Washington, D.C. II. Infection and disease with respect to age, immunologic status, race, and sex. Am J Epidemiol 98:289–300

Patel JA, Kunimoto M, Sim TC, Garofalo R, Eliott T, Baron S, Ruuskanen O, Chonmaitree T, Ogra PL, Schmalstieg F (1995) Interleukin-1 alpha mediates the enhanced expression of intercellular adhesion molecule-1 in pulmonary epithelial cells infected with respiratory syncytial virus. Am J Respir Cell Mol Biol 13:602–609

Prince GA, Hemming VG, Horswood RL, Chanock RM (1985a) Immunoprophylaxis and immunotherapy of respiratory syncytial virus infection in the cotton rat. Virus Res 3:193–206

Prince GA, Horswood RL, Chanock RM (1985b) Quantitative aspects of passive immunity to respiratory syncytial virus infection in infant cotton rats. J Virol 55:517–520

Prince GA, Hemming VG, Horswood RL, Baron PA, Chanock RM (1987) Effectiveness of topically administered neutralizing antibodies in experimental immunotherapy of respiratory syncytial virus infection in cotton rats. J Virol 61:1851–1854

Prince GA, Hemming VG, Horswood RL, Baron PA, Murphy BR, Chanock RM (1990) Mechanism of antibody-mediated viral clearance in immunotherapy of respiratory syncytial virus infection of cotton rats. J Virol 64:3091–3092

Puck JM, Glezen WP, Frank AL, et al. (1980) Protection of infants from infection with influenza A virus by transplacentally acquired antibody. J Infect Dis 142:844–849

Ramphal R, Cogliano RC, Shands JW, Jr., Small PA, Jr. (1979) Serum antibody prevents lethal murine influenza pneumonitis but not tracheitis. Infect Immun 25:992–997

Ramsay AJ, Ruby J, Ramshaw IA (1993) A case for cytokines as effector molecules in the resolution of virus infection. Immunol Today 14:155–157

Ramsay AJ, Husband AJ, Ramshaw IA, Bao S, Matthaei KI, Koehler G, Kopf M (1994) The role of interleukin-6 in mucosal IgA antibody responses in vivo. Science 264:561–563

Reading PC, Hartley CA, Ezekowitz RA, Anders EM (1995) A serum mannose-binding lectin mediates complement-dependent lysis of influenza virus-infected cells. Biochem Biophys Res Commun 217:1128–1136

Renegar KB, Small PA Jr (1991a) Immunoglobulin A mediation of murine nasal anti-influenza virus immunity. J Virol 65:2146–2148

Renegar KB, Small PA Jr (1991b) Passive transfer of local immunity to influenza virus infection by IgA antibody. J Immunol 146:1972–1978

Richman DD, Murphy BR, Baron S, Uhlendorf C (1976) Three strains of influenza A virus (H3N2): interferon sensitivity in vitro and interferon production in volunteers. J Clin Microbiol 3:223–226

Rodriguez WJ, Gruber WC, Welliver RC, Groothuis JR, Simoes EA, Meissner HC, Hemming VG, Hall CB, Lepow ML, Rosas AJ, Robertsen C, Kramer AA (1997) Respiratory syncytial virus (RSV) immune globulin intravenous therapy for RSV lower respiratory tract infection in infants and young children at high risk for severe RSV infections: Respiratory Syncytial Virus Immune Globulin Study Group. Pediatrics 99:454–461

Russi JC, Delfraro A, Borthagaray MD, Velazquez B, Garcia-Barreno B, Hortal M (1993) Evaluation of immunoglobulin E-specific antibodies and viral antigens in nasopharyngeal secretions of children with respiratory syncytial virus infections. J Clin Microbiol 31:819–823

Saito T, Deskin RW, Casola A, Haeberle H, Olszewska B, Ernst PB, Alam R, Ogra PL, Garofalo R (1997) Respiratory syncytial virus induces selective production of the chemokine RANTES by upper airway epithelial cells. J Infect Dis 175:497–504

Sami IR, Piazza FM, Johnson SA, Darnell ME, Ottolini MG, Hemming VG, Prince GA (1995) Systemic immunoprophylaxis of nasal respiratory syncytial virus infection in cotton rats. J Infect Dis 171:440–443

Sarawar SR, Carding SR, Allan W, McMickle A, Fujihashi K, Kiyono H, McGhee JR, Doherty PC (1993) Cytokine profiles of bronchoalveolar lavage cells from mice with influenza pneumonia: consequences of CD4+ and CD8+ T cell depletion. Reg Immunol 5:142–150

Sarawar SR, Sangster M, Coffman RL, Doherty PC (1994) Administration of anti-interferon gamma to β2-microglobulin-deficient mice delays influenza virus clearance but does not switch the response to a T helper cell 2 phenotype. J Immunol 153:1246–1253

Scherle PA, Palladino G, Gerhard W (1992) Mice can recover from pulmonary influenza virus infection in the absence of class I-restricted cytotoxic T cells. J Immunol 148:212–217

Schulman JL, Petrigrow C, Woodruff J (1977) Effects of cell mediated immunity in influenza infections in mice. Dev Biol Stand 39:385–390

Scott R, de Landazuri MO, Gardner PS, Owen JJ (1977) Human antibody-dependent cell-mediated cytotoxicity against target cells infected with respiratory syncytial virus. Clin Exp Immunol 28:19–26

Scott R, Kaul A, Scott M, Chiba Y, Ogra PL (1978) Development of in vitro correlates of cell-mediated immunity to respiratory syncytial virus infection in humans. J Infect Dis 137:810–817

Scott R, Pullan CR, Scott M, McQuillin J (1984) Cell-mediated immunity in respiratory syncytial virus disease. J Med Virol 13:105–114

Sieg S, Xia L, Huang Y, Kaplan D (1995) Specific inhibition of granzyme B by parainfluenza virus type 3. J Virol 69:3538–3541

Sigurs N, Bjarnason R, Sigurbergsson F (1994) Eosinophil cationic protein in nasal secretion and in serum and myeloperoxidase in serum in respiratory syncytial virus bronchiolitis: relation to asthma and atopy. Acta Paediatr 83:1151–1155

Smith TF, McIntosh K, Fishaut M, Henson PM (1981) Activation of complement by cells infected with respiratory syncytial virus. Infect Immun 33:43–48

Smyth MJ, Trapani JA (1998) The relative role of lymphocyte granule exocytosis versus death receptor-mediated cytotoxicity in viral pathophysiology. J Virol 72:1–9

Stark JM, Smith CW, Gruenert DC, Tosi MF (1992) Neutrophil adhesion to parainfluenza virus-infected human airway epithelial cells. Possible contributions of ICAM-1-dependent and ICAM-1-independent mechanisms. Chest 101:40S-41S

Stark JM, Godding V, Sedgwick JB, Busse WW (1996) Respiratory syncytial virus infection enhances neutrophil and eosinophil adhesion to cultured respiratory epithelial cells. Roles of CD18 and intercellular adhesion molecule-1. J Immunol 156:4774–4782

Stein-Streilein J, Guffee J, Fan W (1988) Locally and systemically derived natural killer cells participate in defense against intranasally inoculated influenza virus. Reg Immunol 1:100–105

Tang YW, Graham BS (1994) Anti-IL-4 treatment at immunization modulates cytokine expression, reduces illness, and increases cytotoxic T lymphocyte activity in mice challenged with respiratory syncytial virus. J Clin Invest 94:1953–1958

Taylor G, Stott EJ, Bew M, Fernie BF, Cote PJ, Collins AP, Hughes M, Jebbett J (1984) Monoclonal antibodies protect against respiratory syncytial virus infection in mice. Immunology 52:137–142

Taylor PM, Esquivel F, Askonas BA (1990) Murine CD4+ T cell clones vary in function in vitro and in influenza infection in vivo. Int Immunol 2:323–328

Topham DJ, Tripp RA, Sarawar SR, Sangster MY, Doherty PC (1996) Immune CD4+ T cells promote the clearance of influenza virus from major histocompatibility complex class II -/- respiratory epithelium. J Virol 70:1288–1291

Topham DJ, Doherty PC (1998) Clearance of an influenza A virus by CD4+ T cells is inefficient in the absence of B cells. J Virol 72:882–885

Tosi MF, Stark JM, Hamedani A, Smith CW, Gruenert DC, Huang YT (1992) Intercellular adhesion molecule-1 (ICAM-1)-dependent and ICAM-1-independent adhesive interactions between polymorphonuclear leukocytes and human airway epithelial cells infected with parainfluenza virus type 2. J Immunol 149:3345–3349

Tripp RA, Sarawar SR, Doherty PC (1995) Characteristics of the influenza virus-specific CD8+ T cell response in mice homozygous for disruption of the H-2lAb gene. J Immunol 155:2955–2959

Tsutsumi H, Matsuda K, Yamazaki H, Ogra PL, Chiba S (1995) Different kinetics of antibody responses between IgA and IgG classes in nasopharyngeal secretion in infants and children during primary respiratory syncytial virus infection. Acta Paediatr Jpn 37:464–468

van Haarst JM, de Wit HJ, Drexhage HA, Hoogsteden HC (1994) Distribution and immunophenotype of mononuclear phagocytes and dendritic cells in the human lung. Am J Respir Cell Mol Biol 10:487–492

van Nieuwkerk EB, de Wolf CJ, Kamperdijk EW, van der Baan S (1990) Lymphoid and non-lymphoid cells in the adenoid of children with otitis media with effusion: a comparative study. Clin Exp Immunol 79:233–239

van Nieuwkerk EB, van der Baan S, Richters CD, Kamperdijk EW (1992) Isolation and characterization of dendritic cells from adenoids of children with otitis media with effusion. Clin Exp Immunol 88:345–349

Vasantha S, Coelingh KL, Murphy BR, Dourmashkin RR, Hammer CH, Frank MM, Fries LF (1988) Interactions of a nonneutralizing IgM antibody and complement in parainfluenza virus neutralization. Virology 167:433–441

Virelizier JL (1975) Host defenses against influenza virus: the role of antihemagglutinin antibody. J Immunol 115:434–439

Volovitz B, Welliver RC, De Castro G, Krystofik DA, Ogra PL (1988) The release of leukotrienes in the respiratory tract during infection with respiratory syncytial virus: role in obstructive airway disease. Pediatr Res 24:504–507

Wagner DK, Muelenaer P, Henderson FW, Snyder MH, Reimer CB, Walsh EE, Anderson LJ, Nelson DL, Murphy BR (1989) Serum immunoglobulin G antibody subclass response to respiratory syncytial virus F and G glycoproteins after first, second, and third infections. J Clin Microbiol 27:589–592

Walsh EE, Schlesinger JJ, Brandriss MW (1984) Protection from respiratory syncytial virus infection in cotton rats by passive transfer of monoclonal antibodies. Infect Immun 43:756–758

Watt PJ, Zardis M, Lambden PR (1986) Age related IgG subclass response to respiratory syncytial virus fusion protein in infected infants. Clin Exp Immunol 64:503–509

Welliver RC, Kaul A, Ogra PL (1979) Cell-mediated immune response to respiratory syncytial virus infection: relationship to the development of reactive airway disease. J Pediatr 94:370–375

Welliver RC, Kaul TN, Ogra PL (1980a) The appearance of cell-bound IgE in respiratory-tract epithelium after respiratory-syncytial-virus infection. N Engl J Med 303:1198–1202

Welliver RC, Kaul TN, Putnam TI, Sun M, Riddlesberger K, Ogra PL (1980b) The antibody response to primary and secondary infection with respiratory syncytial virus: kinetics of class-specific responses. J Pediatr 96:808–813

Welliver RC, Wong DT, Sun M, Middleton E, Jr., Vaughan RS, Ogra PL (1981) The development of respiratory syncytial virus-specific IgE and the release of histamine in nasopharyngeal secretions after infection. N Engl J Med 305:841–846

Welliver RC, Sun M, Rinaldo D, Ogra PL (1985) Respiratory syncytial virus-specific IgE responses following infection: evidence for a predominantly mucosal response. Pediatr Res 19:420–424

Welliver RC, Wong DT, Sun M, McCarthy N (1986) Parainfluenza virus bronchiolitis. Epidemiology and pathogenesis. Am J Dis Child 140:34–40

Welliver RC, Sun M, Hildreth SW, Arumugham R, Ogra PL (1989) Respiratory syncytial virus-specific antibody responses in immunoglobulin A and E isotypes to the F and G proteins and to intact virus after natural infection. J Clin Microbiol 27:295–299

Welliver RC, Tristram DA, Batt K, Sun M, Hogerman D, Hildreth S (1994) Respiratory syncytial virus-specific cell-mediated immune responses after vaccination with a purified fusion protein subunit vaccine. J Infect Dis 170:425–428

Wells MA, Ennis FA, Albrecht P (1981) Recovery from a viral respiratory infection. II. Passive transfer of immune spleen cells to mice with influenza pneumonia. J Immunol 126:1042–1046

Whimbey E, Vartivarian SE, Champlin RE, Elting LS, Luna M, Bodey GP (1993) Parainfluenza virus infection in adult bone marrow transplant recipients. Eur J Clin Microbiol Infect Dis 12:699–701

Whimbey E, Elting LS, Couch RB, Lo W, Williams L, Champlin RE, Bodey GP (1994) Influenza A virus infections among hospitalized adult bone marrow transplant recipients. Bone Marrow Transplant 13:437–440

Whimbey E, Couch RB, Englund JA, Andreeff M, Goodrich JM, Raad II, Lewis V, Mirza N, Luna MA, Baxter B (1995) Respiratory syncytial virus pneumonia in hospitalized adult patients with leukemia. Clin Infect Dis 21:376–379

Whimbey E, Champlin RE, Couch RB, Englund JA, Goodrich JM, Raad I, Przepiorka D, Lewis VA, Mirza N, Yousuf H, Tarrand JJ, Bodey GP (1996) Community respiratory virus infections among hospitalized adult bone marrow transplant recipients. Clin Infect Dis 22:778–782

Wong DT, Rosenband M, Hovey K, Ogra PL (1985) Respiratory syncytial virus infection in immunosuppressed animals: implications in human infection. J Med Virol 17:359–370

Wright PF, Murphy BR, Kervina M, Lawrence EM, Phelan MA, Karzon DT (1983) Secretory immunological response after intranasal inactivated influenza A virus vaccinations: evidence for immunoglobulin A memory. Infect Immun 40:1092–1095

Yamazaki H, Tsutsumi H, Matsuda K, Nagai K, Ogra PL, Chiba S (1994) Effect of maternal antibody on IgA antibody response in nasopharyngeal secretion in infants and children during primary respiratory syncytial virus infection. J Gen Virol 75:2115–2119

Yanagihara R, McIntosh K (1980) Secretory immunological response in infants and children to parainfluenza virus types 1 and 2. Infect Immun 30:23–28

Yap KL, Ada GL, McKenzie IF (1978) Transfer of specific cytotoxic T lymphocytes protects mice inoculated with influenza virus. Nature 273:238–239

Yousuf HM, Englund J, Couch R, Rolston K, Luna M, Goodrich J, Lewis V, Mirza NQ, Andreeff M, Koller C, Elting L, Bodey GP, Whimbey E (1997) Influenza among hospitalized adults with leukemia. Clin Infect Dis 24:1095–1099

Bacterial Toxins as Mucosal Adjuvants

L.C. Freytag and J.D. Clements

1 Introduction

The 1995 World Health Organization report of infectious-disease deaths indicated there had been more than 13 million deaths world-wide during that year. The majority of those deaths were caused by organisms that first make contact with and then either colonize or cross mucosal surfaces to infect the host. The overall morbidity caused by these organisms and other pathogens that interact with mucosal surfaces is impossible to calculate.

Traditional vaccine strategies that involve parenteral immunization with inactivated viruses, bacteria or subunits of relevant virulence determinants of those pathogens do not prevent those interactions. In fact, traditional vaccine strategies do not prevent infection, but resolve infection before disease ensues. In some cases once a virus crosses the mucosal surface and enters the host cell, be that a dendritic cell, an epithelial cell or a T-cell, the host-parasite relationship is moved decidedly

Department of Microbiology and Immunology, Tulane University Medical Center, 1430 Tulane Avenue, New Orleans, LA 70112, USA

in favor of the parasite; HIV would be one example. In this case, as in many others, a vaccine strategy that does not prevent the initial infection of the host is unlikely to succeed.

Recently, a great deal of attention has focused on mucosal immunization as a means of inducing secretory immunoglobulin A (S-IgA) antibodies, directed against specific pathogens of mucosal surfaces. The rationale for this is the recognition that S-IgA constitutes greater than 80% of all antibodies produced in mucosal-associated lymphoid tissues in humans and S-IgA may block attachment of bacteria and viruses. This would neutralize bacterial toxins and even inactivate invading viruses inside epithelial cells. In addition, the existence of a Common Mucosal Immune System permits immunization on one mucosal surface to induce secretion of antigen-specific S-IgA at distant mucosal sites. It is now appreciated that mucosal immunization may be an effective means of inducing not only S-IgA, but also systemic antibody and cell-mediated immunity.

The mucosal immune response can be divided into two phases (McGhee and Kiyono 1993). The inductive phase involves antigen presentation and the initiation events which dictate the subsequent immune response. During the initiation events, antigen-specific lymphocytes are primed and migrate from the inductive sites, e.g., Peyer's patches (PP) in the enteric mucosa, through the regional lymph nodes, into the circulation and back to mucosal effector sites, e.g., lamina propria (LP). Once these effector cells have seeded their effector sites, the second phase or effector phase of the mucosal immune response can occur. A significant difference between mucosal immunization and parenteral immunization is that both mucosal and systemic immunity can be induced by mucosal immunization. Parenteral immunization generally results only in systemic responses.

Most studies on the mucosal immune response conducted to date have dealt with the secretory antibody component of the mucosal response and the complex regulatory issues involved with induction of S-IgA following mucosal immunization. These have not dealt with the systemic antibody response or cellular immunity induced by mucosal immunization. Thus, it is important to understand the type of helper T lymphocyte response induced by mucosal immunization, since the type of helper T lymphocyte stimulated by an antigen is one of the most important factors for defining which type of immune response will follow. At least two different types of helper T lymphocytes (Th) which can be distinguished based on cytokine secretion have been identified in mice (Cherwinski et al. 1987; Mosmann and Coffman 1989), humans (Romagnani 1991) and other animal species (Brown et al. 1994). Th1 lymphocytes secrete substantial amounts of interleukin 2 (IL-2) and interferon γ (IFN-γ) and execute cell-mediated immune responses, e.g., delayed-type hypersensitivity and macrophage activation, whereas Th2 lymphocytes secrete IL-4, IL-5, IL-6 and IL-10 and assist in antibody production for humoral immunity. Theoretically, antigenic stimulation of one T helper cell subset and not the other would result in production of a particular set of cytokines which would define the resulting immune response.

The presence of these cytokines coupled with an antigenic stimulus presented by macrophages in the context of class-II MHC molecules can initiate a Th1-type response. The ability of Th1 cells to secrete IL-2 and IFN-γ further amplifies the response by activating Th1 cells in an autocrine fashion and macrophages in a paracrine fashion. These activated leukocytes can release additional cytokines, e.g., IL-6, which may induce the proliferation and differentiation of antigen specific B lymphocytes to secrete antibodies (the effector phase). In this scenario, the predominant isotype secreted by murine B lymphocytes is often IgG2a. In a second scenario (URBAN et al. 1992), antigens such as allergens or parasites can effectively stimulate a Th2-lymphocyte response (the inductive phase). Presentation of such antigens to Th2 cells can result in the production of the lymphokines IL-4 and IL-5 This can induce antigen-specific B lymphocytes to secrete IgE and IgG1 or induce eosinophilia, respectively. Again, this is the effector phase. Furthermore, stimulated Th2 cells can secrete IL-10, which has the ability to specifically inhibit secretion of IL-2 and IFN-γ by Th1 lymphocytes and also inhibit macrophage function.

While these representations are simplistic, it is obvious that the type of T helper cell stimulated affects the resultant cellular immune response as well as the predominant Ig isotype secreted. Specifically, IL-4 stimulates switching to the IgE and IgG1 isotypes, whereas IFN-γ stimulates IgG2a secretion. Numerous studies, predominantly conducted in vitro, have suggested that IL-5, IL-6 and transforming growth factor beta (TGF-β) can cause isotype switching to IgA.

2 Bacterial Enterotoxins as Mucosal Adjuvants

Mucosally administered antigens are frequently not immunogenic. A number of strategies have been developed to facilitate and enhance the immune response obtained after mucosal immunization. Among these strategies are the use of attenuated mutants of bacteria, i.e., *Salmonella* spp., as carriers of heterologous antigens, encapsulation of antigens into microspheres, gelatin capsules, different formulations of liposomes, adsorption onto nanoparticles, use of lipophilic immune-stimulating complexes, and addition of bacterial products with known adjuvant properties. While a number of substances of bacterial origin have been tested as mucosal adjuvants (LOWELL et al. 1997; ROBERTS et al. 1995; VAN DE VERG et al. 1996), it is clear that the two bacterial proteins with the greatest potential to function as mucosal adjuvants are cholera toxin (CT) produced by various strains of *Vibrio cholerae* and the heat-labile enterotoxin (LT) produced by some enterotoxigenic strains of *Escherichia coli* (CLEMENTS et al. 1988; ELSON 1989; LYCKE et al. 1992; XU-AMANO et al. 1993; YAMAMOTO et al. 1996).

Although LT and CT have many features in common, these are clearly distinct molecules with biochemical and immunological differences, which make them unique. Both LT and CT are synthesized as multi-subunit toxins with A and B components. On thiol reduction, the A component dissociates into two smaller polypeptide chains. One of these, the A1 piece, catalyzes the ADP-ribosylation of the stimulatory GTP-binding protein GSα in the adenylate cyclase enzyme complex on the basolateral surface of the epithelial cell, resulting in increasing intracellular levels of cAMP. This increase in cAMP causes secretion of water and electrolytes into the small intestine through interaction with two cAMP-sensitive ion-transport mechanisms involving NaCl cotransport across the brush border of villous epithelial cells and electrogenic Na-dependent Cl secretion by crypt cells (FIELD 1980). The B subunit binds to the host cell-membrane receptor ganglioside GM1 and facilitates the translocation of the A subunit through the cell membrane.

Recent studies have examined the potential of CT and LT as mucosal adjuvants against a variety of bacterial and viral pathogens using whole killed organisms or purified subunits of relevant virulence determinants from these organisms. Representative examples include tetanus toxoid (XU-AMANO et al. 1994; XU-AMANO et al. 1993; YAMAMOTO et al. 1996), inactivated influenza virus (HASHIGUCCI et al. 1996; KATZ et al. 1996, 1997), recombinant urease from *Helicobacter* spp. (LEE et al. 1995; WELTZIN et al. 1997), pneumococcal surface protein A from *Streptococcus pneumoniae* (WU et al. 1997), Norwalk virus capsid protein (MASON et al. 1996), synthetic peptides from measles virus (HATHAWAY et al. 1996) and the HIV-1 C4/V3 peptide T1SP10 MN(A) (STAATS et al. 1996). There are many other examples and it is clear from these studies that both LT and CT have significant potential for use as adjuvants for mucosally administered antigens. This raises the possibility of an effective immunization program against a variety of pathogens involving the mucosal administration of killed or attenuated organisms or relevant virulence determinants of specific agents in conjunction with LT or CT. However, the fact that these toxins stimulate a net luminal secretory response may prevent their use for practical vaccine applications. For instance, it was observed that as little as 5μg of purified CT, administered orally, was sufficient to induce significant diarrhea in human volunteers, while ingestion of 25μg of CT elicited a full 20-l cholera purge (LEVINE et al. 1983)

In recently conducted volunteer studies with LT administered alone or in conjunction with the *V. cholerae* whole-cell/B-subunit vaccine, LT was shown to induce fluid secretion at doses as low as 2.5μg when administered in conjunction with the vaccine, while 25μg of LT elicited up to 6-l of fluid secretion. While the adjuvant effective dose in humans for either of these toxins has not been established, experiments in animals suggest that it may be comparable to the toxic dose. Taken together, these studies suggest that while LT and CT may be attractive as mucosal adjuvants, studies in animals do not reflect the full toxic potential of these molecules in humans, and that toxicity may seriously limit their practical use.

3 Proposed Mechanisms of Adjuvanticity

The phenomenon known as mucosal adjuvanticity is likely to be the final result of a complex series of molecular events that involve several populations of cells and an intricate network of cell-to-cell communication and interactions. The molecular interactions and the cellular targets through which CT and LT mediate their adjuvant properties are not completely understood, although significant efforts have been expended to answer this question (BROMANDER et al. 1993; CEBRA et al. 1986; CLARKE et al. 1991; CLEMENTS et al. 1988; ELSON 1989; ELSON and EALDING 1984a, b; ELSON et al. 1995; HORNQUIST and LYCKE 1993; LYCKE et al. 1991; NEDRUD and SIGMUND 1991; SNIDER et al. 1994; TAKAHASHI et al. 1996; XU-AMANO et al. 1993, 1994). Several models have been proposed, none of which fully accounts for the range of immunological phenomena observed upon mucosal administration of CT or LT.

3.1 Enhanced Luminal Permeability

LYCKE et al. (1991) have proposed that the adjuvanticity of CT and LT is the result of their ability to stimulate hyperabsorption by the luminal enterocytes of the small intestine; a direct consequence of the ability these enterotoxins have to activate adenylate cyclase and stimulate net ion and water secretion into the lumen. In that study, a permeability marker, fluorescent Dextran 3000, was administered orally to mice in conjunction with the protein antigen Keyhole Limpet Hemocyanin (KLH), with or without CT or CT-B as an adjuvant. At various times after oral inoculation, blood was withdrawn and analyzed spectrophotometrically for the presence of FITC-labeled Dextran. An increase in Dextran uptake into the serum was only observed when CT was used as an adjuvant and was correlated with an adjuvant effect as evidenced by significant serum anti-KLH antibody responses and an increased number of anti-KLH spot forming cells (SFCs) in the LP as determined by ELISPOT analysis. These investigators concluded that the adjuvant effect of CT is associated with its ability to increase gut permeability, thereby facilitating access of luminal antigens to the gut mucosal immune system. This effect was attributed to the adenylate cyclase/cAMP activation since CT-B, which does not activate cAMP, did not act as an adjuvant for KLH and was unable to increase uptake of the marker. The augmentation of gut permeability was strongest after the first oral inoculation and considerably lower after the second and third inoculations, suggesting a possible anti-toxin effect. The investigators made it clear that the ability of CT to augment the uptake of a "protein" antigen remain to be determined.

Moreover, small-molecular-weight sugars, such as dextran, are generally nonimmunogenic and are constituents of both host-cell glycoproteins and glycoproteins associated with the normal flora. The biological relevance of dextran's enhanced uptake into the general circulation is not clear. In fact, dextran has been found to access the general circulation by endocytosis–exocytosis of the enterocyte or by a paracellular pathway, the latter of which is thought to be facilitated by CT (VOLKHEIMER and SCHULTZ 1968). However, if this is a component of the adjuvant mechanism of CT, it would be expected that this increased uptake from lumen to serosa would be discriminatory. To this degree, NEDRUD and SIGMUND (1991) found that oral immunization with CT does not enhance uptake of dietary antigens. Rather, the adjuvant effect of CT has been found to be restricted to the specific immunogen with which it is co-administered and not to bystander antigens.

3.2 Selective Induction of a Th2-Mediated Humoral Antibody Response

In contrast to the findings of Lycke et al. (1991), Snider et al. (1994) found that CT did not increase uptake of orally administered hen egg lysozyme (HEL) into the peripheral circulation. In that study, mice were inoculated orally with 200 mg HEL or HEL with CT as an adjuvant and their sera was examined for the presence of intact HEL. Co-administration of CT had no effect in that at no point following oral inoculation did the serum HEL levels exceed 200 ng/ml in either group. These investigators demonstrate that CT stimulates antibody production when given orally with HEL, whereas a similar dose of HEL alone does not. They argue that, since there is no difference in the amount of circulating HEL in either case, CT must be inducing antibody production by a means other than increased HEL uptake. Hence, when a relatively large molecular weight (14,700) immunogenic antigen is used in combination with CT as an adjuvant, adjuvanticity cannot be explained by a simple increased uptake of antigen from the lumen into the general circulation.

In the same study, the authors proposed two distinct mechanisms of adjuvanticity. First, the adjuvant effect could be attributed to a combinatorial effect on B cells and T helper cells or on antigen-presenting cells (APCs), such as macrophages or intestinal epithelial cells themselves. Alternatively, CT may increase antigen uptake preferentially into compartments of the intestinal mucosa, such as the LP. The antigen would be sequestered in this area and, consequently, an estimate of the amount of antigen in the general circulation would not be a precise reflection of the relative amounts of antigen uptake in the presence of CT. This would suggest that antigen in the presence of CT is manipulated differently by the gut-associated lymphoid tissue (GALT) than antigen administered alone. An antigen co-administered with CT or LT follows a different course of antigen presentation, resulting in the generation of a specific immune response in contrast to antigens processed in the absence of an adjuvant.

3.3 Depletion of $CD8^+$ Intraepithelial Lymphocytes

A recent model for adjuvanticity has advanced from an evolving understanding of the effects of CT and LT on the induction of oral tolerance. This model proposes that CT and LT abrogate the induction of oral tolerance by depletion of a "suppressor" T cell population in the GALT. Elson et al. (1995) have proposed that adjuvanticity can be explained in part by the profound effect of CT on regulatory T cells in the mucosal immune system, specifically by depleting the $CD8^+$ intraepithelial lymphocyte (IEL) population upon oral administration. In that study, splenic T cells were cultured in the presence of concanavalin A alone, with CT or with CT-B. The cells were then analyzed by means of flow cytometry for the proportions of $CD3^+$, $CD4^+$ and $CD8^+$ T cells. There was a preferential reduction in the $CD8^+$ cell fraction, although $CD4^+$ cells were also depleted in vitro. Similarly, when $CD4^+$ and $CD8^+$ T-cell subsets were isolated and incubated with phorbol myristic acetate and ionomycin (previously demonstrated to be optimal for mitogenic activity), with and without CT or CT-B, both subsets were significantly inhibited. There was a consistently greater inhibition of $CD8^+$ T cells than $CD4^+$ T cells, as determined by ^{3}H-thymidine incorporation.

Importantly, these investigators examined the relevance of this selective depletion in vivo using an adoptive transfer model. Mice were immunized with ovalbumin (OVA), as an irrelevant antigen or with KLH, both with and without CT as an adjuvant. Splenic T cells were purified from these donor mice and transferred to a naive population of mice. These mice were then immunized with three doses of KLH at biweekly intervals by i.p. inoculation for the first and last inoculation, and by a single intragastric administration separating the two parenteral immunizations. In the control group receiving T cells from OVA-fed mice, there was a strong serum IgG and secretory anti-KLH response, demonstrating that systemic tolerance is an antigen-specific phenomenon. In contrast, the anti-KLH response was significantly reduced in the recipients receiving T cells from the KLH-immunized donors. This demonstrated that suppressor T cells in these donors could be transferred to naive donors and induce tolerance in response to KLH exposure. However, including CT in the oral inoculations inhibited the development of suppressor cells in these mice, such that the donor mice produced an anti-KLH response similar to the control group.

These investigators concluded that both CT and CT-B are able to reduce the IEL numbers in vitro. This suggested that the inhibitory signal to the T cell was neither mediated by the adenylate cyclase pathway, nor the phosphatidylinositol system, both of which require the ADP-ribosylating activity of the holotoxin. Rather, they hypothesized that both CT and CT-B are endocytosed and possibly transcytosed by T cells and that this provides the inhibitory signal required for oral tolerance as well as for adjuvanticity.

3.4 Antigen-Specific T-Cell Response Induced by Cholera Toxin as an Oral Adjuvant

While CT and LT are physicochemically and immunologically related, they have been found to induce different cytokine profiles in vitro and in vivo. Hence, they induce different immune responses to co-administered antigens. In this regard, Xu-Amano et al. (1994) have extensively analyzed the Th1 and Th2 cell responses following oral immunization with CT. In this study, mice were orally inoculated with CT three times at weekly intervals and were sacrificed 1 week after the last immunization. High titers of anti-CT IgG and IgA antibodies were found in serum and fecal extracts, respectively. This correlated with local anti-CT IgA-secreting LP lymphocytes (B cells) as determined by a modified ELISPOT. To examine the effect of CT on Th-cell responses during this peak, PP and spleens (SP) were collected from these mice and the CD4+ T-cell fractions purified and analyzed for specific cytokine expression.

Using an in vitro re-stimulation assay, CD4+ T cells were incubated with CT-B-coated latex microspheres for 1 day, 3 days and 6 days, and then analyzed by ELISPOT for IL-2, IFN-γ, IL-4 and IL-5 secretion. In the CT-immunized mice, high levels of IL-4 and IL-5 SFCs were observed in CD4+ T cells isolated from PP and SP, while IL-2 and IFN-γ SFCs were scarce. These findings were confirmed by reverse-transcribed polymerase chain reaction (RT-PCR) of total cellular RNA collected from the CD4+ T-cell fractions of PP and SP cells of immunized mice re-stimulated in vitro with CT-B-coated microspheres. Following re-stimulation, high levels of IL-4 and IL-5 mRNA were detectable, while low to undetectable levels of IL-2 and IFN-γ mRNA were observed. This indicated that Th2-type responses induced by oral immunization were due to de novo synthesis of cytokines and that induced cytokine response was due to CT-specific CD4+ T cells, both in the PP and in the SP. These findings are significant in that they suggest Th2 cell regulation of S-IgA response in vivo and may, in part, explain how CT acts as an adjuvant. It is important to determine whether the same antibody and cytokine profile is initiated when CT is used as an adjuvant for a given antigen. Xu-Amano et al. (1993) also investigated the Th cell subsets induced upon oral immunization with tetanus toxoid (TT) as an antigen and with CT as an adjuvant. Similarly, they found that an antigen-specific Th2-type response was generated using similar techniques of re-stimulation assays with TT-coated microspheres and cytokine-specific Northern blots of RT-PCR products from re-stimulated CD4+ T cells isolated from PP and SP of orally immunized mice.

In contrast to the above findings, Hornquist and Lycke (1993) have observed that CT greatly promotes antigen priming of both Th1 and Th2 cells. In this study, mice were administered a single oral priming of KLH, with or without CT as an adjuvant. One week later, SP, mesenteric lymph nodes (MLN), PP and LP T cells were cultured in vitro with KLH and peritoneal macrophages from naive syngeneic mice as APCs. Cellular proliferation, in response to the recall antigen, was determined by incorporation of ^{3}H-thymidine into cellular DNA. CT was found to promote antigen-specific priming of T cells derived from both mucosal tissues (LP and PP) and systemic organs (MLN and SP). This was in contrast to KLH-primed T cells which responded poorly to re-stimulation with KLH. Further, MLN, LP and SP T cells from KLH- and CT-inoculated mice produced elevated amounts of both Th1 cytokines (IL-2 and IFN-γ) and Th2 cytokines (IL-4, 5 and 6) on re-stimulation compared with KLH-primed mice in which all tested cytokines were low or undetectable. Finally, it was found that KLH-specific CD8+ T cells did not contribute significantly to the CT-enhanced proliferation; CT caused no significant change in the frequencies of either CD4+ or CD8+ T cells in the SP or MLN, relative to the distribution observed in KLH-primed mice, as determined by fluorescence-activated cell sorter (FACS) analysis.

This study differed from previous studies in that the investigators purposefully separated the antigen-primed T cells from CT-exposed APCs, including both macrophages and B cells, prior to re-exposure of T cells to the recall antigen in vitro. This is significant because previous studies have indicated that CT preferentially induces a Th2 profile upon oral immunization (CLARKE et al. 1991; WILSON et al. 1991). Other groups did not separate immune T cells from the primed APCs in their in vitro re-stimulation assays. It is important to note that in vitro re-stimulation assays are useful in determining the effect of CT and LT as adjuvants on single, isolated lymphocyte populations. However, this assay system is artificial in the sense that lymphocytes are removed from the immunological milieu in which they are initially primed. Hence, the influence of other antigen-primed cell populations, such as B cells, macrophages, dendritic cells, IEL and epithelial cells on T-cell responses are not present as would be in a natural in vivo situation.

3.5 Role of Enterocytes in the Adjvuanticity of CT and LT

A role for intestinal epithelial cells in antigen presentation has recently been proposed to the extent that they constitute the vast majority of cells in the small intestine. In contrast, M cells overlying the PP constitute only a small percentage of cells in the small bowel. Current dogma suggests that the M cells are the predominant antigen-sampling cells of the gut which then pass an internalized antigen to the macrophage cells of the PP. These then serve as the major APCs that initiate T and B cell priming in response to oral immunization.

An alternative pathway of antigen uptake and presentation, provided by the absorptive epithelial cells of the intestinal mucosa, has been suggested by BROMANDER et al. (1993). These investigators analyzed the effect of CT exposure on alloantigen presentation by cultured intestinal epithelial cells. The crypt, small intestinal epithelial cell line, IEC-17, derived from the duodenum of SPRAGUE-DAWLEY rats was cultured with IFN-γ to induce optimal expression of MHC II in the presence or absence of CT (QUARONI and ISSELBACHRE 1981). After 48 h, the IEC-17 cells were washed, treated with mitomycin C and incubated with histoincompatible spleen cells. Alloantigen-stimulated CD4+ T-cell proliferation was then determined by ^{3}H-thymidine incorporation. CT enhanced the antigen presentation of IEC-17 cells. This enhancement was not due to an upregulation of MHC-II expression, but rather a dose-dependent increase in IL-1 and IL-6 secretion by IEC-17 cells. These data suggest that the potent adjuvanticity of CT may be attributed to its ability to enhance the co-stimulating ability of APCs, such as the intestinal epithelial cells of the mucosal immune system. However, it was noted that intestinal epithelial cells have only been found to process and present antigens in vitro and that whether these cells are capable of functioning as APCs in vivo remains to be examined.

Two limitations of the above study are the relevance of using a crypt cell line (IEC-17) rather than a non-crypt intestinal cell line, such as Henle-407 cells, and the nature of the antigen that is presented. Specifically, alloantigen presentation is an atypical form of antigen presentation that does not require presentation in the context of self MHC. Therefore, the enhanced "presentation" observed in this study might not accurately reflect the conventional antigen processing and presenting capability of intestinal epithelial cells in vitro.

3.6 Combined Effect of More Than One Mechanism

Rather than a single defined mechanism, adjuvanticity should be viewed as an outcome and not an event. It is likely to be some combination of effects that collectively results in the outcome observed as enhanced immunity or adjuvanticity. It is also important to note that most studies attempting to define the mechanism of adjuvanticity of CT and LT focus on induction of S-IgA and Th2 events as the only or most relevant outcomes. These studies ignore the Th1 induction potential of LT which is likely to be important for protection against intracellular bacterial pathogens and viruses. Moreover, any attempt to define the mechanism of action of CT or LT outside the context of the antigen with which it is applied is likely to be unproductive. Clearly, a gram-negative bacterium, such as *Salmonella* or *Shigella* makes a contribution to the immunological outcome by virtue of its lipopolysaccharide (LPS) content. This is different than the contribution to the outcome made by a viral glycoprotein which may have inherent immunomodulating properties of its own.

4 Development of a Safe, Effective Mucosal Adjuvant

A number of attempts have been made to alter the toxicity of LT and CT, most of which have focused on eliminating enzymatic activity of the A-subunit associated with enterotoxicity. This has involved the use of site-directed mutagenesis to change amino acids associated with the crevice where nicotinamide adenine dinucleotide (NAD) binding and catalysis is thought to occur. Recently, a model for NAD binding and catalysis was proposed (Domenighini et al. 1994; Pizza et al. 1994), based on computer analysis of the crystallographic structure of LT (Sixma et al. 1991, 1993). Replacement of any amino acid in CT or LT involved in NAD-binding and catalysis by site-directed mutagenesis has been shown to alter ADP-ribosyltransferase activity with a corresponding loss of toxicity in a variety of biological assay systems (Burnette et al. 1991; Fontana et al. 1995; Harford et al. 1989; Haese et al. 1994; Lobet et al. 1991; Lycke et al. 1992; Merritt et al. 1995; Moss et al. 1993; Pizza et al. 1994; Tsuji et al. 1990, 1991; Yamamoto et al. 1997a,b). The adjuvanticity potential of some of these mutants has been tested on animal models using a variety of co-administered antigens (DiTommaso et al. 1996; Lycke et al. 1992; Partidos et al. 1996; Yamamoto et al. 1997a,b). In addition, it has been shown that exchanging K for E112 in LT not only removes ADP-ribosylating enzymatic activity, but cAMP activation and adjuvant activity as well (Lycke et al.

1992). A logical conclusion from the LYCKE et al. (1992) studies was that ADP-ribosylation and induction of cAMP are essential for the adjuvant activity of these molecules. As a result, a causal linkage was established between adjuvanticity and enterotoxicity. That is, the accumulation of cAMP responsible for net ion and fluid secretion into the gut lumen was thought to be a requisite to adjuvanticity. Recent studies by a number of laboratories have challenged that linkage.

DICKINSON and CLEMENTS (1995) explored an alternative approach to dissociation of enterotoxicity from adjuvanticity. Like other bacterial toxins that are members of the A-B toxin family, both CT and LT require proteolysis of a trypsin sensitive bond to become fully active. In these two enterotoxins, that trypsin-sensitive peptide is subtended by a disulfide interchange that joins the A1 and A2 pieces of the A-subunit. In theory, if the A1 and A2 pieces cannot separate, A1 may not be able to find its target (adenylate cyclase) on the basolateral surface or may not assume the conformation necessary to bind or hydrolyze NAD.

A mutant of LT was constructed using site-directed mutagenesis to create a single amino acid substitution within the disulfide subtended region of the A subunit separating A1 from A2. This single amino acid change altered the proteolytically sensitive site within this region, rendering the mutant insensitive to trypsin activation. The physical characteristics of this mutant were examined by SDS-PAGE, its biological activity was examined on mouse Y-1 adrenal tumor cells and Caco-2 cells, its enzymatic properties determined in an in vitro NAD–agmatine ADP-ribosyltransferase assay, and its immunogenicity and immunomodulating capabilities determined by testing for the retention of immunogenicity and adjuvanticity. This mutant LT, designated LT(R192G), has been shown to be an effective mucosal adjuvant and has recently been evaluated in two phase-I safety studies (OPLIGER et al. 1997; TRIBBLE et al. 1997). The properties of LT(R192G) are shown in Table 1.

TSUJI et al. (1997) recently demonstrated that an equivalent protease-site deletion mutant LT(Δ192–194) also lacks in vitro ADP-ribosylagmatine activity, has a 10-fold reduction in enterotoxicity in rabbit ligated ileal loops and a 50% reduction and delayed onset of cAMP induction in cultured myeloma cells. LT(Δ192–194) was shown to have increased adjuvant activity for induction of serum IgG and mucosal IgA against measles virus when compared with native LT, LT-B or LT(E112 K). LT(Δ192–194) was effective when administered intranasally, subcutaneously, intraperitoneally or orally, although mucosal IgA responses were only demonstrated following mucosal administration. These investigators also demonstrated increased adjuvant activity for mucosally administered LT(Δ192–194) in conjunction with KLH, BCG and OVA. These findings substantiate the findings with LT(R192G).

Efficacy studies for LT(R192G) in appropriate animal models have demonstrated that LT(R192G) can function as an effective mucosal adjuvant for killed whole bacteria (*Salmonella* spp., *Vibrio cholerae*, *Escherichia coli*, *Shigella* spp., *Campylobacter* and *Helicobacter*) and a number of inactivated viruses of importance to both human and veterinary medicine (Sendai, Influenza, Herpes and TGEV).

Table 1. Properties of LT(R192G)

100 to 1000-fold less active than cholera toxin or native heat-labile enterotoxin on mouse Y-1 adrenal tumor cells
Not sensitive to proteolytic activation
Does not possess detectable in vitro NAD:agmatine ADP-ribosyltransferase activity
Does not increase production of cAMP in cultured Caco-2 cells
Reduced enterotoxicity in the patent mouse intestinal challenge model
Promotes the development of both humoral (antibody) and cell-mediated immune responses against co-administered antigens in both the systemic and mucosal compartments
Functions as an effective adjuvant when administered mucosally (i.e., orally, intranasally) or parenterally (i.e., subcutaneously)
Lacks enterotoxicity in humans at adjuvant-effective doses

NAD, nicotinamide adenine dinucleotide; *ADP*, adenosine diphosphate; *cAMP*, cyclic adenosine monophosphate

5 Use of LT(R192G) as a Mucosal Adjuvant for Protection Against Lethal Oral Challenge by *Salmonella*

A series of experiments in mice were conducted to determine the level of protection against lethal oral challenge by *Salmonella* spp. following oral immunization with killed whole bacteria in conjunction with LT(R192G). For these studies, mice were immunized orally with viable, attenuated *Salmonella dublin* (SL1438) or with killed *S. dublin*, and their humoral and cellular immune responses during the effector phase of the immune response examined. In addition, some groups of mice were immunized orally with killed *S. dublin* in conjunction with LT(R192G). Following two oral immunizations, mice were challenged with approximately 100 LD_{50} of the virulent wild-type parent strain. Results from these experiments showed that animals orally immunized with viable, attenuated *S. dublin* were protected against lethal oral challenge with approximately 100 LD_{50} of the virulent wild-type parent strain, while animals orally immunized with killed *S. dublin* were not protected. When LT(R192G) was included as an adjuvant with killed *S. dublin*, animals were solidly protected against lethal oral challenge (Fig. 1.).

We then examined both the humoral and cellular immunological parameters associated with this immunization strategy. Systemic and mucosal anti-*Salmonella* antibody responses were examined by ELISA and mononuclear cells from MLN- and SP were examined for production of Th1 and Th2-type cytokines in an in vitro antigen re-stimulation assay.

As shown in Table 2, animals immunized orally with viable *S. dublin* SL1438 had a fourfold or greater increase in both serum anti-LPS IgG and fecal anti-LPS IgA compared with animals immunized orally with killed SL1438 in conjunction with LT(R192G) or with LT(R192G) alone. Animals immunized orally with killed SL1438 in conjunction with LT(R192G) demonstrated a fourfold increase in serum anti-LPS IgG, but not in fecal anti-LPS IgA when compared with animals immunized orally with killed SL1438 or with LT(R192G) alone. These findings indicate that there is a correlation between serum anti-LPS IgG antibody responses

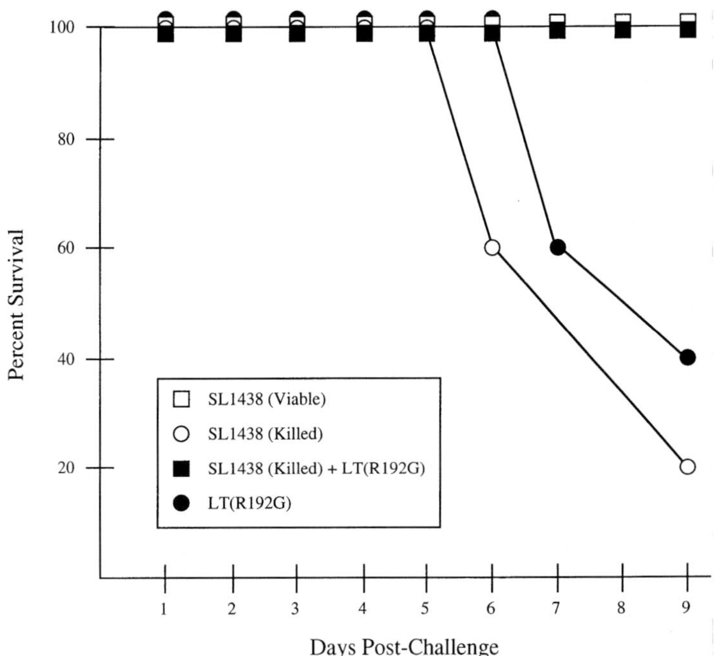

Fig. 1. Survival of BALB/c mice orally immunized with killed SL1438 in conjunction with LT(R192G). Mice were immunized orally with 1×10^{10} cells of either viable SL1438, ethanol-killed SL1438 or ethanol-killed SL1438 in conjunction with 25µg LT(R192G) or 25µg LT(R192G) alone, on day 0 and day 7. Mice were challenged orally with 1×10^7 CFU of *S. dublin* SL1363 (approximately 100 LD_{50}) on day 14, and their survival was monitored. Each group consisted of five mice [from Chong et al. (1998)]

Table 2. Serum and fecal antibody response to *S. dublin* LPS following oral immunization

Immunogen	GMT[a] Serum anti-LPS IgG	GMT Fecal anti-LPS IgA
Viable SL1438	512	32
Killed SL1438	32	4
Killed SL1438 + LT(R192G)	128	8
LT(R192G)	<32	4

[a]Antibody levels are expressed as Geometric Mean Titer (GMT) for an optical density of 0.1, calculated from a plot of the mean responses of 5–6 animals through eight doubling dilutions [from Chong et al. (1998)]

and adjuvant-induced protection in orally immunized mice. Specifically, animals orally immunized with killed organisms in conjunction with LT(R192G) were protected against lethal challenge and had greater levels of serum anti-LPS IgG, but not fecal IgA than animals immunized orally with killed organisms. We next investigated potential cellular correlates of adjuvant-induced protection by examining antigen-induced mononuclear cell production of cytokines.

As seen in Figs. 2A and 3A, significant antigen-induced IL-2 was detected only in culture supernatants of mononuclear cells from the MLN (Fig. 2A) and SP (Fig. 3A) of mice orally immunized with killed SL1438 in conjunction with LT(R192G). There was little or no antigen-induced IL-2 detected in culture supernatants of mononuclear cells from animals immunized with viable SL1438 or killed SL1438 or with LT(R192G) alone. Likewise, there was no antigen-induced IL-4 detected in culture supernatants of mononuclear cells from any immunization group (data not shown).

Mononuclear cells from the MLN of mice orally immunized with viable SL1438 produced higher amounts of antigen-induced IFN-γ (Fig. 2B) than mononuclear cells of mice orally immunized with killed SL1438. However, mononuclear cells from the MLN of mice orally immunized with killed SL1438 in conjunction with LT(R192G) had 5- and 14-fold higher levels of antigen-induced IFN-γ compared with mononuclear cells of mice orally immunized with viable SL1438 or killed SL1438, respectively. Mononuclear cells from the MLN of mice orally immunized with LT(R192G) alone produced antigen-induced IFN-γ at levels similar to those of mononuclear cells from mice orally immunized with killed SL1438. As shown in Fig. 2C, mononuclear cells from the MLN of mice orally immunized with viable SL1438, killed SL1438 in conjunction with LT(R192G) or LT(R192G) alone produced similar amounts of antigen-induced IL-10.

As with MLN cells, the levels of antigen-induced IFN-γ from mononuclear cells from the SP (Fig. 3B) of mice orally immunized with viable SL1438 were higher than those in mice orally immunized with killed SL1438. Likewise, mononuclear cells of mice orally immunized with killed SL1438 in conjunction with LT(R192G) had higher levels of antigen-induced IFN-γ (Fig. 3B) than mice orally immunized with killed SL1438. The levels of antigen-induced IL-10 (Fig. 3C) in mononuclear cells were highly variable among groups of mice. There was a marked increase in production of antigen-induced IL-10 by mononuclear cells from animals immunized orally with LT(R192G) alone.

Since IFN-γ and IL-10 are able to cross-regulate the development of helper T cells, i.e., Th1 and Th2, we also examined the ratio of antigen-induced IFN-γ to IL-10 from cells isolated from the MLN and spleen. Mononuclear cells isolated from the MLN and SP of mice orally immunized with viable SL1438 and killed SL1438 in conjunction with LT(R192G) produced higher IFN-γ to IL-10 ratios than mononuclear cells of mice orally immunized with killed SL1438 or LT(R192G) alone (Fig. 4).

Fig. 2a–c. Production of cytokines by mononuclear cells isolated from the mesenteric lymph nodes (MLN) of orally immunized mice. BALB/c mice were orally immunized on day 0 and day 7 as in Fig. 1. On day 14, the mice were sacrificed and mononuclear cells from the MLN were isolated and pooled by group. The pooled cells were used in an in vitro antigen-re-stimulation assay and were cultured with killed SL1438 as the antigen (Ag), or with complete medium as a negative control in the presence of peritoneal macrophages as antigen presenting cells. Culture supernatants were collected on day 1 (*striped bars*) and 3 (*open bars*) post-culture and murine cytokines analyzed by ELISA. The data are representative of three experiments. Each group consisted of eight mice [from Chong et al. (1998)]

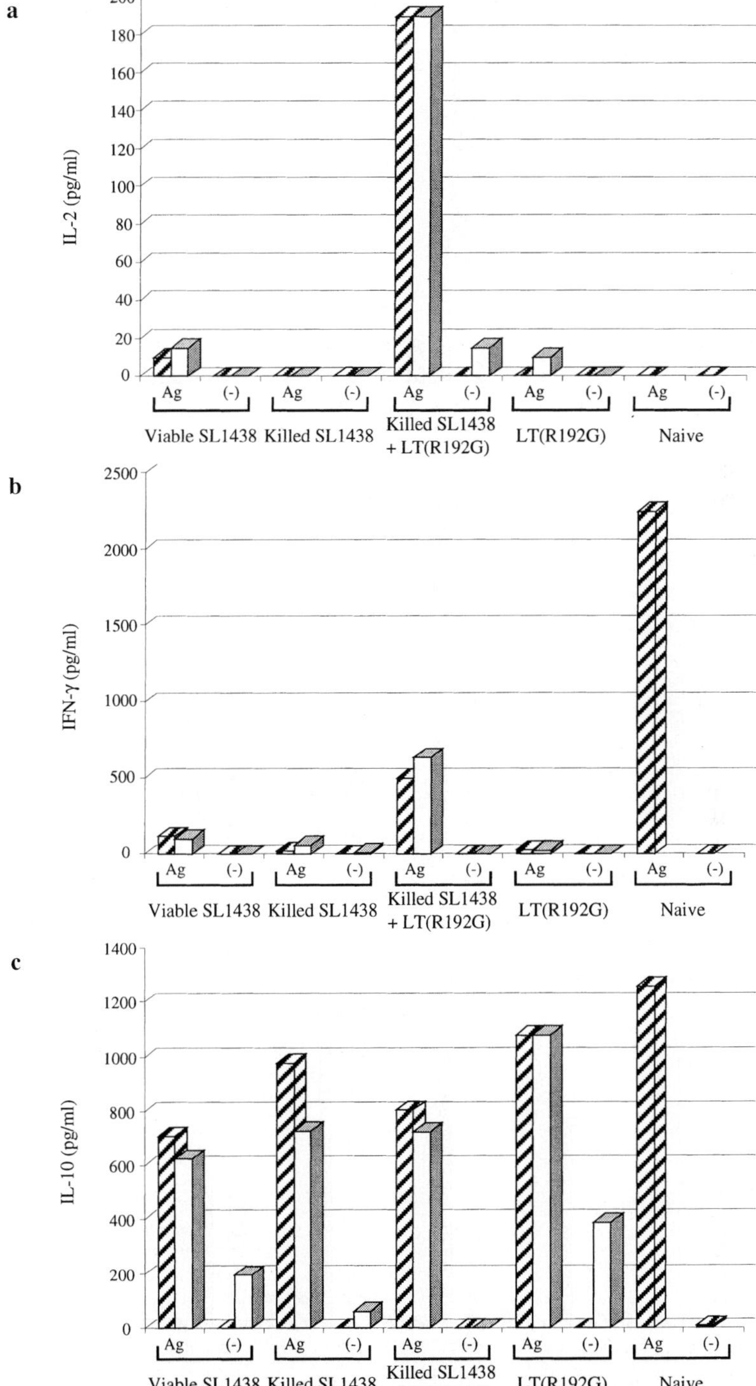
a
IL-2 (pg/ml)
200
180
160
140
120
100
80
60
40
20
0
Ag
(-)
Viable SL1438
Killed SL1438
Killed SL1438 + LT(R192G)
LT(R192G)
Naive
b
IFN-γ (pg/ml)
2500
2000
1500
1000
500
0
c
IL-10 (pg/ml)
1400
1200
1000
800
600
400
200
0

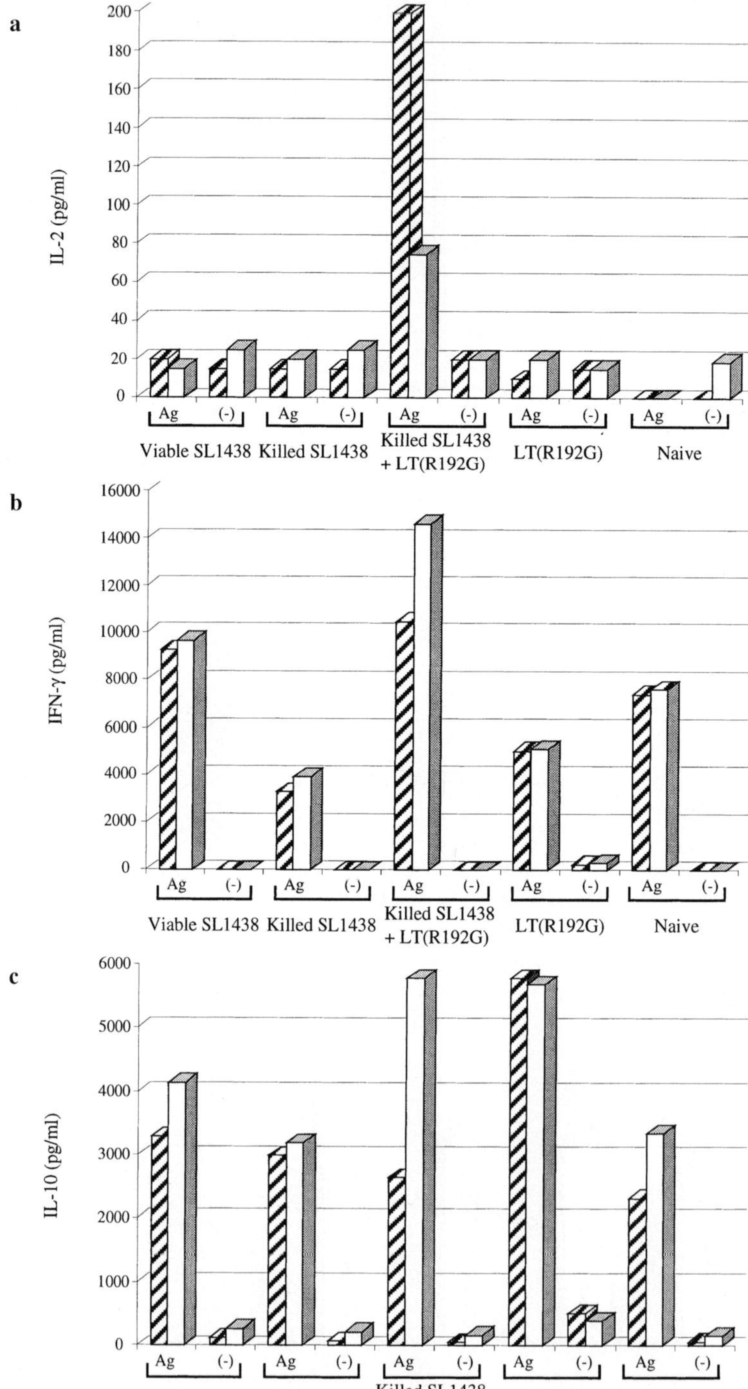
a
IL-2 (pg/ml)
200
180
160
140
120
100
80
60
40
20
0
Ag
(-)
Viable SL1438
Killed SL1438
Killed SL1438
+ LT(R192G)
LT(R192G)
Naive
b
IFN-γ (pg/ml)
16000
14000
12000
10000
8000
6000
4000
2000
0
c
IL-10 (pg/ml)
6000
5000
4000
3000
2000
1000
0

Fig. 3a–c. Production of cytokines by mononuclear cells isolated from the spleens (SP) of orally immunized mice. Mice were orally immunized on day 0 and day 7 as in Fig. 1. On day 14, the mice were sacrificed and mononuclear cells from the SP were isolated and pooled by group. The pooled cells were used in an in vitro antigen-re-stimulation assay and were cultured with killed SL1438 as the antigen (Ag) or with complete medium (–) as a negative control in the presence of peritoneal macrophages as antigen presenting cells. Culture supernatants were collected on day 1 (*striped bars*), day 3 (*open bars*), day 5 and day 7 post-culture and murine cytokines analyzed by ELISA. The data are representative of three experiments. Each group consisted of 10 mice. Results obtained from day 5 and day 7 are indistinguishable from day-3 results and are not shown [from Chong et al. (1998)]

This study demonstrates that the function of LT(R192G) in protection against typhoid-like disease is to enhance the Th1 arm of the immune response against killed organisms. Specifically, mice orally immunized with killed *S. dublin* in conjunction with LT(R192G) were protected against lethal challenge and had higher IFN-γ, IL-2 and IgG responses than unprotected mice orally immunized with killed *S. dublin* alone. This is also the first demonstration of a role for IFN-γ in protection against a challenge following oral immunization with viable attenuated *Salmonella* spp.

6 Summary

The use of mucosally administered killed bacteria or viruses as vaccines has a number of attractive features over the use of viable attenuated organisms, including safety, cost, storage and ease of delivery. Unfortunately, mucosally administered killed organisms are not usually effective as vaccines. The use of LT(R192G), a genetically detoxified derivative of LT, as a mucosal adjuvant enables the use of killed bacteria or viruses as vaccines by enhancing the overall humoral and cellular host immune response to these organisms, especially the Th1 arm of the immune response. With this adjuvant, protective responses equivalent to those elicited by live attenuated organisms can be achieved with killed organisms without the potential side effects. These findings have significant implications for vaccine development and further support the potential of LT(R192G) to function as a safe, effective adjuvant for mucosally administered vaccines.

There are a number of unresolved issues regarding the use of LT and CT mutants as mucosal adjuvants. Both active-site and protease-site mutants of LT and CT have been constructed and adjuvanticity reported for these molecules in various animal models and with different antigens. There needs to be a side-by-side comparison of CT, LT, active-site mutants, protease-site mutants and recombinant B subunits regarding the ability to induce specific, targeted immunological outcomes as a function of route of immunization and nature of the co-administered antigen. Those side-by-side comparisons have not been carried out and there is a substantial body of evidence indicating that the outcomes may very well be different. With that information, vaccine strategies could be designed employing the

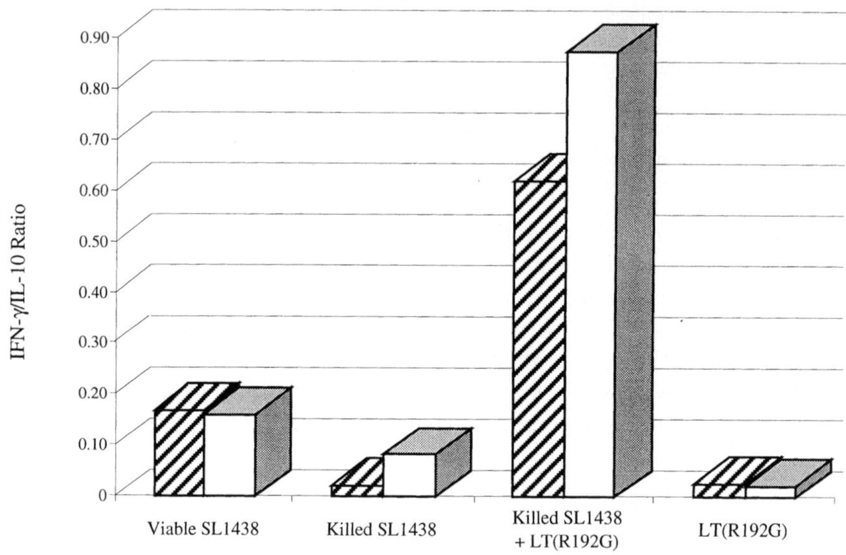

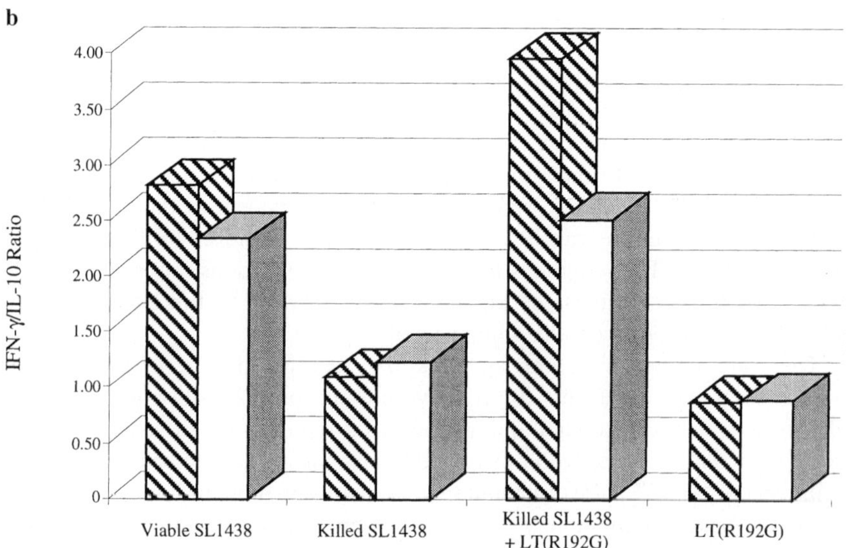

Fig. 4a,b. Ratio of IFN-γ to IL-10 from mononuclear cells isolated from the MLN (**a**) and spleen (**b**). Comparison of IFN-γ and IL-10 from day 1 (*striped bars*) and day 3 (*open bars*). Data from Figs. 2 and 3 [from Chong et al. (1998)]

optimum adjuvant/antigen formulation and route of administration for a variety of bacterial and viral pathogens. Also lacking is an understanding of the underlying cellular and intracellular signaling pathways activated by these different molecules and an understanding of the mechanisms of adjuvanticity at the cellular level. These are important issues because they take us beyond the phenomenological observations of "enhanced immunity" to a more clear understanding of the mechanisms of adjuvant activity.

Acknowledgments. This work was supported by Public Health Service Grants AI28835 and AI36519 from the National Institute of Allergy and Infectious Diseases and by a grant from SmithKline Beecham Biologics.

References

Bromander AK, Kjerrulf M, Holmgren J, Lycke N (1993) Cholera toxin enhances alloantigen presentation by cultured intestinal epithelial cells. Scand J Immunol 37:452–458

Brown W, Woods V, Chitko-McKown C, Hash S, Rice-Ficht A (1994) Interleukin-10 is expressed by bovine type 1 helper, type 2 helper,and unrestricted parasite-specific T-cell clones and inhibits proliferation of all three subsets in an accessory-cell-dependent manner. Infect Immun 62:4697–4708

Burnette WN, Mar VL, Platler BW, Schlotterbeck JD, McGinley MD, Stoney KS, Rhode MF, Kaslow HR (1991) Site-specific mutagenesis of the catalytic subunit of cholera toxin: substituting lysine for arginine 7 causes loss of activity. Infect Immun 59:4266–4270

Cebra JJ, Fuhrman JA, Lebman DA, London SD (1986) Effective gut mucosal stimulation of IgA-committed B cells by antigen. In: Brown F, Channok RM, Lerner RA (eds) Vaccines 86: new approaches to immunization. Developing vaccines against parasitic, bacterial, and viral diseases. Cold Spring Harbor, New York, pp 129–133

Cherwinski HM, Schumacher JH, Brown KD, Mosmann TR (1987) Two types of mouse helper T cell clones. J Exp Med 166:1229–1244

Chong C, Friberg M, Clements JD (1998) LT(R192G), a non-toxic mutant of the heat-labile enterotoxin of Escherichia coli, elicits enhanced humoral and cellular immune responses associated with protection against lethal oral challenge with Salmonella spp. Vaccine 16:732–740

Clarke CJ, Wilson AD, Williams NA, Stokes CR (1991) Mucosal priming of T-lymphocyte responses to fed protein antigens using cholera toxin as an adjuvant. Immunology 72:323–328

Clements JD, Hartzog NM, Lyon FL (1988) Adjuvant activity of Escherichia coli heat-labile enterotoxin and effect on the induction of oral tolerance in mice to unrelated protein antigens. Vaccine 6:269–277

Dickinson BL, Clements JD (1995) Dissociation of Escherichia coli heat-labile enterotoxin adjuvanticity from ADP-ribosyltransferase activity. Infect Immun 63:1617–1623

DiTommaso A, Saletti G, Pizza M, Rappuoli R, Dougan G, Abrignani S, Douce G, DeMagistris MT (1996) Induction of antigen-specific antibodies in vaginal secretions by using a nontoxic mutant of heat-labile enterotoxin as a mucosal adjuvant. Infect Immun 64:974–979

Domenighini M, Magagnoli C, Pizza M, Rappuoli R (1994) Common features of the NAD-binding and catalytic site of ADP-ribosylating toxins. Mol Microbiol 14:41–50

Elson CO (1989) Cholera toxin and its subunits as potential oral adjuvants. Immunol Today 146:29–33

Elson CO, Ealding W (1984a) Cholera toxin feeding did not induce oral tolerance in mice and abrogated oral tolerance to an unrelated protein antigen. J Immunol 133:2892–2897

Elson CO, Ealding W (1984b) Generalized systemic and mucosal immunity in mice after mucosal stimulation with cholera toxin. J Immunol 132:2736–2741

Elson CO, Holland SP, Dertzbaugh MT, Cuff CF, Anderson AO (1995) Morphologic and functional alterations of mucosal T cells by cholera toxin and its B subunit. J Immunol 154:1032–1040

Field M (1980) Regulation of small intestinal ion transport by cyclic nucleotides and calcium. In: Field M, Fordtran JS, Schultz SG (eds) Secretory diarrhea. Waverly, Baltimore pp 21–30

Fontana MR, Manetti R, Giannelli V, Magagnoli C, Marchini A, Olivieri R, Domenighini M, Rappuoli R, Pizza M (1995) Construction of nontoxic derivatives of cholera toxin and characterization of the immunological response against the A subunit. Infect Immun 63:2356–2360

Harford S, Dykes CW, Hobden AN, Read MJ, Halliday IJ (1989) Inactivation of the Escherichia coli heat-labile enterotoxin by in vitro mutagenesis of the A-subunit gene. Eur J Biochem 183:311–316

Häse CC, Thai LS, Boesman-Finkelstein M, Mar VL, Burnette WN, Kaslow HR, Stevens LA, Moss J, Finkelstein RA (1994) Construction and characterization of recombinant Vibrio cholerae strains producing inactive cholera toxin analogs. Infect Immun 62:3051–3057

Hashigucci K, Ogawa H, Ishidate T, Yamashita R, Kamiya H, Watanabe K, Hattori N, Sato T, Suzuki Y, Nagamine T, Aizawa C, Tamura S, Kurata T, Oya A (1996) Antibody responses in volunteers induced by nasal influenza vaccine combined with Escherichia coli heat-labile enterotoxin B subunit containing a trace amount of the holotoxin. Vaccine 14:113–119

Hathaway LJ, Partidos CD, Vohra P, Steward MW (1995) Induction of systemic immune responses to measles virus synthetic peptides administered intranasally. Vaccine 13:1495–1500

Hornquist E, Lycke N (1993) Cholera toxin adjuvant greatly promotes antigen priming of T cells. Eur J Immunol 23:2136–2143

Katz JM, Lu X, Galphin JC, Clements JD (1996) Heat-labile enterotoxin from Escherichia coli as an adjuvant for oral influenza vaccination. In: Brown LE, Hampson AW, Webster RG (eds) Options for the control of influenza. III. Elsevier, New York, pp 292–297

Katz JM, Lu X, Young SA, Galphin JC (1997) Adjuvant activity of the heat-labile enterotoxin from enterotoxigenic Escherichia coli for oral administration of inactivated influenza virus vaccine. J Infect Dis 175:352–363

Lee CK, Weltzin R, Thomas WD, Kleanthous H, Ermak TH, Soman G, Hill JE, Ackerman SK, Monath TP (1995) Oral immunization with recombinant Helicobacter pylori urease induces secretory IgA antibodies and protects mice from challenge with Helicobacter felis. J Infect Dis 172:161–171

Levine MM, Kaper JB, Black RE, Clements ML (1983) New knowledge on pathogenesis of bacterial enteric infections as applied to vaccine development. Microbiol Rev 47:510–550

Lobet Y, Cluff CW, Cieplak W, Jr. (1991) Effect of site-directed mutagenic alterations on ADP-ribosyltransferase activity of the A subunit of Escherichia coli heat-labile enterotoxin. Infect Immun 59:2870–2879

Lowell G, Kaminski R, VanCott T, Slike B, Kersey K, Zawoznik E, Loomis-Price L, Smith G, Redfield R, Amselem S, Brix D (1997) Proteosomes, emulsions, and cholera toxin B improve nasal immunogenicity of human immunodeficiency virus gp160 in mice: induction of serum, intestinal, vaginal, and lung IgA and IgG. J Infect Dis 175:292–301

Lycke N, Karlsson U, Sjolander A, Magnusson K-E (1991) The adjuvant effect of cholera toxin is associated with an increased intestinal permeability for luminal antigens. Scand J Immunol 33:691–698

Lycke N, Tsuji T, Holmgren J (1992) The adjuvant effect of Vibrio cholerae and Escherichia coli heat-labile enterotoxins is linked to their ADP-ribosyltransferase activity. Eur J Immunol 22:2277–2281

Mason HS, Ball JM, Shi J-J, Jiang X, Estes MK, Arntzen CJ (1996) Expression of Norwalk virus capsid protein in transgenic tobacco and potato plants and its oral immunogenicity in mice. Proc Natl Acad Sci USA 93:5335–5340

McGhee JR, Kiyono H (1993) New perspectives in vaccine development: mucosal immunity to infections. Infect Agents Dis 12:55–73

Merritt EA, Sarfaty S, Pizza M, Domenighini M, Rappuoli R, Hol WG (1995) Mutation of a buried residue causes loss of activity but no conformational change in the heat-labile enterotoxin of Escherichia coli. Nat Struct Biol 2:269–272

Mosmann TR, Coffman RL (1989) TH1 and TH2 cells: different patterns of lymphokine secretion lead to different functional properties. Ann Rev Immunol 7:145–173

Moss J, Stanley SJ, Vaughan M, Tsuji T (1993) Interaction of ADP-ribosylation factor with Escherichia coli enterotoxin that contains an inactivating lysine 112 substitution. J Biol Chem 268:6383–6387

Nedrud JG, Sigmund N (1991) Cholera toxin as a mucosal adjuvant. III. Antibody responses to nontarget dietary antigens are not increased. Reg Immunol 3:217–222

Oplinger ML, Baqar S, Trofa AF, Clements JD, Gibbs P, Pazzaglia G, Bourgeois AL, Scott. DA (1997) Safety and immunogenicity in volunteers of a new candidate oral mucosal adjuvant, LT(R192G). In: Program and abstracts of the 37th Interscience Conference on Antimicrobial Agents and Chemotherapy. Am Soc Microbiol, Washington, DC

Partidos CD, Pizza M, Rappuoli R, Steward MW (1996) The adjuvant effect of a non-toxic mutant of heat-labile enterotoxin of Escherichia coli for the induction of measles virus-specific CTL responses after intranasal co-immunization with a synthetic peptide. Immunology 89:483–487

Pizza M, Domenighini M, Hol W, Giannelli V, Fontana MR, Giuliani MM, Magagnoli C, Peppoloni S, Manetti R, Rappuoli R (1994) Probing the structure-activity relationship of Escherichia coli LT-A by site-directed mutagenesis. Mol Microbiol 14:51–60

Quaroni A, Isselbachre K (1981) Cytotoxic effects and metabolism of benzo(a)pyrene and 7, 12-diamethylbenzene (a)-anthrazene in duodenal and ileal epithelial cell cultures. J Natl Cancer Inst 67:1353–1359

Roberts M, Bacon A, Rappuoli R, Pizza M, Cropley I, Douce G, Dougan G, Marinaro M, McGhee J, Chatfield S (1995) A mutant pertussis toxin molecule that lacks ADP-ribosyltransferase activity, PT-9K/129G, is an effective mucosal adjuvant for intranasally delivered protein. Infect Immun 63:2100–2108

Romagnani S (1991) Human Th1 and Th2 subsets: doubt no more. Immunol Today 12:256–257

Sixma TK, Pronk SE, Kalk KH, Wartna ES, van Zanten BAM, Witholt B, Hol WGJ (1991) Crystal structure of a cholera toxin-related heat-labile enterotoxin from E. coli. Nature (London) 351:371–377

Sixma TK, Kalk KH, van Zanten BAM, Dauter Z, Kingma J, Witholt B, Hol WGJ (1993) Refined structure of Escherichia coli heat-labile enterotoxin, a close relative of cholera toxin. J Mol Biol 230:890–918

Snider DP, Marshall JS, Perdue MH, Liang H (1994) Production of IgE antibody and allergic sensitization of intestinal and peripheral tissues after oral immunization with protein Ag and cholera toxin. J Immunol 53:647–657

Staats HF, Nichols WG, Palker TJ (1996) Systemic and vaginal antibody responses after intranasal immunization with HIV-1 C4/V3 peptide T1SP10 MN(A). J Immunol 157:462–472

Takahashi I, Marinaro M, Kiyono H, Jackson RJ, Nakagawa I, Fujihashi K, Hamada S, Clements JD, Bost KL, McGhee JR (1996) Mechanisms for mucosal immunogenicity and adjuvanticity of Escherichia coli labile enterotoxin. J Infect Dis 173:627–635

Tribble DR, Baqar S, Oplinger ML, Bourgeois AL, Clements JD, Pazzaglia G, Pace J, Walker RI, Gibbs P, Scott DA (1997) Safety and enhanced immunogenicity in volunteers of an oral, inactivated whole cell Campylobacter vaccine co-administered with a modified E. coli heat-labile enterotoxin adjuvant – LT(R192G). In: Program and abstracts of the 37th Interscience Conference on Antimicrobial Agents and Chemotherapy. Am Soc Microbiol, Washington, DC

Tsuji T, Inoue T, Miyama A, Okamoto K, Honda T, Miwatani T (1990) A single amino acid substitution in the A subunit of Escherichia coli enterotoxin results in loss of its toxic activitiy. J Biol Chem 265:22520–22525

Tsuji T, Inoue T, Miyama A, Noda M (1991) Glutamic acid-112 of the A subunit of heat-labile enterotoxin from enterotoxigenic Escherichia coli is important for ADP-ribosyltransferase activity. FEBS Lett 291:319–321

Tsuji T, Yokochi T, Kamiya H, Kawamoto Y, Miyama A, Asano Y (1997) Relationship between a low toxicity of the mutant A subunit of enterotoxigenic Escherichia coli enterotoxin and its strong adjuvant action. Immunology 90:176–182

Urban JF, Madden KB, Svetic A, Cheever A, Trotta PP, Gause WC, Katona IM, Finkelman FD (1992) The importance of TH2 cytokines in protective immunity to nematodes. Immunol Rev 127:205–220

Van de Verg L, Hartman A, Bhattacharjee A, Tall B, Yuan L, Sasala K, Hadfield T, Zollinger W, Hoover D, Warren R (1996) Outer membrane protein of Neisseria meningitidis as a mucosal adjuvant for lipopolysaccharide of Brucella melitensis in mouse and guinea pig intranasal immunization models. Infect Immun 64:5263–5268

Volkheimer G, Schulz FH (1968) The phenomenon of persorption. Digestion 1:213–218

Weltzin R, Kleanthous H, Guirakhoo F, Monath TP, Lee CK (1997) Novel intranasal immunization techniques for antibody induction and protection of mice against gastric Helicobacter felis infection. Vaccine 4:370–376

Wilson AD, Bailey M, Williams NA, Stokes CR (1991) The in vitro production of cytokines by mucosal lymphocytes immunized by oral administration of keyhole limpet hemocyanin using cholera toxin as an adjuvant. Eur J Immunol 21:2333–2339

Wu Y-Y, Nahm MH, Guo Y, Russell MW, Briles DE (1997) Intranasal immunization of mice with PspA (pneumococcal surface protein A) can prevent intranasal carriage, pulmonary infection, and sepsis with Streptococcus pneumoniae. J Infect Dis 175:839–846

Xu-Amano J, Kiyono H, Jackson RJ, Staats HF, Fujihashi K, Burrows PD, Elson CO, Pillai S, McGhee JR (1993) Helper T cell subsets for immunoglobulin A responses: oral immunization with tetanus toxoid and cholera toxin as adjuvant selectively induces Th2 cells in mucosal associated tissues. J Exp Med 178:1309–1320

Xu-Amano J, Jackson RJ, Fujihashi K, Kiyono H, Staats HF, McGhee JR (1994) Helper Th1 and Th2 cell responses following mucosal or systemic immunization with cholera toxin. Vaccine 12:903–911

Yamamoto M, Vancott JL, Okahashi N, Marimanr M, Kiyono H, Fujihashi K, Jackson RJ, Chatfield SN, Bluethmann H, McGhee JR (1996) The role of Th1 and Th2 cells for mucosal IgA responses. Ann NY Acad Sci 778:64–71

Yamamoto S, Kiyono H, Yamamoto M, Imaoka K, Yamamoto M, Fujihashi K, VanGinkel FW, Noda M, Takeda Y, McGhee JR (1997a) A nontoxic mutant of cholera toxin elicits Th-2 type responses for enhanced mucosal immunity. Proc Natl Acad Sci USA 94:5267–5272

Yamamoto S, Takeda Y, Yamamoto M, Kurazono H, Imaoka K, Yamamoto M, Fujihashi K, Noda M, Kiyono H, McGhee JR (1997b) Mutants in the ADP-ribosyltransferase cleft of cholera toxin lack diarrheagenicity but retain adjuvanticity. J Exp Med 185:1203–1210

Bacterial Mucosal Vaccines: *Vibrio cholerae* as a Live Attenuated Vaccine/Vector Paradigm

K. KILLEEN[1], D. SPRIGGS[1], and J. MEKALANOS[2]

1 Introduction

Immunization is the most effective public health tool used to control infectious disease. Moreover, immunization is extremely cost effective given that disease treatment is far more expensive than prevention of disease. The cost of vaccines and their administration from birth to age 16 is estimated by the Centers for Disease Control (CDC) to be US $500. Each US $1 spent on vaccinations saves US $16 in avoiding costly drug therapies and hospitalizations (FETTNER 1994), ultimately

[1] Virus Research Institute, Cambridge, USA
[2] Department of Microbiology and Molecular Genetics, Harvard Medical School and the Shipley Institute of Medicine, 200 Longwood Avenue, Boston, MA 02215, USA

saving approximately US $7500 per vaccinated individual. Furthermore, phenomena such as herd immunity can provide protection to a community, even when only a minority of the total population has been vaccinated. Ideally, vaccination leads to the total eradication of an infectious agent that has no alternative hosts or environmental reservoirs, e.g., smallpox and, in the near future, polio.

Although commercialized vaccines are generally quite safe, examples of adverse reactions have been reported for every vaccine that has been widely used. In some cases, such as the diphtheria, tetanus and pertussis (DTP) vaccine, widespread fears concerning the potential harmful side-effects of vaccines have resulted in declining use and a resultant resurgence of disease (Brennan et al. 1992).

During the past decade, recombinant DNA technology has made it possible to design a new generation of vaccines. Although the first recombinant killed vaccine, i.e., hepatitis B, has been a tremendous success, we are only beginning to see the fruits of biotechnology in the arena of live attenuated vaccines. A series of live attenuated recombinant vaccines for cholera, typhoid, and shigellosis have entered advanced clinical trials. We are now at an appropriate stage to consider the safety and environmental impact of these new live bacterial mucosal vaccines and vaccine vectors. In this review, we will briefly discuss these issues, with particular emphasis on recent efforts to construct safe live cholera vaccines.

2 Types of Vaccines

Three basic types of vaccines exist: antigen subunits, inactivated agents, and live attenuated organisms. Vaccines composed of subunits, i.e., purified inactivated proteins/peptides, carbohydrate capsules or cross-linked conjugates of these, are generally considered safest because they are used in small quantities, are chemically well defined and do not replicate. Thus, they cannot spread to the environment or other non-vaccinated or immunocompromised individuals. However, subunit vaccines can be expensive to manufacture and have a limited ability to induce immune responses, particularly mucosal responses. Therefore, these vaccines frequently require adjuvants and multiple doses.

Inactivated agents, i.e., bacteria/viruses killed by heat or chemical treatment, are considered safe. However, these suffer from other drawbacks, such as ill-defined molecular characteristics and poor immunogenicity, i.e., large quantities of vaccine are required to elicit protective immune responses. Inactivated vaccines are relatively easy to manufacture and can possess inherent immunomodulatory activity, e.g., the whole cell pertussis component of DTP, although the spectrum of immune response induced is usually limited to the humoral arm of the immune system.

In contrast, live attenuated vaccines, i.e., living viruses and bacteria that carry mutations rendering them avirulent or greatly reduced in virulence, offer significant advantages in terms of manufacture and immunogenicity. For example, a single inoculation of live vaccine at a modest dose will replicate in vivo to a very large immunogenic dose and, during the course of replication, express the majority of immunogens seen during natural disease. Moreover, the processing and presentation of these antigens is more "natural", mimicking convalescent immune responses. Live vaccines can induce mucosal immune responses, which are not elicited by systemically administered vaccines and the mode of delivery of live vaccines can be quite simple, e.g., ingestion. Finally, because attenuated vaccines are living organisms, they may also be used as vectors by genetically engineering them to express heterologous antigens and, thus, provide protection from more than one disease.

Of the 21 licensed vaccines in the United States, 11 are directed against bacterial and 10 against viral pathogens. There are 17 improved vaccine candidates and 45 novel vaccine candidates in research and development at assorted vaccine companies throughout the world (Jordan Report 1996). Of the 62 total vaccine candidates, 28 are bacterial antigens, 24 are viral antigens and 10 are fungal or parasitic antigens. Optimal delivery of these future vaccines is crucial in order to minimize the number of immunizations required to render them commercially viable. Thus, combination parenteral and mucosal vaccines, as well as multivalent live vaccine delivery have central roles in the vaccines of the 21st century.

2.1 Attenuated Live Bacterial Vaccines

The most intensely investigated live attenuated bacterial vaccines have been derived from the genera *Mycobacterium*, *Salmonella*, *Shigella* and *Vibrio*. Vaccines composed of attenuated mutants of these bacteria have been evaluated for safety and efficacy in humans for prevention of tuberculosis, typhoid fever, shigellosis or cholera, respectively. Moreover, many of these attenuated bacterial strains have also been examined for their capacity to deliver foreign antigens as multivalent vaccine vectors in extensive pre-clinical studies.

The modes of attenuation of these bacteria can be grouped into two general categories: undefined mutation or defined mutation. *Salmonella typhi* (*S. typhi*) strain Ty21a and *Mycobacterium bovis* strain Bacille, Calmette, Guerin (BCG) are examples of vaccines that are attenuated by an undefined mutation(s). Ironically, these two strains represent the most widely used attenuated bacterial vaccines worldwide. In contrast, numerous examples exist of attenuated strains of *Vibrio*, *Salmonella*, and *Shigella* that carry defined attenuating mutations. Defined genetic attenuation can be further dissected into two functional subclasses: (1) gene(s) disruption affecting metabolism or regulation and (2) gene(s) disruption affecting virulence.

2.1.1 Undefined Genetic Attenuation

BCG is widely used throughout the world as a live vaccine for tuberculosis and in the treatment of certain cancers. BCG is the result of successive *Mycobacterium bovis* in vitro passages, culminating in undefined spontaneous genetic lesions.

Despite BCG's unknown cause(s) of attenuation, it is extremely safe; it has been administered to more human subjects than any other vaccine and is effective in combatting tuberculosis.

S. typhi Ty21a is the sole example of a chemically-induced attenuation process successfully employed in a licensed bacterial vaccine product. Nitrosoquanidine was used as a chemical mutagen to induce random genetic lesions with selection of a galE (encoding UDP galactose 4-epimerase) mutation, presumably resulting in galactose sensitivity in vivo. Retrospectively, a galE mutant of *S. typhi* was constructed and found to be virulent, indicating that the reason for attenuation of Ty21a is associated with other undefined mutations (Hone et al. 1988). Ty21a is administered in three or four oral doses and is very well tolerated in humans. The immune response elicited is both humoral and cell-mediated and confers significant, but incomplete, protection from typhoid fever (Wahdan et al. 1982; Levine et al. 1987).

2.1.2 Defined Genetic Attenuation

In direct contrast to BCG and Ty21a; attenuated *S. typhi*, *Shigella* sp. and *Vibrio cholerae* are the result of defined genetic mutations. Each bacteria and its mode of genetic attenuation is described in Sect. 2.1.2.1 and Sect. 2.1.2.2.

2.1.2.1 Metabolic/Regulatory Attenuation

The elucidation of bacterial metabolic pathways, coupled with the identification of the genes that encode metabolic enzymes, enabled the concept of generating metabolically attenuated live bacteria. Disruption of genes encoding essential metabolic functions was the first rational mode of genetic attenuation applied to live bacteria. *S. typhi* aroA$^-$, purA$^-$, genes, which encode essential biosynthetic enzymes in the aromatic amino acid pathway and DNA biosynthesis, respectively, termed 541Ty, were constructed (Edwards and Stocker 1984). 541Ty was evaluated in human clinical trials and, although well tolerated, was less immunogenic than Ty21a (Levine et al. 1987) and was not considered for further clinical development. CVD908, an aroC/D$^-$ strain attenuated due to aromatic amino-acid biosynthesis dysfunction, was constructed by the Center for Vaccine Development (CVD), and evaluated in volunteers. At the single, oral dose level required for immunogenicity, CVD908 caused transient bacteremias impeding this strain's clinical development (Tacket et al. 1992 and Sztein et al. 1994). Consequently, several other groups attempted to attenuate *S. typhi* aroA$^-$ strains further by the addition of other mutations. *S. typhi* TyLH445, a combined aroA$^-$, phoP/Q$^-$ strain, which encodes virulence regulation, was evaluated in human volunteers and determined to be hyper-attenuated. Although the vaccine was safe, it was ineffective at provoking significant immune responses after two oral doses (Hohmann et al. 1996). CVD908 htrA$^-$, a gene encoding a bifunctional stress protein/serine protease, was constructed and evaluated in human volunteers. The strain was well tolerated in the dose ranges tested (10^7–10^9cfu) and immunogenic, but significantly compromised in immunogenicity compared with its progenitor CVD908 (Tacket et al. 1997). Exploiting metabolic/regulatory genes, Curtiss and colleagues constructed a cya$^-$, crp$^-$ strain of *S. typhi*, Chi3927 (corresponding to genes encoding adenylate cyclase and cAMP receptor proteins, respectively). Chi3927 caused unacceptable levels of high fever and bacteremia, thwarting its clinical development (Tacket et al. 1992).

Historically, the metabolic attenuation strategy for *Shigella* sp. strongly parallels that of *S. typhi*. Initially, an *aroD* deletion was constructed in *Shigella flexneri* and evaluated in human volunteers. This strain was inadequately attenuated and, thus, was not well tolerated (Li et al. 1992). In vitro studies have since defined many important virulence factors of this invasive pathogen, and mutations in these properties are actively being combined with auxotrophic lesions to produce many new vaccine candidates (Fontaine et al. 1990).

2.1.2.2 Virulence Attenuation

The major premise of virulence attenuation is that virulence determinants are often non-essential functions of bacterial pathogens. Thus, in direct contrast to metabolic attenuation, the attenuation of virulence has no direct manifestation on replication/colonization in vivo, yielding a more fit live vaccine candidate. Attenuated *V. cholerae* marked a new era for the construction of live bacterial vaccine strains which are metabolically intact and attenuated solely for virulence. Initially, a rational process was used, in which a deletion of the *cholera toxin A* subunit gene, (*ctxA*), which encodes the enzymatically active subunit of the toxin, was introduced as the sole attenuating mutation (Mekalanos et al. 1983; Kaper et al. 1984). In volunteer studies, these early vaccines were shown to be "reactogenic", causing symptoms in volunteers that ranged from moderate diarrhea and fever, to simple malaise and loose stools. Notably, these first generation, live, attenuated, recombinant cholera vaccines could, in a single oral dose, induce long-lasting immunity to experimental wild-type cholera challenge (Levine et al. 1988; Herrington et al. 1988). Until recently, only one recombinant strain, *V. cholerae* CVD103-HgR, has been sufficiently tolerated by those vaccinated for it to be considered for further development (Levine et al. 1988). Apart from the *ctxA* deletion, the mutations attenuating this strain are largely undefined. This vaccine has been tested for efficacy in a large field trial in Indonesia (the results of which have yet to be reported) and based on challenge data, has been licensed for use as a traveler's cholera vaccine in several countries in Europe. In experimental challenge models, volunteers receiving CVD 103-HgR show 87% protection against challenge with homologous classical biotype strains, but only 62% against El Tor strains (Levine et al. 1988).

Prompted by the need to develop cholera vaccines that provide better protection against the El Tor biotype of *V. cholerae* as well as recently emerged 0139 strains, a second generation of cholera vaccines have been constructed and are undergoing evaluation. For example, vaccine candidates Peru-15 and Bengal-15 are examples of genetically defined, live *Vibrio* strains which have, so far, performed well in initial safety and efficacy tests in volunteers. These strains are deleted for the entire genome of the recently described CTX phage (WALDOR and MEKALANOS 1996b), and additionally carry mutations in *recA*, eliminating homologous recombination, and a gene required for flagellar assembly and motility. Thus, these two vaccine strains are metabolically wild type and highly competent for replication in the human host. Peru-15 and Bengal-15 were proven safe, immunogenic, and protective as live vaccines against O1 and O139 cholera, respectively, in human-volunteer challenge studies (KENNER et al. 1995; COSTER et al. 1995). The success of these live attenuated *V. cholerae* vaccines has encouraged other investigators to explore attenuation strategies based on deletion of genes encoding virulence factors or the regulatory genes controlling their expression.

The attenuated *Salmonella* and *Shigella* sp. vaccine fields have migrated from the largely unsuccessful, metabolic/regulatory-attenuated candidates, to the successful, virulence-attenuated vaccine strains. Also, regulatory mutations that affect virulence of these organisms have been explored with considerable success.

As previously described in Sect. 2.1.2.1, *S. typhi* TyLH445 was overly attenuated. Thus, an aroA$^+$, phoP/Q$^-$, Q$^-$ *S. typhi* strain, Ty800, was constructed by HOHMANN and colleagues (1996). Ty800 is the first example of an attenuated *S. typhi* strain that is metabolically intact and attenuated solely for virulence. Ty800 was evaluated in human studies and was discovered to be safe and immunogenic after single, oral doses ranging from 10^7cfu to 10^{10}cfu (HOHMANN et al. 1996). Another attenuated *S. typhi* strain, Chi4073, is a derivative of *crp cyc* defective mutant Chi3927 which carries an additional mutation in *cdt*$^-$ (a gene that participates in dissemination of *S. typhi* from gut-associated lymphoid tissue to deeper organs of the reticuloendothelial system). Chi4073 is, thus, a combination regulatory/virulence-attenuated strain. In volunteer studies, it was found to be reasonably well tolerated and immunogenic at doses of $10^7/10^8$cfu (TACKET et al. 1997), thereby substantially improving the safety profile of its progenitor Chi3927.

Improvements in the attenuation of *Shigella* live vaccines have also been realized by combining mutations affecting virulence in vaccine strains: for example, *S. flexneri* 2a strain, SC602, a complex mutant that is virG$^-$, icsA$^-$ (genes that encode intra- and intercellular migration) and iuc$^-$, iut$^-$ (genes that encode iron acquisition). This strain was tested in volunteers and was determined to be safe and immunogenic following a single oral dose (World Health Organization 1996).

In conclusion, metabolic attenuation generally compromises a live bacteria's ability to produce essential metabolites, thereby reducing replication in vivo, resulting in decreased immunogenicity. Moreover, the bioavailability of essential metabolic substrates in vivo is very difficult to predict during the colonization process in humans. Oral immunization with *S. typhi* aroC/D$^-$ (CVD908), which results in transient bacteremias, supports the conclusion that metabolic attenuation alone will seldom be an adequate means for attenuation of live vaccine destined for human use. In direct contrast, some virulence genes encode functions that are essential for deep tissue invasion, or disease symptoms, but play little or no role in in vivo replication per se. Construction of vaccines using virulence attenuation can thus produce safer vaccines, while minimizing any negative impact that attenuation has on immunogenicity. As additional genes encoding virulence functions are elucidated, safer and more effective live bacterial vaccines will continue to be constructed by laboratories interested in vaccinology.

3 Live Bacterial Vectors

Successful genetic attenuation of live bacteria yielding safe, immunogenic, and protective vaccines enabled the development of safe vaccine vectors. Pre-clinical data support the utility of bacterial vectors for assorted viral-, bacterial-, and parasitic-derived antigens. The nature of the immune response engendered by a given bacterial vector with respect to Th1 versus Th2 and humoral versus cellular immunity can be considerably different. Thus, the type of immune response desired against the vectored antigen should be carefully considered with regard to the bacterial vector to be employed.

4 Pre-clinical Studies

BCG elicits strong humoral and cell-mediated immune responses in humans. In addition, it has vectored several bacterial and viral antigens from *Borrelia*, pneumococci, simian immunodeficiency virus (SIV) and human immunodeficiency virus (HIV); eliciting immune responses against each antigen when used to immunize mice. Specifically, BCG-vectored outer-surface protein A (OspA) and pneumococcal-surface protein A (PspA) conferred protection against *Borrelia* and *S. pneumoniae*, respectively, in murine challenge models. The use of BCG as a delivery system for recombinant antigens was recently reviewed elsewhere (Stover et al. 1994).

Attenuated *Salmonella* evokes humoral, secretory, and cell-mediated immune responses in man. It is the most intensely studied bacterial vector, having proven capable of delivering more than 30 bacterial, 10 viral, and 10 parasitic antigens in pre-clinical murine studies (Kaufmann 1996). Examples of protective efficacy were demonstrated with bacterial challenge (*Y. pestis*, *L. monocytogenes*, and *B. pertussis*), viral challenge (Herpes simplex virus and influenza) and parasitic challenge (*L. major*, and *S. mansoni)*, to name but a few (Kaufmann 1996).

Shigella elicits humoral and mucosal immune responses in man. Live-attenuated *Shigella* vaccine candidates that are well tolerated in humans have only recently emerged, delaying *Shigella*'s development as a vector. In pre-clinical studies, attenuated *Shigella* has proven capable of delivering a native antigen, IpaC, a potential "protein carrier", and provoking an anti-IpaC antibody response, following combination systemic/intranasal administrations (Barzu et al. 1996). Attenuated *Shigella* vectoring a plasmid that harbors a eukaryotic promoter controlling the expression of an antigen has been reported to induce mucosal immune responses (Sizemore et al. 1997). For example, such a delivery system for fragment C (FC) of tetanus toxoid provoked significant anti-FC antibodies following intranasal administrations to guinea pigs (Anderson et al. 1997).

Attenuated *V. cholerae* elicits strong serum and mucosal immune responses (Waldor and Mekalanos 1996). There have been several recent publications demonstrating attenuated *Vibrio's* use as a live vector. Immunogenicity was demonstrated with the following vectored antigens: Shiga-like toxin-B subunit (SltB) (Butterton et al. 1995), *S. sonnei O*-antigen (Haider et al. 1995), *E. histolytica* dodecapeptide repeat unit of serine-rich *Entamoeba histolytica* protein (SREHP) (Ryan et al. 1997), and *C. difficile* toxoidA (Ryan et al. 1997). In pre-clinical models where challenge was possible, SltB, *S. sonnei* and *C. difficile* toxin A protective immunity was engendered.

5 Human Clinical Studies

Despite the vast number of pre-clinical successes of live bacterial vectors, to date, none has been effective at provoking reproducible immune responses to vectored antigens in human clinical trials. There have been several studies evaluating metabolically attenuated *Salmonella* to deliver Csp, the malarial circumsporozoite antigen (Gonzalez et al. 1994) and a modified hepatitis B surface antigen (Tacket et al. 1997). Each clinical study was unimpressive, failing to demonstrate significant levels of vectored antigen immune response in human volunteers. Despite BCG's numerous pre-clinical successes as a vector, a recent clinical study showed that human volunteers immunized with BCG–rOspA failed to seroconvert against OspA (Mucosal Vaccine Meeting Abstracts, NIH 1997). Clearly, improvements in vector fitness, antigen expression levels, antigen stability and antigen localization need to be thoroughly re-examined if bacterial vectors are to become commercially viable in the near future.

6 Next Generation Bacterial Vectors

The first generation bacterial vectors employed strains with undefined mutations (BCG and Ty21a) or metabolically-attenuated strains (CVD908 and Chi4073) to express and deliver heterologous antigens. The inability to provoke an immune response may have been due to the nature of attenuating metabolic defects, compromising the vector's replication in vivo. Thus, virulence-attenuated live vectors are more likely to deliver a vectored antigen and reproducibly provoke an immune response in humans.

Two criteria for attenuated bacteria to elicit effective immune responses against vectored antigens, are high-level antigen expression and extracellular antigen localization. In order to augment expression of vectored antigens, "balanced-lethal" plasmid-based foreign-antigen expression systems have been constructed in attenuated *Salmonella* (NAKAYAMA et al. 1988; HOHMANN 1997) and *Vibrio* vaccine strains (KOCHI et al. unpublished observations). Conceptually, the balanced lethal expression system consists of an essential gene deleted from the chromosome, then complemented on a high-copy plasmid that also drives expression of antigen. Thus, endogenous selective pressure for the essential plasmid-based gene results in maintenance of the plasmid without the usual requirement of antibiotic resistance selection. Balanced lethal systems have been constructed exploiting different essential genes, such as *asd* and *purB* in *Salmonella* and *glnA* in *Vibrio*. Heterologous antigen expression in both attenuated *Salmonella* and *Vibrio* was significantly enhanced when engineered into the balanced lethal plasmid-expression system.

Antigen localization is another important parameter for optimizing the immune response against a vectored antigen. The *E. coli* alpha-hemolysin secretion system (HSS), consisting of HlyB, HlyD, and TolC, has been demonstrated to extracellularly export protein antigens that have been fused with the HlyA carboxy terminus (GENTSCHEV et al. 1990). There are several attractive features of HSS. First, antigens are secreted, and are therefore more readily processed by APCs throughout in vivo expression. Second, there is no "antigen burden" to the vector. That is, heterologous antigens may form inclusion bodies or localize to inappropriate intracellular compartments of bacterial vectors, negatively impacting vector fitness. Heterologous antigens that are secreted should not compromise a live vector during its course of replication and colonization within the human host, enhancing the probability of seroconversion. HSS was applied to attenuated *Salmonella* to enhance heterologous antigen expression and antigen-presenting cell (APC) processing. Listeriolysin, secreted from *Salmonella* via HSS, was superior to cell-associated listeriolysin in conferring protection against L. monocytogene challenge (Hess et al. 1996). Further, *L. monocytogenes* superoxide dismutase (SOD), a non-protective antigen, could be converted to a protective antigen once secreted via HSS (HESS et al. 1997). HSS has also been adapted to attenuated *Vibrio* vectors expressing *C. difficile* toxoid-A fusions. The resulting constructs were found to be secreted, immunogenic and protective in pre-clinical studies (RYAN et al. 1997). Likewise, attenuated *Shigella dysenteriae* was shown to express and secrete Shiga toxin-B subunit using HSS (TZASCHASHEL et al. 1996).

7 Special Issues with Live Bacterial Vaccines

Although the development and testing of live vaccines follows the conventional strategies for defining clinical safety, immunogenicity and efficacy, there are a special set of issues that must be considered. These issues focus on concerns about how these modified organisms might alter the ecology of similar organisms in the environment, and what impact this might have on human health. Not surprisingly, this has already been given a great deal of attention, particularly in the context of releasing recombinant organisms for pollution control and agricultural applications (Käppeli and Auberson 1997; Wilson and Lindow 1993). We will outline in the following paragraph how similar principles can be applied to live bacterial vaccines, and develop various arguments, using *V. cholerae* vaccines as an example.

First, we propose the following five questions as an outline to begin framing any discussion: (1) will the vaccine organisms be released into the environment; (2) will the organisms survive; (3) will the organism proliferate; (4) will the organisms cause ecological disruption where they spread; and (5) will they transfer genes to or acquire genes from other organisms in the environment? These questions have been modified from a recent review on genetic exchange in natural microbial communities (Veal et al. 1992).

Before we begin to think about how to answer these questions, we need to sort out two very important concepts that have been discussed by Käppeli and Auberson (1997) in their review of environmental safety issues: safety assessment versus risk analysis. In their model, safety assessment is the objective and scientifically based approach used to define potential hazards and perform experiments to determine – to the extent possible – whether such events could occur. For example, if one is concerned about a live vaccine strain reacquiring a virulence trait from wild-type organisms, controlled laboratory experiments can be conducted to test various conditions under which this might occur. Parallel to this prospective safety assessment, it is important to establish, retrospectively, all possible independent events that might lead to the same outcome: the emergence of a new pathogenic variant of the organism. These analyses will yield very different results, depending upon the organism being studied.

The second concept, risk analysis, is more difficult. Although a variety of studies can be performed to assess the safety of recombinant organisms, this risk cannot be reliably defined. Risk analysis is therefore a subjective, value-based exercise that must be kept distinct from the science-based safety assessment. This distinction can be seen by the example of polio immunization. It is well known that the live, attenuated polio vaccine can revert to become virulent during replication in the intestinal tract of those vaccinated. Indeed, in the U.S., there are four to six cases of vaccine-associated paralytic disease each year. In this case, the safety assessment was quite clear. The risk analysis was, however, a different issue. Public health officials felt that the risk of vaccine-associated disease was worth the benefit of protecting the majority from wild-type disease. With the elimination of polio from this hemisphere, however, the risk has been reassessed, and these same officials have now decided that this risk is no longer acceptable (CDC 1997).

8 Safety Assessment for Live Cholera Vaccines

Cholera is a severe and sometimes fatal diarrheal disease caused by the bacterium *V. cholerae*. It has been extensively studied, not only because of its importance as a human disease, but also because it is an important source of information about how bacteria respond to environmental signals (HERRINGTON et al. 1988; PARSOT et al. 1991; DIRITA et al. 1996). For example, while its attributes as a human pathogen are well described, *V. cholerae* can also can exist as an autochthonous inhabitant of brackish water and estuarine ecosystems, where it colonizes copepods and plankton and grows to high densities (COLWELL 1996). The ecology of *V. cholerae* is therefore best viewed as a cycle of growth in environment, infection of humans, and reseeding of the environment.

To begin assessing the safety issues associated with the release of a live, attenuated *V. cholerae* into the environment, one must frame the studies in this ecological context. We will discuss two prototype live, oral cholera vaccines (Peru-15 and CVD-103HgR) and define how one would assess safety in the context of the five questions outlined in Sect. 8. We will then discuss how these studies might inform the risk analysis.

8.1 Will the Organisms Be Released into the Environment?

Both the Peru-15 and CVD103-HgR vaccines are live, attenuated vaccines that are given orally. They actively multiply in the intestinal tract and are shed in the stools. In the case of CVD 103-HgR, the strain is shed at comparably low levels (10^2cfu/g of stool) for up to 3 days, while Peru-15 is shed at higher levels ($>10^3$cfu/g of stool) for up to 10 days. The answer to this first question is therefore "yes."

8.2 Will the Organisms Survive?

The answer to this question depends upon the location of release. In the U.S. and other industrialized countries, the process of waste treatment, i.e., filtration, chlorination, or irradiation, virtually eliminated water-borne infectious diseases, including cholera. These non-specific factors would also kill the vaccine strains. In developing countries, the situation may be different; although one study in Indonesia suggests that CVD-103 HgR cannot survive in raw sewage (SUHARYONO et al. 1992). There are of course a variety of reconstruction studies that could be designed to assess the survival and die-off kinetics of these vaccine strains to answer this question.

8.3 Will the Organisms Proliferate?

To date, studies in reconstructed ecosystems have shown that the attenuated strains grow poorly, if at all, in various media such as estuarine water and soil (CRYZ et al.

1995, unpublished observation). These laboratory findings were validated by environmental sampling during a clinical trial of CVD-103HgR in Indonesia (Suharyono et al. 1992). In this trial, the investigators failed to detect the vaccine strain in open sewers even though non-O1 *Vibrio* were detected from virtually every house that was being monitored.

8.4 Will the Organisms Cause Ecological Disruption?

For these vaccine strains, it seems highly unlikely, given their lack of fitness for autochthonous growth outside of the laboratory. The ecological balance among various *V. cholerae* is, however, a complex phenomenon, as judged by the shifts in bacterial biotypes and serotypes that cause cholera in endemic areas. The most recent example of this phenomenon was seen in 1992, when a new serotype of *V. cholerae* (O139) appeared in India and Bangladesh (Cholera Working Group 1993). It rapidly displaced the existing O1 serotype and appeared poised to become the next pandemic strain. Surprisingly, however, the O1 serotype reappeared and has made a dramatic comeback (Faruque et al. 1997). Currently, both O1 and O139 serogroups co-exist in the cholera endemic areas of South Asia. The reasons for these shifts are not known although they clearly support genetic exchange as a contributor to the evolution of pathogenic *V. cholerae*. (see Sect. 8.5).

8.5 Will They Transfer to or Acquire Genes from Other Organisms?

This is the question that has received the most attention from vaccine developers and regulatory authorities. Genetic exchange between bacteria is a well-described phenomenon, and its role in human disease, e.g., the spread of antibiotic resistance genes, is clear. In *V. cholerae*, the emergence of the O139 strain was traced to the introduction of novel cluster of genes into an El Tor O1 strain (Stroeher et al. 1995; Mooi and Bik 1997). Although the origin of these sequences is not known, they were probably acquired by an El Tor O1 strain via conjugation, transduction, or transformation from an unknown donor bacterium. Although *V. cholerae* does not appear to have Hfr-like conjugation elements, some strains do contain conjugative *P* plasmids which can mobilize chromosomal markers to *P*-negative cells at relatively low frequency. For example, transfer of the cholera toxin genes (marked with a chromosomal antibiotic-resistance marker) occurred at a frequency of less than 2×10^{-8} events (Cryz et al. 1995). A recent study identified a novel conjugative transposon that encodes antibiotic resistance, but this element apparently does not mediate transfer of unlinked chromosomal markers (Waldor et al. 1996a). Because *V. cholerae* is not a naturally competent organism, transformation is probably the least likely mechanism for this organism to acquire or donate DNA. Nonetheless, a recent review pointed out that extracellular DNA can be detected in significant quantities in a variety of aquatic and terrestrial environments (Lorenz and Wackernagel 1994). For example, estuarine water can contain 10–20µg/l

DNA, and the half-life of this DNA, particularly in soil, can be up to 28h. Intergeneric natural transformation between *E. coli* and non-*V. cholerae*, *Vibrio* sp. in an aquatic microenvironment has been reported (Lorenz and Wackernagel 1994).

In contrast to conjugation and transformation, transduction has recently been recognized as an important mechanism of gene transfer between different *V. cholerae* strains. We now know that the structural genes of cholera toxin are actually encoded by a lysogenic filamentous bacteriophage (Waldor and Mekalanos 1996b). The CTX phage uses another virulence factor, the toxin co-regulated pilus (TCP), as its receptor. Because most strains of *V. cholerae* only express TCP within the gastrointestinal tract, it has been proposed that the emergence of toxigenic *V. cholerae* involves a transduction event that must occur in vivo. Such in vivo CTX phage transduction events have been observed in infected animals and, basically, do not occur with El Tor strains under in vitro conditions (Waldor and Mekalanos 1996).

While the existence of the CTX phage was not appreciated during the early development of several cholera vaccines, its DNA was recognized by some investigators as a potentially mobile genetic element encoding a site-specific recombination system (Pearson et al. 1993). Accordingly, vaccine strains Peru-15 and Bengal-15 carry deletions of the entire phage genome together with its chromosomal attRS1 attachment site (Kenner et al. 1995; Sack et al. 1997a,b). Recent experiments have shown that without its chromosomal attachment site, the CTX phage cannot stably maintain its DNA with such vaccine strains without artificial antibiotic pressure (Waldor and Mekalanos, unpublished results). Thus, reacquisition of the cholera toxin genes by vaccine strains Peru-15 and Bengal-15 would require a minimum of four low-probably events:

1. The vaccine strain would have to be co-infecting, with a toxigenic CTX phage donor strain, the gastrointestinal tract of the patient vaccinated.
2. Because toxigenic *V. cholerae* strains are stable, CTX phage lysogens, the donor strain would need to be induced in vivo by DNA damage in order to produce infectious CTX phage particles.
3. CTX phage particles produced in vivo by the donor strain would have to find and infect the TCP piliated vaccine strain in vivo.
4. The CTX phage DNA would have to be inserted into the chromosome of the vaccine strain by an illegitimate recombination event since its usual attachment site would be deleted.

With regard to these hypothesized steps, it important to note that step 4 has not been demonstrated to occur under laboratory conditions designed to detect these rare integration events.

In addition to the deletion of the entire CTX phage, Peru-15 and Bengal-15 have a deletion in the *recA* gene, which encodes an enzyme essential for homologous recombination (Waldor and Mekalanos 1996b; Mekalanos 1994). This deletion therefore greatly decreases the possibility of Peru-15 or Bengal-15 from stably acquiring any new genes at any other locus by homologous recombination (Kenner et al. 1995). Surprisingly, the *recA* deletion also greatly impairs the ability of Bengal-15 to donate DNA during conjugation (K. Haider and K. Killeen, unpublished observations). Finally, these *recA–* strains are also less environmentally fit since they have an increased sensitivity to ultraviolet light (Mekalanos et al. 1995).

9 Risk Analysis

As mentioned above, the objective, scientific studies carried out for the safety assessment should be followed by the subjective analysis of risk. In this context, the potential use of a vaccine depends to a large part on how one answers questions about tolerable level of risk, always recognizing that there is never a "zero-risk" option.

Although we don't have a good understanding of the forces that shape the ecological behavior of *V. cholerae*, we do know that these organisms are well adapted for genetic exchange and evolution. Serotypes and biotypes wax and wane in the environment and, occasionally, new variants appear and cause widespread disease (Albert et al. 1997; Faruque et al. 1997; Sharma et al. 1997). While this may not seem to be encouraging from the perspective of releasing attenuated strains into the environment, it is nonetheless important to remember that the population biology of these organisms is complex and that it is unlikely that adding attenuated vaccine strains to this ecological mix – even if they can survive – will, in any way, influence this ongoing process.

To explore this further, let's consider one of the major concerns that has been raised about these vaccines: the possibility that they will reacquire the cholera-toxin gene and, thereby, become toxigenic, virulent organisms. While one might be able to demonstrate the phenomenon of gene transfer in the laboratory, this is not the same as showing that this organism is virulent. A toxin-positive organism that could not colonize the intestine, would not be a virulent organism (Herrington et al. 1988), and it would still be unable to compete in natural ecosystems. If a vaccinated individual was colonized with the vaccine strain, and this strain acquired the cholera toxin gene in vivo, one would have to construct a fairly tenuous argument to explain how this event poses more of a threat than the vaccinated individual normally faces from the wild-type *V. cholerae* donor strain that was the source of the gene. Given the relative inability of the vaccine strains to survive in the environment, the risk issue is even less significant in this context.

Another series of concerns have been raised about the possibility of altering the gene pool by introducing these vaccine strains. By analogy with influenza epidemiology, we can observe antigenic "drifts" in which the biotype or serotype of the circulating strains changes from year to year. For *V. cholerae*, this has been seen in changes from Inaba to Ogawa serotype or from Classical to El Tor biotypes (Manning et al. 1994); the former is due to genetic events and the latter is due to unknown environmental influences. The equivalent of the influenza antigenic "shift" occurred in 1992, when a new serogroup (O139) appeared and caused a major epidemic (Cholera Working Group 1993). These events clearly occur, and they support the notion of genetic plasticity in these organisms. The demonstration of these phenomena, however, does not really address the risk issue.

Here again, we have to envision a situation in which the vaccine strain is part of the biomass and participates equally as a player in the ongoing evolution of these strains. The fact that a new, virulent strain can arise by recombination – as happened with the O139 serogroup – should not lead to the assumption of significant risk. Indeed, the observation that this serotype arose independently at two different sites suggests that the risk from natural processes is significant (Mooi and Bik 1997).

In all these scenarios, the ultimate decision about risk must be made on the basis of our understanding of the molecular biology, pathogenesis, and ecology of the organisms being targeted. The issues that are relevant to *V. cholerae* are not unique to this organism. Indeed, in the case of *E. coli*, toxin converting phages exist for Shiga toxin and enterotoxin encoding plasmids that are conjugative have also been found. Similarly, *Salmonella* is a genus that is well known for its transducing and converting phages. Just as decisions about vaccine use are often based on issues beyond the clinical utility of the vaccine, so too should risk-based decisions be based on a matrix of information and need. Inevitably, these will be handled on a case-by-case basis, and we should focus on providing the best data possible to help this process.

References

Albert MJ, Bhuiyan NA, Talukder KA, Faruque ASG, Nahar S, Faruque SM, Ansaruzzaman M, Rahman M (1997) Phenotypic and genotypic changes in Vibrio cholerae O139 Bengal. J Clin Microbiol 35:2588–2592

Anderson RJ, Pasetti MF, Sztein MB, Levine MM, Noriega FR (1997) Immune response induced by a Shigella vaccine strain harboring a eukaryotic expression vector. ASM 97th General Meeting, Miami Beach, FL, May 4–8

Barzu S, Fontaine A, Sansonetti P, Phalipon A (1996) Induction of a local anti-IpaC antibody response in mice by use of a Shigella flexneri 2a vaccine candidate: implications for use of IpaC as a protein carrier. Infect Immum 64:1190–1196

Brennan MJ, Burns DL, Meade BD, Shahin RD, Manclark CR (1992) Recent development in the development of pertussis vaccines. In: Ellis RW (ed) Vaccines: new approaches to immunological problems. Butterworth-Heinemann, Boston, pp 23–52

Butterton JR, Beattie DT, Gardel CL, Carroll PA, Hyman T, Killeen KP, Mekalanos JJ, Calderwood SB (1995) Heterologous antigen expression in Vibrio cholerae vector strains. Infect Immun 63:2689–2696

Centers for Disease Control (1997) Poliomyelitis prevention in the United States: introduction of a sequential schedule of inactivated poliovirus vaccine followed by oral poliovirus vaccine. 46:RR-3

Cholera Working Group (1993) Large epidemic of cholera-like disease in Bangladesh caused by Vibrio cholerae O139 synonym Bengal. Lancet 342:387–390

Colwell RC (1996) Global climate and infectious disease: the cholera paradigm. Science 274:2025–2031

Coster TS, Killeen KP, Waldor MK, Beattie DT, Spriggs DR, Kenner JR, Trofa A, Sadoff JC, Mekalanos JJ, Taylor DN (1995) Safety, immunogenicity, and efficacy of live attenuated Vibrio cholerae 0139 vaccine prototype. Lancet 345:959–952

Cryz SJ, Kaper J, Tacket C, Nataro J, Levine MM (1995) Vibrio cholerae CVD103-HgR live oral attenuated vaccine: construction, safety, immunogenicity, excretion, and non-target effects. Dev Biol Stand 84:237–244

DiRita VJ, Neely M, Taylor RK, Bruss PM (1996) Differential expression of the ToxR regulon in classical and El Tor biotypes of Vibrio cholerae is due to biotype-specific control over toxT expression. Proc Natl Acad Sci USA 93:7991–7995

Edwards MF, Stocker BAD (1984) Construction of ΔaroA his ΔpurA strains of Salmonella typhi. J Bacteriol 170:3991–3995

Faruque SM, Ahmed KM, Abdul Alim ARM, Qadri F, Siddique AK, Albert J (1997) Emergence of a new clone of toxigenic Vibrio cholerae O1 biotype El Tor displacing V. cholerae O139 Bengal in Bangladesh. J Clin Microbiol 35:624–630

Faruque SM, Ahmed SM, Siddique AK, Zaman K, Abdul Amin ARM, Albert J (1997) Molecular analysis of toxigenic Vibrio cholerae O139 Bengal strains isolated in Bangladesh between 1993 and 1996: evidence for emergence of a new clone of the Bengal vibrios. J Clin Microbiol 35:2299–2306

Fettner A (1994) A Crack in the shield – our unvaccinated children. A report on the colloquium. Cold Spring Harbor Laboratory, December 11–14

Fontaine A, Arondel J, Sansonetti PJ (1990) Construction and evaluation of live attenuated vaccine strains of Shigella flexneri and Shigella dysenteriae 1. Res Microbiol 141:907–912

Gentschev K, Hess J, Goebel W (1990) Change in the cellular localization of alkaline phosphatase by alteration of its carboxy-terminal sequence. Mol Gen Genet 222:211–216

Gonzalez C, Hone D, Noriega F, et al. (1994) *S. typhi* vaccine strain CVD 908 expressing circumsporozoite protein of Plasmodium falciparum: strain construction and safety and immunogenicity in humans. J Infect Dis 169:927–931

Haider K, Metcalfe KA, Beattie DT, Killeen KP (1995) Vibrio-vectored Shigella sonnei O-antigen biosynthetic genes. 95th General Meeting of the American Society of Micriobiology

Herrington DA, Hall RH, Losonsky G, Mekalanos JJ, Taylor RK, Levine MM (1988) Toxin, toxin-coregulated pili, and the toxR regulon are essential for Vibrio cholerae pathogenesis in humans. J Exp Med 168:1487–1492

Hess J, Gentschev I, Miko D, Welzel M, Ladel C, Goebel W, Kaufmann SHE (1996) Superior efficacy of secreted over somatic antigen display in recombinant Salmonella vaccine induced protection against Listeriosis. Proc Natl Acad Sci USA 93:1458–1463

Hess J, Dietrich G, Gentschev I, Miko D, Goebel W, Kaufmann SHE (1997) Protection against murine Listeriosis by an attenuated recombinant Salmonella typhimurium vaccine strain that secretes the naturally somatic antigen superoxide dismutase. Infect Immun 65:1286–1292

Hohmann EL, Oletta CA, Miller SI (1996) Evaluation of a phoP/phoQ-deleted, aroA-deleted Live Oral *S. typhi* vaccine strain in human volunteers. Vaccine 14:19–24

Hohmann EL, Oletta CA, Killeen KP, Miller SI (1996) phoP/phoQ-deleted *S. typhi* (Ty800) is a safe and immunogenic single-dose typhoid fever vaccine in volunteers. J Infect Dis 173:1408–1414

Hone DM, Attridge SR, Forrest B, Morona R, Daniels D, LaBrooy JT, Bartholomeusz RZ, Shearman DJ, Hackett J (1988) A galE via (Vi antigen-negative) mutant of *S. typhi* Ty2 retains virulence in humans. Infect Immun 56:1326–1333

Jordan Report (1996) Accelerated Development of Vaccines. Division of Microbiology and Infectious Diseases. National Institute of Allergy and Infectious Disease. National Institutes of Health

Kaper JB, Lockman H, Balini M, Levine MM (1984) Recombinant nontoxinogenic Vibrio cholerae strains as attenuated cholera vaccine candidates. Nature 308:655–658

Käppeli O, Auberson L (1997) The science and intricacy of environmental safety evaluations. Tibtech 15:342–349

Karolis DKR, Lan R, Reeves PR (1995) The sixth and seventh cholera pandemics are due to independent clones separately derived from environmental, nontoxigenic, non-O1 Vibrio cholerae. J Bacteriol 177:3191–3198

Kaufmann SHE (1996) Concepts in vaccine development. de Gruyter, Berlin

Kenner JR, Coster TS, Taylor DN, Trofa AF, Barrera-Oro M, Hyman T, Adams JM, Beattie DT, Killeen KP, Spriggs DR, Mekalanos JJ, Sadoff JC (1995) Peru-15, an improved live attenuated oral vaccine candidate for Vibrio cholerae 01. J Infect Dis 172:1126–1129

Levine MM, Ferreccio C, Black RE, Germanier R (1987) Large-scale field trial of Ty21a live oral typhoid vaccine in enteric-coated capsule formulation. Lancet 1:1049–1052

Levine MM, Herrington D, Murphy JR, et al. (1987) Safety, infectivity, immunogenicity and in vivo stability of two attenuated auxotrophic mutant strains of *S. typhi* 541Ty and 543Ty, as live oral vaccines in man. J Clin Invest 79:888–902

Levine MM, Kaper JB, Herrington D, Losonsky G, Morris JG, Clements M, Black RE, Tall B, Hall R (1988) Volunteer studies of deletion mutants of Vibrio cholerae 01 prepared by recombinant techniques. Infect Immun 56:161–167

Levine MM, Kaper JB, Herrington D, Ketley J, Losonsky G, Tacket CO, Tall B, Cryz R (1988) Safety, immunogenicity, and efficacy of recombinant live oral cholera vaccines, CVD 103 and CVD 103-HgR. Lancet ii:467–470

Li A, Tibor P, Forsum U, Lindberg AA (1992) Safety and immunogenicity of live oral auxotrophic Shigella flexneri SFK24 in volunteers. Vaccine 10:395–404

Lorenz MG, Wackernagel W (1994) Bacterial gene transfer by natural genetic transformation in the environment. Microbiol Rev 58:563–602

Manning PA, Stroeher UH, Moron a R (1994) Molecular basis for O-antigen biosynthesis in Vibrio cholerae O1: Ogawa-Inaba switching. In: Wacksmuth IK, Blake PA, Olsvik O (eds) Vibrio cholerae and cholera. ASM Press, Washington, DC, pp 77–94

Mekalanos JJ, Swartz DJ, Pearson GD, Harford N, Groyne F, deWilde M (1983) Cholera toxin genes: nucleotide sequence, deletion analysis and vaccine development. Nature 306:551–557

Mekalanos JJ, Waldor MK, Gardel CL, Coster TR, Kenner J, Killeen KP, Beattie DT, Trofa A, Taylor DN, Sadoff JC (1995) Live cholera vaccines: perspectives on their construction and safety. Bull Inst Pasteur 93:255–262

Mekalanos JJ (1994) Live bacterial vaccines: environmental aspects. Curr Opin Biotech 5:312–319

Mooi RF, Bik EM (1997) The evolution of epidemic Vibrio cholerae strains. Trends Microbiol 5:161–165

Nakayama K, Kelly SM, Curtiss III R (1988) Construction of an ASD+ expression-cloning vector: stable maintenance and high level expression of cloned genes in a Salmonella vaccine strain. Biotechnology 6:693–697

Parsot C, Taxman E, Mekalanos JJ (1991) ToxT regulated the production of lipoproteins and the expression of serum resistance in Vibrio cholerae. Proc Natl Acad Sci USA, 88:1641–1645

Pearson GDN, Woods A, Chaing S Mekalanos JJ (1993) CTX genetic element encodes a site-specific recombination system and an intestinal colonization factor. Proc Natl Acad Sci USA 90:3750–3754

Ryan ET, Butterton JR, Zhang T, Baker MA, Stanley, Jr. SL, Calderwood SB (1997) Oral immunization with attenuated vaccine strains of Vibrio cholerae expressing a dodecapeptide repeat of the serine-rich entamoeba histolytica protein fused to the cholera toxin B subunit induces systemic and mucosal antiamebic and anti-V. cholerae antibody responses in mice. Infect Immun 65:3118–3125

Ryan ET, Butterton JR, Smith RN, Carroll PA, Crean TI, Calderwood SB (1997) Protective immunity against clostridium difficile toxin A induced by oral immunization with a live, attenuated Vibrio cholerae vector strain. Infect Immun 65:2941–2949

Sack DA, Sack RB, Shimko J, Gomes G, O'Sullivan D, Metcalfe K, and Spriggs D (1997a) Evaluation of Peru-15, a new live oral vaccine for cholera, in volunteers. J Infect Dis 176:201–205

Sack DA, Shimko J, Sack RB, Gomes G, MacLeod K, O'Sullivan D, Spriggs D (1997b) Comparison of alternative buffers for use with a new live oral cholera vaccine, Peru-15, in outpatient volunteers. Infect Immun 65:2107–2111

Sharma C, Balakrish Nair N, Mukhopadhyay AK, Bhattacharya SK, Ghosh RK, Ghosh A (1997) Molecular characterization of Vibrio cholerae O1 biotype El Tor strains isolated between 1992 and 1995 in Calcutta, India: evidence for the emergence of a new clone of the El Tor biotype. J Infect Dis 175:1134–1141

Sizemore, DR, Branstrom, AA, Sadoff, JC (1997) Attenuated bacteria as a DNA delivery vehicle for DNA-mediated immunization. Vaccine 15:804–807

Stover CK (1994) Recombinant vaccine delivery systems and encoded vaccines. Curr Opin Immunol 6:568–571

Stroeher UH, Jedani KE, Dredge BK, Morona R, Brown MH, Karageorgos LE, Albert MJ, Manning PA (1995) Genetic rearrangements in the rfb regions of Vibrio cholerae O1 and O139. Proc Natl Acad Sci USA 92:10374–10378

Suharyono C, Simanjuntak N, Witham N, Punjabi K, Heppner G, Losonsky G, Totsusdirjo H, Rifai A, Clemens J, Lim Y, Burr D, Wasserman S, Kaper J, Sorenson S, Cryz S, Levine M (1992) Safety and immunogenicity of single-dose live oral cholera vaccine CVD 103-HgR in 5–9 year-old Indonesian children. Lancet 340:689–694

Sztein MB, Wasserman SS, Tacket CO, et al. (1994) Cytokine production patterns and lymphoproliferative responses in volunteers orally immunized with attenuated vaccine strains of Salmonella typhi. J Infect Dis 170:1508–1517

Tacket CO, Hone DM, Losonsky GA, Guers L, Edelman R, Levine MM (1992) Clinical acceptability and immunogenicity of CVD 908 *S. typhi* vaccine strain. Vaccine 10:443–446

Tacket CO, Hone DM, Curtiss R, III, et al. (1992) Comparison of the safety and immunogenicity of ΔaroC ΔaroD and ΔcyaΔcrp *S. typhi* strains in adult volunteers. Infect Immun 60:536–541

Tacket CO, Kelly SA, Schodel F, Losonsky G, Nataro JP, Edelman R, Levine MM, Curtiss III R (1997) Safety and immunogenicity in humans of an attenuated *S. typhi* vaccine vector strain expressing plasmid-encoded hepatitis B antigens stabilized by the Asd-balanced lethal vector system. Infect Immun 65:33881–33885

Tacket CO, Sztein MB, Losonsky GA, et al. (1997) Safety and immune response in humans of live oral *S. typhi* vaccine strains deleted in htrA and aroC, aroD. Infect Immun 65:452–456

Tacket CO, Kelly SM, Schodel F, et al. (1997) Safety and immunogenicity in humans of an attenuated *S. typhi* vaccine vector strain expressing plasmid-encoded hepatitis B antigens stabilized by the ASD balanced lethal system. Infect Immun 65:3381–3385

Tzschaschel BD, Klee SR, de Lorenzo V, Timmis KN, Guzmán CA (1996) Towards a vaccine candidate against Shigella dysenteriae 1: expression of the Shiga toxin B-subunit in an attenuated Shigella flexneri aroD carrier strain. Micro Path 21:277–288

Veal DA, Stokes HW, Daggard G (1992) Genetic exchange in natural microbial communities. Adv Microb Ecol 12:383–430

Wahdan MH, Serie C, Cerisier Y, Sallam S, Germanier R (1982) A controlled field trial of live *S. typhi* strain Ty21a oral vaccine against typhoid: three year results. J Infect Dis 145:292–296

Waldor MK, Mekalanos JJ (1996a) Vibrio cholerae : molecular pathogenesis, immune response, and vaccine development. In: Paradise LJ (ed) Enteric infections and immunity. Plenum, New York, pp 37–56

Waldor MK, Mekalanos JJ (1996b) Lysogenic conversion by a filamentous phage encoding cholera toxin. Science 272:1910–1914

Waldor MK, Tschäpe H, Mekalanos JJ (1996) A new type of conjugative transposon encodes resistance to sulfamethoxazole, trimethoprim, and streptomycin in Vibrio cholerae O139. J Bacteriol 178: 4157–4165

Wilson M, Lindow SE (1993) Release of recombinant microorganisms. Annu Rev Microbiol 47:913–944

World Health Organization (1996) Report of the meeting on new strategies of accelerating Shigella vaccine development. 25–26 November, Geneva

Recombinant Viruses as Vectors for Mucosal Immunity

C.D. Morrow, M.J. Novak, D.C. Ansardi, D.C. Porter, and Z. Moldoveanu

1 Introduction

Vaccines represent one of the most effective ways to control the spread of infectious diseases. The design of new vaccines that are effective against complex infectious agents presents several formidable challenges. Viral, bacterial and parasitic agents, which cause diseases of the respiratory, intestinal, and genital tract, enter through and can sometimes infect mucosal surfaces. Thus, a vaccine designed to protect against infection would need to stimulate both systemic and mucosal immune responses. Development of new vaccines designed to stimulate both humoral and mucosal immunity faces several hurdles. To stimulate mucosal immunity, the antigens need to be delivered to immunoreactive sites, such as the small intestine,

Department of Microbiology, University of Alabama at Birmingham, Birmingham, AL 35294, USA

nasal tissue, genital tract or rectum, where discrete lymphoid follicles are found (McGhee et al. 1992; Mestecky 1988; Ogra et al. 1973). Antigen stimulation at these mucosal sites results in the generation of large numbers of plasma-cell precursors, which have the capacity to migrate to distinct mucosal sites. This results in the appearance of antibodies in the corresponding secretions (Mestecky et al. 1987). The delivery of antigens to mucosal sites must overcome many difficulties. For example, the low pH environment of the stomach, as well as the presence of proteolytic enzymes in the stomach and intestines, precludes the oral administration of many antigens (Mestecky 1988). In many instances, mucosal surfaces also have physical as well as biochemical barriers which make the administration of these vaccines difficult.

1.1 Viral Infections at Mucosal Surfaces

The natural transmission of numerous viruses is initiated at a mucosal surface (Table 1). As such, viruses have evolved numerous strategies to circumvent the physical and biochemical barriers. For example, picornaviruses and reoviruses are transmitted via the fecal or oral route and rely on a virion that is naturally resistant to the low pH and the numerous proteolytic enzymes (Fields et al. 1996). In

Table 1. Major viruses that infect via mucosa. Viruses that have their major route of infection at a mucosal surface. The most prominent members of each virus family is listed. The three major routes of transmission that were considered include fecal/oral, respiratory (upper and lower), and sexual transmission

RNA viruses	Route of immunization
Picornaviruses	
Enteroviruses	Fecal/Oral
Poliovirus	
Coxsackievirus	
Rhinoviruses	Nasal/Upper respiratory
Calicivirus	Fecal/Oral
Norwalk group	
Orthomyxovirus	
Influenza virus	Upper/Lower respiratory
Paramyxovirus	Respiratory
Parainfluenza virus	
Mumps	
Measles	
Respiratory syncytial virus	
Reovirus	
Rotavirus	Fecal/Oral
Retrovirus	
HIV	Sexual transmission
DNA Viruses	
Adenoviruses	Upper/Lower respiratory
Herpes simplex virus	Sexual transmission
Cytomegalovirus	Sexual transmission/respiratory
Papillomavirus	Sexual transmission/respiratory (?)

contrast, adenoviruses and orthomyxoviruses (influenza) replicate in the upper and lower respiratory tract, respectively, and are transmitted mainly by aerosolization (FIELDS et al. 1996). Finally, viruses such as human immunodeficiency virus (HIV) and herpes simplex virus, which are rapidly inactivated in the environment, have evolved to be sexually transmitted (FIELDS et al. 1996).

Viruses naturally transmitted through mucosal surfaces pose a challenging problem for the development of vaccines targeted to prevent infection or disease. The most effective vaccine for these viruses would probably be expected to target a mucosal surface. One of the first and still most effective vaccines in this regard is the oral poliovirus vaccine (SABIN et al. 1973). This vaccine, termed the Sabin vaccine, is composed of the attenuated strains of the three different serotypes of poliovirus. Oral administration of these vaccine strains in infants results in the production of mucosal and systemic immunity and prevents spread of the wild-type virus to the central nervous system, which results in poliomyelitis (HANSON et al. 1984; OGRA 1984; OGRA et al. 1971; OGRA et al. 1968; SANDERS et al. 1974).

The administration of a live virus vaccine at the mucosal surface, based on studies with the poliovirus vaccine, affords many attractive features. First, immunization results in stimulation of humoral and cell-mediated responses in both mucosal and systemic compartments (GRAHAM et al. 1993; HANSON et al. 1984; OGRA and KARZON 1971; SIMMONS et al. 1993). Second, the immunity generated from the vaccination is usually long lasting. Oral immunization with the Sabin strains of poliovirus result in lifelong protection against poliomyelitis (HANSON et al. 1984; OGRA 1984; OGRA and KARZON 1971; OGRA et al. 1968; SANDERS and CRAMBLETT 1974). Oral vaccine strategies are also under development for influenza, parainfluenza and rotavirus (BELSHE et al. 1996; CONNOR et al. 1996; COUCH et al. 1996). The effectiveness of each strategy, compared with the oral Sabin vaccine for prevention of the disease poliomyelitis, will await evaluation.

An alternative strategy to oral delivery for generation of mucosal immunity involving administration via the respiratory tract has several potential advantages. First, the preparation can be given intranasally in a local and confined area. Second, the conditions within the nasal and upper respiratory tract are generally viewed as less harsh than the intestinal environment. Recent studies have established that administration via the nasal tract results in an effective mucosal immune response, not only in the respiratory tract, but also in secondary sites such as the female reproductive organ (GALLICHAN and ROSENTHAL 1995).

1.2 Viral Vaccine Vectors

The ability of certain viruses to infect mucosal surfaces as well as stimulate both the systemic and mucosal immune systems forms the foundation for the development and use of these viruses as vaccine vectors (Table 2). The development of these novel recombinant vectors has been facilitated by rapid advancements in molecular biology related to complementary DNA (cDNA) cloning of large DNA inserts and the rapid advancement of polymerase chain reaction (PCR)-based technology.

Table 2. Viral vector systems for delivery of antigens to mucosal surfaces. Viruses discussed in this review that have been developed as vaccine vectors and have potential uses as mucosal vaccines

RNA Viruses
1. Picornaviruses
Poliovirus
2. Alphaviruses
a. Sindbis virus/Semliki Forest virus
b. Venezuelan equine encephalitis virus
3. Orthomyxoviruses
a. Influenza (A, B, C)
4. Paramyxoviruses
a. Respiratory syncytial viruses
DNA Viruses
1. Poxviruses
2. Adenoviruses

Experimental systems are now available in which infectious viruses can be generated following transfection of suitable cDNA clones into tissue-culture cells. Using a molecular biological approach, these cDNAs can be manipulated as vaccine vectors to encode target foreign proteins of specified pathogens.

Three general strategies have been employed by investigators using infectious clones of viruses. In the first, investigators used recombinant DNA techniques to display regions of the antigenic proteins on the virion surface. This approach involved the genetic grafting of epitopes of viral proteins on the outer surface of the virus. For example, the influenza hemagglutinin (HA) protein can be genetically engineered to display foreign epitopes. The altered HA can then be incorporated onto an influenza virion. Immunization with these recombinant influenza virions results in an immune response against both the influenza and the target epitope (Muester et al. 1995). The information from the three-dimensional structure of several picornaviruses has provided the opportunity to design recombinant viruses that display foreign epitopes. Both poliovirus and, more recently, human rhinoviruses have been constructed which display foreign epitopes substituted for the amino acids constituting the major neutralization site (Almond and Burke 1990; Ferstandig-Arnold et al. 1994). Immunization with these viruses resulted in production of antibodies specific for foreign peptides displayed on their surface. A second approach is to modify the infectious clone of the designated virus to encode foreign genes. This vector system relies on the capacity of some of these viruses to encapsidate genomes that are larger in size than the wild-type genome. This strategy has been used for both RNA and DNA viruses and will be discussed later in this chapter. The third strategy is to substitute specific genes of viruses with those encoding foreign proteins, resulting in viruses that are naturally defective and can undergo only a single round of infection. Growth of these viruses is facilitated by supplying the substituted genes from an external source in *trans*. This strategy has been successfully used for numerous RNA viruses, including picornaviruses and alphaviruses and DNA viruses, such as adenoviruses and poxviruses.

In the following sections, recombinant vaccine strategies using RNA or DNA viruses are presented. The use of RNA viruses as vaccine vectors initially lagged behind the use of DNA viruses due to the relative ease of manipulating the DNA genomes with molecular biological methods. Given the explosion of recombinant DNA technology, including PCR, the development of RNA viruses as vaccine vectors has now caught up with the DNA viruses.

2 Vaccines Based on Recombinant RNA Viruses

RNA viruses are categorized based on the polarity of their genomes (Fields et al. 1996). Viruses in which the virion RNA is the same coding sense as a messenger RNA (mRNA) are designated as plus-strand viruses. Distinguishing features of plus-strand RNA viruses include the fact that the virion RNA is infectious upon transfection and the virion does not contain a polymerase. The exception to this would be retroviruses. By virtue of the fact that the cDNAs for plus-strand RNA viruses are infectious when introduced into cells, the development of plus-strand RNA viruses as vectors has preceded that of the negative-strand RNA viruses. However, recent advances in understanding the biology of negative-strand RNA viruses have led to the development of new vector systems based on some of these viruses. In Sect. 2.1, a summary of the most common vector systems based on RNA viruses and their unique properties that point to applications as mucosal vaccine vectors will be discussed.

2.1 Negative-Strand RNA Viruses

The port of entry for many negative-strand RNA viruses used as vectors is mainly restricted to the respiratory tract. Viruses such as influenza (A, B, and C), parainfluenza, respiratory syncytial virus, measles and mumps infect the host respiratory tract via aerosolization (Fields et al. 1996). For the most part, the general virion structure of these viruses is very similar. The RNA genome is protected by a helical nucleocapsid (NP) which is surrounded by a host-derived membrane (envelope). Viral proteins that determine host range and cell tropism are included in this envelope.

The development of infectious cDNAs for negative-strand RNA viruses was slowed due to inherent features of virus replication in common with all negative-strand RNA viruses. Most importantly, the virion RNA from a negative-strand virus is not infectious. The first step in replication following entry of these viruses is the transcription of the negative-strand RNA genome to generate plus-strand RNAs, which can then be translated to produce viral proteins. These viral proteins, in turn, replicate the negative-strand RNA which interacts with viral proteins prior to release, in the form of a NP or infectious particle. Since the negative-strand RNA itself is not infectious, generation of encapsidated negative-strand RNA was required to facilitate development of an infectious clone. Success in this area has now been achieved for several negative-strand RNA viruses which will undoubtedly lead to further development as mucosal vaccine vectors.

2.2 Segmented Negative-Strand RNA Viruses

In the last 5 years, considerable efforts have been directed toward the development of suitable molecular biological methods to engineer segmented RNA virus genomes and, in particular, influenza. The initial strategies, pioneered by Palese and co-workers (1996), relied on reconstituting a biologically active ribonucleic–protein complex consisting of synthetic RNA and purified nucleoprotein/polymerase proteins in vitro. This complex was then transfected back into helper virus (influenza)-infected cells. The helper virus in this system provided in *trans* those proteins necessary for the replication and encapsidation of the synthetic gene. The selection of the virus is achieved by utilizing a host range or temperature-sensitive mutant as the helper virus. Using this procedure, Palese and co-workers (1996) have demonstrated that six of the eight genes of influenza viruses can be altered using genetic-engineering methodologies (Fig. 1).

A second approach, which has been developed recently, is to reconstitute the biologically active influenza-virus ribonucleic–protein complex within the cell. To do this, plasmids encoding the different influenza virus proteins are co-transfected with a plasmid encoding the synthetic gene. To derive a transcript with the necessary 3′ termini, a hepatitis delta ribozyme was positioned 3′ to the synthetic gene; the 5′ termini is generated via initiation of transcription from a DNA promoter. This system also requires the use of a helper virus which provides the genetic backbone into which the plasmid-derived genes can be introduced. The advantage of this system is that it eliminates the need to isolate the ribonucleic–protein complexes (Palese et al. 1996).

The development of molecular biological techniques for manipulation of the influenza genome has also resulted in the development as an expression vector. For the most part, these studies have been directed at "epitope grafting", in which small defined epitopes substitute for regions in the outer surface proteins of influenza HA and/or neuraminidase (NA). Epitopes derived from HIV, plasmodia or lymphocytic choriomenengitis virus proteins have been successfully expressed when used to substitute for regions of the HA or NA from different influenza viruses (Li et al. 1993; Muester et al. 1995). Administration of these recombinant influenza viruses resulted in a potent B-cell and or T-cell response against the foreign epitope. A companion approach in which bicistronic mRNAs, encoding the HA/NA and a second foreign gene of interest, are generated has also been developed. To allow expression of the second gene, an element containing the encephalomyocarditis virus internal ribozome-entry site was added between the genes. This strategy affords the possibility of expression of larger and more complex genes (Palese et al. 1996). These viruses offer the exciting possibility of the generation of a mucosal

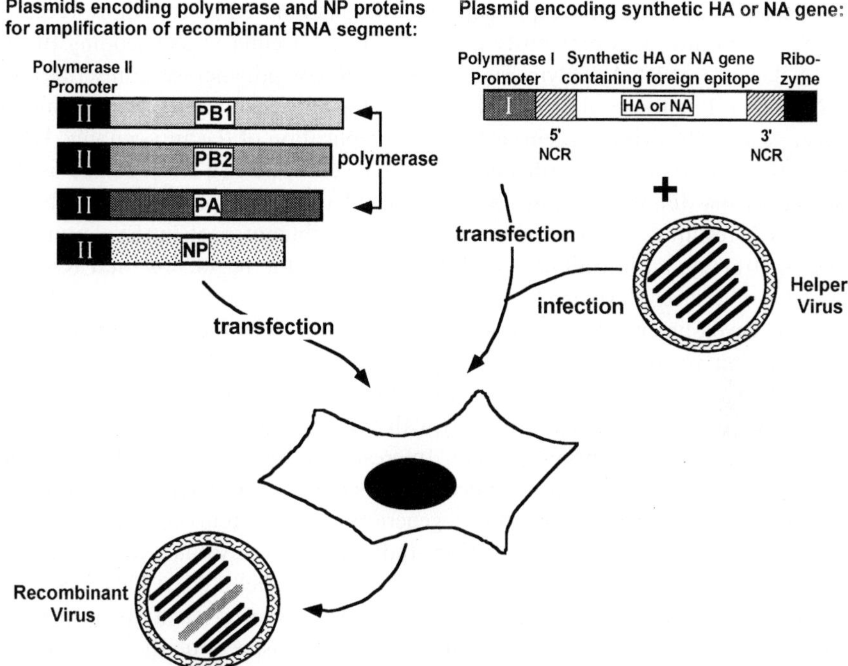

Fig. 1. Expression of foreign genes by recombinant negative-strand segmented viruses. Plasmid-based expression systems for segmented negative-strand RNA viruses (such as Influenza) have been developed. In the schematic shown, influenza replication proteins (PB1, PB2, PA, and NP) required for amplification of viral RNA segments are expressed from plasmids containing cDNA copies of the individual genes under control of a pol II promoter. These plasmids are cotransfected into cells along with a plasmid cDNA encoding a synthetic influenza gene (HA or NA encoding a foreign epitope) flanked by cDNA copies of influenza 5′ and 3′ non-coding regions. A synthetic ribozyme sequence is positioned after the 3′ non-coding region to generate an authentic 3′ viral RNA end through autocatalytic cleavage. These non-coding regions contain the *cis*-sequences required for replication and amplification of the viral RNA segment upon expression from the pol I promoter. This amplification is mediated by the influenza replication proteins provided in trans from the co-transfected cDNAs, resulting in high level expression of the synthetic HA or NA. If the transfected cells are co-infected with a helper virus, a recombinant influenza virion containing the synthetic influenza RNA segment can be generated (Adapted from Palese et al. 1996)

immune response, since they are naturally tropic for the upper and lower respiratory tracts.

2.3 Non-segmented Negative-Strand RNA Viruses

Several recent developments in the study of non-segmented negative-strand RNA viruses have resulted in simultaneous development of expression systems for vesicular stomatitis virus, (VSV), Sendai virus, respiratory syncytial virus, and parainfluenza virus type 3. The general strategy for each of these viruses is similar. Plasmids encoding viral capsid proteins, NP and polymerase (L) proteins are co-transfected with a plasmid expressing the anti-genomic RNA (or, more recently, naked RNA) into cells previously infected with a vaccinia virus encoding the T7 DNA-dependent RNA polymerase. The T7 RNA polymerase transcribes the mRNA encoding the viral proteins, as well as the anti-genomic RNA encoding the target protein. The viral proteins catalyze the replication of the anti-genomic RNA, which is encapsidated into authentic virions. The encapsidated RNA can be used to infect cells and, following transcription, the RNA produced (plus-strand mRNA) is translated, resulting in the expression of the recombinant protein. Exploiting this general strategy, an infectious cDNA for several of the negative-strand RNA viruses has now been developed.

The development of an infectious clone has prompted the use of these viruses as expression vectors. In a recent study, SCHNELL et al. (1996) generated recombinant VSV which encodes the genes for HIV envelope or measles hemagglutinin-neuraminidase (HN). Infection of cells with the recombinant VSV resulted in the expression of the appropriate protein. Interestingly, encapsiration of the mutant VSV genome resulted in an extended bullet-shaped virus. Whether this unique feature of the VSV biology will allow a generation of recombinant viruses encoding even larger proteins will require further study. Studies to evaluate the immunogenicity of these recombinant viruses are ongoing. Parallel studies have been performed with other non-segmented RNA virus genomes, such as respiratory syncytial virus and parainfluenza virus. The unique aspects of these viruses, including their port of entry being the respiratory tract, point to the exciting possibility of development of these viruses as mucosal vaccines.

3 Plus-Strand RNA Viruses

The development of expression systems based on plus-strand RNA viruses has been greatly facilitated by the fact that the RNA genome, upon transfection into tissue-culture cells, results in the production of infectious viruses. A major difference in the replication strategy between plus- and negative-strand RNA viruses is that the first step in the replication of plus-strand RNA viruses is translation of the incoming viral genome. In contrast, transcription of negative-strand RNA genomes must occur prior to translation. This feature of plus-strand RNA viruses resulted in the early development of infectious cDNAs for poliovirus and, later, several different plus-strand RNA viruses. In this section, the vector systems derived from two plus-strand RNA virus families, the alphavirus and picornaviruses, will be discussed.

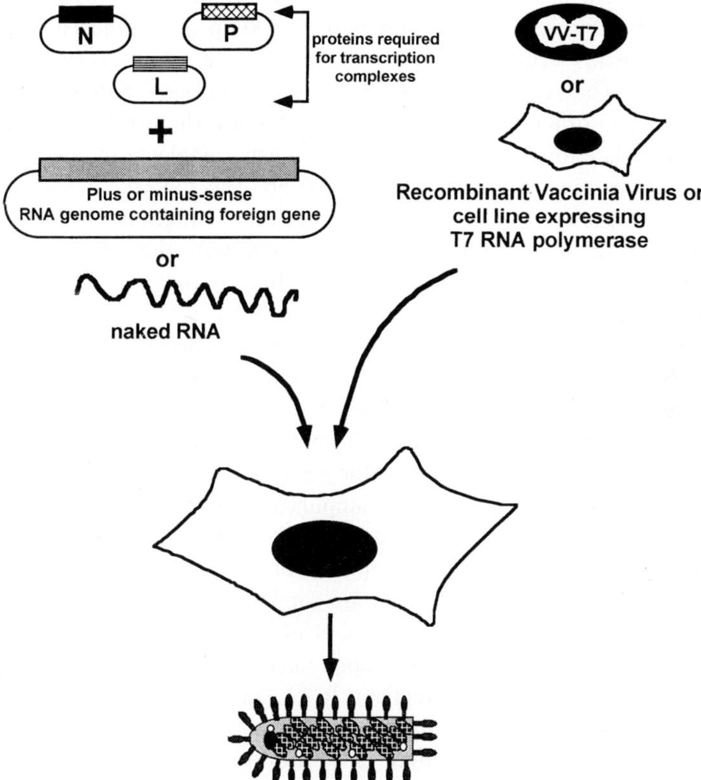

Fig. 2. Expression of foreign genes by non-segmented negative-strand RNA viruses. Reverse genetic systems for amplification of recombinant non-segmented negative-strand RNA viruses (such as VSV) are based on transfection of plasmids encoding three viral proteins in conjuction with a plasmid encoding a cDNA copy of the viral genome (positive or negative sense). Negative strand RNA viruses require viral proteins to initiate the transcription/replication processes, since the genomic RNA cannot be directly translated. In the plasmid-based systems for expression of recombinant negative-strand viral genomes (such as VSV-based systems depicted in this diagram), the required proteins N, P, and L are expressed from plasmids. N, P, L are obligatory for assembly of transcription complexes. The recombinant viral RNA genome is expressed from an additional plasmid or, alternatively, naked RNA may be introducted into the cells. Expression of the viral proteins and/or the recombinant RNA genome is under control of a promoter for T7 RNA polymerase. Expression of these genes/RNA is amplified upon transfection into cel lines that express T7 RNA polymerase or cells that have been infected with a recombinant vaccinia virus that expresses T7 RNA polymerase. Upon expression of these components, replication of the recombinant RNA genome ensues. Assembly of infectious, recombinant viral particles occurs if all of the wild-type viral genes are maintained within the RNA genome, and the foreign gene is inserted as an additional transcriptional unit. These recombinant RNA genomes are packaged into helical virions that are longer than the wild-type virus as a result of the longer RNA genome (Adapted from PALESE et al. 1996)

3.1 Alphavirus Vectors

The alphaviruses are enveloped plus-strand RNA viruses. Viral glycoproteins that specify the tissue tropism are embedded within this envelope. The plus-strand RNA genomes of alphaviruses are approximately 11,000 bp in length. The expression of viral proteins occurs first with the production of non-structural (replication) proteins. The replication proteins copy the plus-strand RNA, resulting in the production of a complete negative-strand RNA; the negative-strand RNA serves as a template for the synthesis of either a new complete genome or a sub-genomic RNA consisting of approximately 4500 bp. The sub-genomic RNA encodes the structural proteins for alphaviruses. Since more structural than non-structural proteins are required during the alphavirus infection, the sub-genomic RNAs are produced in a greater quantity than the full-length genomic RNAs (Fields et al. 1996).

Sindbis virus and Semliki Forest viruses have received considerable attention in the development of vaccine vectors (Frolow et al. 1996; Liljestrom et al. 1991). These viruses are unique in their ability to infect a wide range of vertebrate and invertebrate cells and initiate a prolific cytoplasmic replication cycle. Previous studies have demonstrated that substitution of the genes encoding the sub-genomic mRNA with those of foreign genes allows the amplification and expression of the foreign protein upon introduction into the designated target cell. Two different strategies have been employed for the development of expression systems using alphaviruses. In the first, the structural genes are supplied in *trans* from a complementing cDNA (Frolow et al. 1996; Roden et al. 1996). This results in the production of an encapsidated genome which, upon infection, can undergo a single round of infection/amplification of the replicon RNA (Liljestrom and Garoff 1991). A second approach is to take advantage of the fact that the alphavirus capsid can accommodate larger than genome length RNA. In this case, the heterologous genes are cloned in tandem with the structural genes. This results in an infectious virus that, upon infection, expresses the target heterologous genes. This feature has recently been exploited in the development of recombinant vectors based on Venezuelan equine encephalitis virus (VEE) (Davis et al. 1996). Replication-competent VEE vectors expressing the influenza virus HA gene were found to be potent vaccine vectors with respect to their capacity to induce both systemic and mucosal immune responses in animals. VEE has an interesting feature in that it first replicates in the lymph nodes draining the site of inoculation, thus resulting in potent systemic and mucosal immune responses. Coupled with the fact that most human and animal populations are not already immune to VEE, this virus affords exciting future potential as a mucosal vaccine vector.

3.2 Picornavirus Vectors

Picornaviruses encompass a diverse group of small plus-strand RNA viruses. The viruses are non-enveloped. The enteroviruses (poliovirus, coxsackievirus) are transmitted mainly via the fecal or oral route. Thus, they are inherently stable to the harsh environmental conditions found in the intestinal tract. A second group of picornaviruses, rhinoviruses, infect the nasal epithelium and upper respiratory tract. In contrast to enteroviruses, these viruses are not as stable and replicate best at 33°C (Fields et al. 1996).

Poliovirus affords many attractive features for development and use as a recombinant vaccine vector. The attenuated strains of poliovirus have been routinely administered to infants over the last 30 years as a safe and effective vaccine against poliomyelitis. Numerous studies have confirmed that the use of the oral Sabin vaccine results in the generation of both systemic and mucosal immune responses (Hanson et al. 1984; Ogra and Karzon 1971; Sabin and Boulger 1973). Since studies have demonstrated that poliovirus has the propensity to interact with mucosal M cells, which are important for the antigen sampling of the environment required to generate a mucosal immune response (Neutra et al. 1996), the development of poliovirus, in particular, as a mucosal vaccine vector is attractive.

One approach for the development of poliovirus as a vaccine vector involves the insertion of foreign genes into the complete poliovirus infectious cDNA clone. (Racaniello et al. 1981). This approach, though, results in increasing the size of the poliovirus's genome to greater than the 7500 base pairs of the wild-type genome. Recombinant polioviruses generated using this strategy are replication competent and express the designated foreign protein upon infection. Using this strategy, recombinant polioviruses that express antigenic fragments of hepatitis B, HIV or simian immunodeficiency virus (SIV), group antigen (Gag), HIV or SIV envelope, or cholera toxin B subunit have been generated (Altmeyer et al. 1994; Andino et al. 1994; Mattion et al. 1995). Immunization of transgenic mice that carry the human receptor for poliovirus resulted in an immune response to the foreign protein. A potential drawback of this approach, however is the deletion of the foreign gene following successive replication cycles in the vaccinated individuals.

3.3 Poliovirus Replicons

A second strategy for the expression of foreign proteins from poliovirus genomes has been developed within this laboratory. (Morrow et al. 1994; Porter et al. 1993; Porter et al. 1995). This approach relies on the fact that regions within the poliovirus P1 gene, which encodes the capsids, can be deleted without compromising the overall capacity of the RNA to undergo self-amplification when transfected into cells. These deleted RNA genomes, termed "replicons", encode the necessary P2 and P3 region proteins required for self-amplification upon introduction into cells (Fig. 3). The foreign gene is used to substitute for the poliovirus genes encoding VP2-VP3-VP1 capsid proteins. The foreign gene is positioned such that the translational reading frame is maintained between the VP4 foreign gene and remaining P2-P3 region proteins. Upon transfection into cells, the replicon RNA is translated. A viral-encoded protease, $2A^{Pro}$, autocatalytically releases the

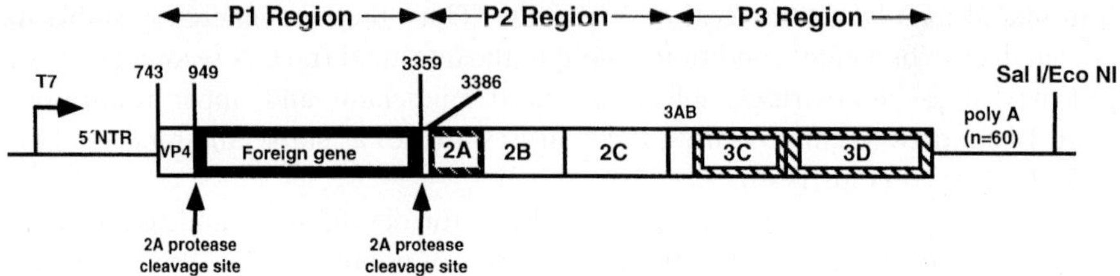

Fig. 3. Prototype replicon genome. Complementary DNA copies of replicon RNA genomes are positioned downstream of the promoter for bacteriophage T7 RNA polymerase, which allows in vitro transcription of full-length RNA from templates linearized at the *Sal I* (or *EcoNI*) site at the 3′ end. Replicon genomes are derived from type-1 poliovirus genomes in which most of the capsid gene has been replaced with sequences encoding a foreign gene that maintains the translational reading frame with the downstream viral genes. The encoded proteins are translated on transfection of the RNA into tissue-culture cells, first as a long polyprotein precursor. This polyprotein is cleaved by virus-encoded proteases, 2A and 3C. The foreign protein encoded by the replicon is released from the polyprotein upon cleavage by 2A at processing sites located at the amino and carboxy termini. In addition to the proteases, the proteins 2B, 2C, 3AB, and 3D (RNA-dependent RNA polymerase) are required for replication of the RNA genome. The 3CD protease (a fusion protein consisting of 3C and 3D) cleaves the viral capsid precursor protein (designated P1) into individual capsid subunits which assemble to form a complete capsid shell. In the case of replicons, this protease cleaves the P1 capsid precursor expressed by the recombinant vaccinia virus, VV-P1, initiating the process of *trans*-encapsidation of the replicon RNA genomes

VP4 foreign protein from the P2-P3 region proteins. Subsequent processing of the P2-P3 region proteins by a second viral protease, $3C^{Pro}$, releases the proteins required for RNA replication. We have engineered a second 2A cleavage site immediately 5′ to the foreign gene. Cleavage of the VP4 foreign protein by $2A^{Pro}$ in *trans* results in the production of a protein with only minimal additional amino acids at the amino terminus (two extra amino acids) and at the carboxy terminus (eight extra amino acids). Using this replicon strategy, we have expressed a wide array of proteins including HIV and SIV Gag and polymerase (Pol), SIV Nef, carcinoembryonic antigen (CEA), HER-2/neu oncogene, firefly luciferase, and the C-fragment of tetanus toxin.

Since the replicons do not encode capsid proteins, they do not have the capacity to spread from cell to cell. The replicons, though, can become encapsidated in poliovirions if the capsid proteins are provided in *trans*. To accomplish the encapsidation of replicons, we have derived a recombinant vaccinia virus that encodes the capsid proteins for poliovirus (Fig. 4) (Ansardi and Morrow 1993). Infection of cells with this recombinant vaccinia virus, (designated as VV P1), followed by transfection of the replicon RNA results in the production of P1 from the vaccinia virus. The P1 capsid protein encapsidates the replicon which is released from cells following successive freeze-thaw cycles. Reinfection of cells with the encapsidated replicon along with VV P1 results in the subsequent encapsidation of the new replicon genomes. Repetition of this process results in the generation of encapsidated stocks of replicon. Using this process, we have derived large-scale stocks of encapsidated replicons encoding numerous foreign genes.

The immunogenicity of the encapsidated replicons has been analyzed using several model systems. In one study, replicons encoding HIV-1 Gag were passaged

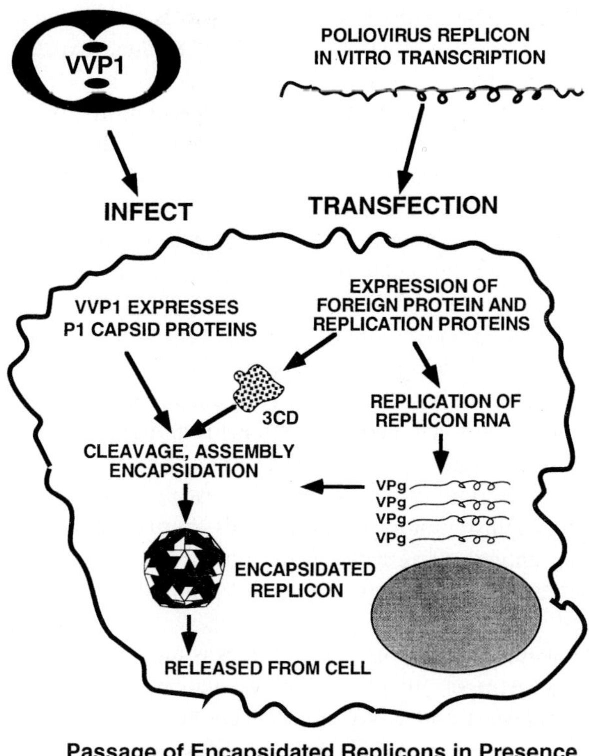

Fig. 4. Encapsidation system for recombinant replicons based on poliovirus. Replicon RNA genomes are transcribed in vitro using bacteriophage T7 RNA polymerase. The in vitro transcribed RNA is transfected into cells that have been infected previously with VVP1, the recombinant vaccinia virus that expresses the Poliovirus P1 capsid precursor protein. Upon transfection, the replicon RNA expresses replication proteins that self-amplify the RNA genome; the foreign protein encoded by the replicon is also expressed. A protease encoded by the replicon, 3CD, cleaves the P1 capsid precursor resulting in assembly and encapsidation of progeny replicon RNA genomes. The encapsidated replicons are released from the cell upon lysis

with poliovirus type 2, under conditions of high multiplicity of infection (MOI) reinfection. This process allowed the replicons to become encapsidated into type-2 capsids (Moldoveanu et al. 1995; Porter et al. 1993). Since mice are susceptible to type-2 poliovirus, we used the type-2/replicon preparation to immunize mice. Inoculation of mice (Balb/c strain) with the type-2/replicon preparation via intramuscular, intragastric or intrarectal routes resulted in the production of antibodies to poliovirus and HIV-1 Gag in both serum and secretions (Moldoveanu et al. 1995; Porter et al. 1993). In a different study, we used transgenic mice that express the human receptor for poliovirus (Mendelsohn et al. 1989). Infection of these mice with wild type poliovirus results in infection and paralysis (Ren et al. 1990). It was found that immunization of these mice with the encapsidated replicons encoding the C-fragment of tetanus toxin resulted in the production of serum antibodies against tetanus toxin. These antibodies were found to be protective in that they allowed the animals to resist a lethal challenge with tetanus toxin. In follow-up studies, it was found that the administration of the replicons via mucosal routes, i.e., gastric or nasal, resulted in the production of low levels of antibodies in secretions (Porter et al. 1997). In a recent study, we constructed a replicon that encodes the B subunit of *Helicobacter pylori* urease. Infection of cells with this encapsidated replicon encoding *ureB* resulted in the production of a protein that was immunoprecipitated with rabbit anti-urease antibodies. Immunization of transgenic mice with these replicons via the intramuscular route resulted in antibodies that reacted against the urease protein from a pathogenic strain of *H. pylori*. At present, it is being determined whether prior immunization with replicons encoding *ureB* will protect against challenge with *H. pylori* in a murine model system.

Studies have also been carried out with encapsidated replicons to determine whether immunization would result in a cell-mediated immune response. Replicons encoding the tumor antigen, CEA, were used to immunize mice (Ansardi et al. 1994). Challenge of these mice with tumors expressing CEA revealed that the immunized mice were protected against tumors, while animals immunized with a replicon encoding an irrelevant antigen all presented with tumors. A direct analysis of the cell-mediated immune response generated as a result of immunization with the replicons is being pursued.

In summary, the development of RNA viruses as vectors is an active area of experimentation. From advancements in molecular biology and expression technology, both negative- and positive-strand RNA viruses have been developed as vaccine vectors. The unique properties of some of these viruses, with respect to the capacity to replicate at mucosal surfaces, points to future applications as mucosally targeted vaccine vectors.

4 Vaccines Based on Recombinant DNA Viruses

DNA viruses have been used as experimental tools to understand basic elements of eukaryotic cell replication, transcription and translation. As a result of these studies, several DNA viruses have also been extensively exploited for use as expression vectors. In general, DNA viruses differ from RNA viruses in that certain DNA viruses can accommodate larger genes and are less prone to generating mistakes following replication due to proofreading mechanisms during replication. A feature of several DNA viruses is that many genes encoded by these viruses were not required for growth in tissue culture. For example, the thymidine kinase gene of vaccinia virus is not required for growth and can be substituted with a gene encoding a foreign protein. In other cases, genes essential for virus growth, e.g., E1A of adenovirus, can be expressed in continuous cell lines. Recombinant adenoviruses with foreign genes inserted into the E1A region could be propagated in this cell line. These features of poxviruses and adenoviruses have been exploited for the development of different DNA viruses as vaccine vectors.

4.1 Poxvirus-Based Vectors

The refinement of techniques in molecular biology, as well as the development of systems with which to manipulate large viral genomes, has led to the use of larger and considerably more complex DNA viruses as expression vectors. Most prominently, during the last 15 years, considerable progress has been made in the development of poxvirus as a recombinant vaccine vector (Moss 1996; Paoletti 1996). Poxviruses infect a wide variety of cells and replicate in the cytoplasm of infected cells. Through the efforts of several investigators, numerous transfer plasmids have been developed that allow the insertion of foreign genes behind poxvirus promoters, resulting in high-level expression of the recombinant protein. The promoter/foreign genes are positioned within genes of the poxvirus that are dispensable for growth in tissue culture. For example, the gene encoding poxvirus thymidine-kinase can be eliminated from the genome without compromising the capacity of this virus to replicate in tissue-culture cells. An important feature of vaccinia virus is the high recombination frequency that occurs between transfected plasmids and the virus genome in pox virus-infected cells. Thus, the generation of recombinant vaccinia viruses is, for the most part, relatively straightforward. The recombinant vaccinia viruses can also be purified and grown to high titers.

Recombinant vaccinia viruses have been used in numerous studies in small animals as well as in limited human studies to demonstrate that these viruses are immunogenic when given via routes typical for vaccinia immunization (scarification). Limited studies are available that describe the use of vaccinia vectors for mucosal routes of immunization. In a notable study, Small and colleagues demonstrated that oral administration of recombinant vaccinia viruses encoding influenza HA and NA resulted in the development of anti-influenza antibodies, both in the serum and secretions (Meitin et al. 1994). However, the use of replication-competent vaccinia viruses as a mucosal vaccine vector has several potential drawbacks. For example, the application with a replication-competent virus to mucosal tissues might lead to scarification. Also, administration of replication-competent viruses to immunosuppressed patients could lead to disease. Recent studies have attempted to circumvent the problems associated with live viruses by using vaccinia viruses that are defective for replication and assembly in mammalian cells. Vector systems have been developed based on either the avipox virus, which undergoes productive infection in avian cells, or the Ankara strain of vaccinia virus (MVA), which has been serially passaged in chicken embryo fibroblasts, resulting in large deletions in the genome, thus compromising the capacity of the virus to replicate in mammalian cells, and allowing single round infections within those cells (Moss 1996; Paoletti 1996). The enhanced safety properties of the new defective vaccinia virus vectors also point to their future use as mucosal vaccine vectors.

4.2 Adenovirus-Based Vectors

Human adenoviruses were isolated over four decades ago. At present, there are over 47 distinct human adenovirus serotypes, which are classified into six subgroups (A–F), based on a number of criteria. The different serotypes can infect and replicate in various locations in the body, including the upper respiratory tract and gastrointestinal tract (Fields et al. 1996). Infection with adenovirus usually results in a mild respiratory illness, with symptoms including rhinorrhea, nasal congestion, sneezing, or conjunctivitis. The use of adenovirus as a mucosal vaccine vector is attractive because serotypes 4 and 7 have been proven safe and effective when administered to millions of military recruits over the last 30 years. These viruses can be administered to mucosal surfaces. Oral administration with enteric coated capsules containing the live serotypes produces an asymptomatic intestinal infection and protection against acute respiratory disease.

The development of adenovirus-based vectors has been facilitated by the immense scientific knowledge about virus replication and expression of viral proteins. Adenoviruses are non-enveloped icosahedral viruses, containing a linear double-stranded DNA. The life cycle has been arbitrarily divided into two phases, early and late, corresponding to the events before and after initiation of viral DNA replication. Two different strategies have been utilized for the development of recombinant adenoviruses. The first takes advantage of the fact that adenoviruses can package approximately 105% of their genomes, which corresponds to approximately 2 kb of additional foreign DNA. To insert larger fragments of DNA, deletions within the adenovirus genome must be created, which can then be compensated by proteins provided in *trans*. Most of these regions are located within the E1 and E3 regions of the adenovirus genome which can be complemented in *trans*, using 293 cells which contain and express the left end of the adenovirus genome. The use of these deletions results in an increase of the capacity for foreign genes up to 8 kb.

Adenovirus vectors have been used to express a variety of key genes involved in recombinant vaccines. Included in this list are those for hepatitis B surface antigen, VSV glycoproteins, HIV *gag*, *pol* and *env*, Herpes simplex virus glycoprotein B, respiratory syncytial virus F and G glycoproteins, rabies virus glycoproteins and parainfluenza type III glycoproteins F and HN (Davis et al. 1985; Dewar et al. 1989; Prevec et al. 1991; Schneider et al. 1989). Studies using these recombinant viruses in animal model systems have demonstrated their immunogenicity. Most importantly, adenoviruses expressing HIV Gag and Env proteins have been shown to be immunogenic in primates, including chimpanzees. In a recent study, Buge and co-workers demonstrated that oral and intranasal immunization of rhesus macaques with a recombinant adenovirus expressing SIV envelope resulted in the generation of both humoral and cellular, as well as mucosal immune responses (Buge et al. 1997). Furthermore, challenge of these animals with infectious SIV via the vaginal route resulted in a decreased viral burden. Thus, the possibility exists that the further development of recombinant adenoviruses as mucosal vaccine vectors could be used for mucosally targeted vaccines.

5 Summary and Conclusions

The development and characterization of viral based vaccine vectors is an extremely active research field. Much of this work has been facilitated by developments in molecular biology that allow work with large plasmid-based vectors, as well as the use of PCR. Several different vector systems are now available using RNA viruses and DNA viruses. Each vector system has its own strengths and weaknesses. Due to the differences and diversity between the viruses used as vectors, it is doubtful that a single system will be useful for all desired vaccines. However, the further development of existing, as well as potentially new systems, will provide a repertoire for vaccinologists to design the recombinant vaccine which will generate an optimal humoral and immune response for protection against infection or disease caused by pathogens that infect via mucosal surfaces.

Acknowledgements. We thank Dee Martin for preparation of the manuscript. The research was supported by grants from the NIH (AI 25005, AI 28147) and the Department of Defense (DAMD17-J-4403).

References

Almond JW, Burke KL (1990) Poliovirus as a vector for the presentation of foreign antigens. Semin Virol 1:11–20

Altmeyer R, Escriou N, Girard M, Palmenberg A, van der Werf S (1994) Attenuated Mengo virus as a vector for immunogenic human immunodeficiency virus type 1 glycoprotein 120. Proc Natl Acad Sci USA 91:9775–9779

Andino R, Silvera D, Suggett SD, Achacoso PL, Miller CJ, Baltimore D, Feinberg MB (1994) Engineering poliovirus as vaccine vector for the expression of diverse antigens. Science 265:1448–1451

Ansardi DC, Moldoveanu Z, Porter DC, Walker DE, Conry RM, LoBuglio AF, McPherson S, Morrow CD (1994) Characterization of poliovirus replicons encoding carcinoembryonic antigens. Cancer Res 54:6359–6363

Ansardi DC, Morrow CD (1993) Poliovirus capsid proteins derived from P1 precursors with glutamine-valine cleavages sites have defects in assembly and RNA encapsidation. J. Virol., 67:7284–7297

Belshe RB, Newman FK, Ray R (1996) Parainfluenza Virus Vaccines. In: H Kiyono, PL Orga, JR McGhee, (eds) Mucosal Vaccines. Academic Press, San Diego, pp 311–319

Buge SL, Richardson E, Alipanah S, Markham P, Cheng S, Kalyan N, Miller CJ, Lubeck M, Udem S, Eldridge J, Guroff MR (1997) An adenovirus-simian immunodeficiency virus *env* vaccine elicits humoral, cellular, and mucosal immune responses in Rhesus Macaques and decreases viral burden following vaginal challenge. J Virol 71:8531–8541

Connor ME, Estes MK, Offit PA, Clark HF, Franco M, Feng Jr. N, Greenberg Jr. H (1996) Development of a mucosal rotavirus vaccine. In: H Kiyono, PL Orga, JR McGhee (eds) Mucosal Vaccines. Academic Press, San Diego, pp 325–339

Couch RB, Cate TR, Keitel WA (1996) Oral immunization with influenca virus vaccines. In: H Kiyono, PL Orga, JR McGhee (eds). Mucosal Vaccines. Academic Press, San Diego, pp 303–308

Davis AR, Kostek B, Mason BB, Hsiao CL, Morin J, Dheer SK, Hung PP (1985) Expression of hepatitis B surface antigen with a recombinant adenovirus. Proc Natl Acad Sci USA 82:7560–7564

Davis NL, Brown KW, Johnston RE (1996) A viral vaccine vector that expresses foreign genes in lymph nodes and protects against mucosal challenge. J Virol 70:3781–3787

Dewar RL, Natarajan V, Vasudevachari MB, Sakzman NP (1989) Synthesis and processing of human immunodeficiency virus type 1 envelope proteins encoded by a recombinant human adenovirus. J Virol 63:129–136

Ferstandig-Arnold G, Resnick DA, Li Y, Zhang A, Smith AD, Geisler SC, Jacobo-Molina A, Lee W-M, Webster RG, Arnold E (1994) Design and construction of rhinovirus chimeras incorporating immunogens from polio, influenza, and human immunodeficiency viruses. Virology 198:703–708

Fields BN, Knipe DM, Howley PM (1996). In: Fields (ed) Virology, 3rd edn. Lippincott-Raven, Philadelphia, pp 609–654

Frolow I, Hoffman TA, Pragal BM, Dryga SA, Huang HV, Schlesinger S, Rice CM (1996) Alphaviruses-based expression vectors: strategies and applications. Proc Natl Acad USA 93:11371–11377

Gallichan WS and Rosenthal KL (1995) Specific secretory immune responses in the female genital tract following intra-nasal immunization with a recombinant adenovirus expressing glycoprotein β of Herpes simplex virus. Vaccine 13(16):1589–1595

Graham S, Wang ECY, Jenkins O, Borysiewicz LK (1993) Analysis of the human T-cell response to picornaviruses:identification of T-cell epitopes close to B-cell epitopes in poliovirus. J Virol 67:1627–1637

Hanson LA, Carlsson B, Jalil F, Lindblad BS, Khan SR, van Wezel AL (1984) Different secretory IgA antibody responses after immunzation with inactivated and live poliovirus vaccines. Rev Infect Dis 6:S356–S360

Li S, Rodrigues M, Rodriguez D, Rodriguez HR, Esteban M, Palese P, Nussenzweig RS, Xavala F (1993) Priming with recombinant influenza virus followed by administration of recombinant vaccinia virus induces CD8+ T cell-mediated immunity against malaria. Proc Natl Acad Sci USA 90:5214–5218

Liljestrom P, Garoff H (1991) A new generation of animal cell expression vectors based on the Semliki Forest virus replicon. Bio/Technology 9:1356–1361

Mattion NM, Reilly PA, Camposano E, Wu S-L, DiMichele SJ, Ishizaka T, Fantini SE, Crowley JC, Weeks-Levy C (1995) Charaterization of recombinant polioviruses expressing regions of rotavirus VP4, hepatitis B surface antigen, and herpes simplex virus type 2 glycoprotein D.J. Virology 69:5132–5137

McGhee JR, Mestecky J (1992) The mucosal immune system in HIV infection and prospects for mucosal immunity to AIDS. AIDS Res Rev 2:289–312

Meitin CA, Bender BS, Small Jr. PA (1994) Enteric immunization of mice against influenza with recombinant vaccinia. Proc Natl Acad Sci USA 91:11187–11191

Mendelsohn CL, Wimmer E, Racaniello VR (1989) Cellular receptor for poliovirus: molecular cloning, nucleotide sequence, and expression of a new member of the immunoglobulin superfamily. Cell 56:855–865

Mestecky J (1988) The common mucosal immune system and current strategies for induction of immune responses in external secretions. J Clin Immunol 7:265–276

Mestecky J, McGhee JR (1987) Immunoglobulin A (IgA): molecular and cellular interactions involved in IgA biosynthesis and immune response. Adv Immunol 40:153–245

Moldoveanu Z, Porter DC, Lu A, McPherson S, Morrow CD (1995) Immune responses induced by administration of encapsidated poliovirus replicons which express HIV-1 gag and envelope proteins. Vaccine 13:1013–1022

Morrow CD, Porter DC, Ansardi DC, Moldoveanu Z, Fultz PN (1994) New approaches for mucosal vaccines for AIDS: Encapsidation and serial passage of poliovirus replicons that express HIV-1 proteins on infection. AIDS Res Hum Retroviruses 10:S61–S66

Moss B (1996) Genetically engineered poxviruses for recombinant gene expression, vaccination, and safety. Proc Natl Acad Sci USA 93:11341–11348

Muester T, Freko B, Klima A, Purtscher M, Trkola A, Schultz P, Grassauer A, Engelhardt OG, Garcia-Sastre A, Palese P, Katinger J (1995) Mucosal model of immunization against human immunodeficiency virus type 1 with a chimeric influenza virus. J Virol 69:6678–6686

Neutra MR, Frey A, Kraehenbuhl J-P (1996) Epithelial M cells: gateway for mucosal infection and immunization. Cell 86:345–348

Ogra PL (1984) Mucosal immune response to poliovirus vaccines in childhood. Rev Infect Dis 6:S361–S368

Ogra PL, Karzon DT (1971) Formation and function of poliovirus antibody in different tissues. Prog Med Virol 13:56–193

Ogra PL, Karzon DT, Righthand F, MacHillivray M (1968) Immunoglobulin response in serum and secretions afer immunization with live and inactivated poliovaccine and natural infection. New Engl J Med 279:893–900

Ogra PL, Ogra SS (1973) Local antibody response to poliovaccine in the human female genital tract. J Immunol 110:1307–1311

Palese P, Zheng H, Engelhardt OG, Pleschika S, Garcia-Sastre A (1996) Negative-strand RNA viruses: genetic engineering and applications. Proc Natl Acad Sci USA 93:11354–11358

Paoletti E (1996) Applications of poxvirus vectors to vaccination: an update. Proc Natl Acad Sci USA 93:11349–11353

Porter DC, Ansardi DC, Choi WS, Morrow CD (1993) Encapsidation of genetically engineered poliovirus mini-replicons which express HIV-1 *gag* and *pol* proteins upon infection. J Virol 67:3712–3719

Porter DC, Ansardi DC, Morrow CD (1995) Encapsidation of poliovirus replicons encoding the complete human immunodeficiency virus type 1 *gag* gene using a complementation system which provides the P1 capsid protein *in trans*. J Virol 69:1548–1555

Porter DC, Wang J, Moldoveanu Z, McPherson S, Morrow CD (1997) Immunization of mice with poliovirus replicons expressing the C-fragment of tetanus toxin protects against lethal challenge with tetanus toxin. Vaccine 15:257–264

Prevec L, Christie BS, Laurie KE, Bailey (Smith) MB, Graham FL, Rosenthal KL (1991) Immune response to HIV-1 gag antigens induced by recombinant adenovirus vectors in mice and Rhesus Macaque monkeys. J Acquir Immune Defic Syndr 4:568–576

Racaniello VR, Baltimore D (1981) Cloned poliovirus complementary DNA is infectious in mammalian cells. Science 214:916–919

Ren R, Costantini FC, Gorgacz EJ, Lee JJ, Racaniello VR (1990) Transgenic mice expressing a human poliovirus receptor: a new model for poliomyelitis. Cell 63:353–362

Roden RBS, Greenstone HL, Kirnbauer R, Booy FP, Jessie J, Lowy DR, Schiller JT (1996) In vitro generation and type-specific neutralization of a human papillomavirus type 16 virion pseudotype. J Virol 70:5875–5883

Sabin AB, Boulger LR (1973) History of Sabin attenuated poliovirus oral live vaccine strains. J Biol Stand 1:115–118

Sanders DY, Cramblett HG (1974) Antibody titers to polioviruses in patients ten years after immunization with Sabin vaccine. J Pediatr 84:406–408

Schneider M, Graham FL, Prevec L (1989) Expression of the glycoprotein of vesicular stomatitis virus by infectious adenovirus vectors. J Gene Virol 70:417–427

Schnell MJ, Bluonocore L, Kretzschmar E, Johnson E, Rose JK (1996) Foreign glycoproteins expressed from recombinant vesicular stomatitis viruses are incorporated efficiently into virus particles. Proc Natl Acad Sci USA 93:11359–11365

Simmons J, Kutubuddin M, Chow M (1993) Characterization of poliovirus-specific T lymphocytes in the peripheral blood of Sabin-vaccinated humans. J Virol 67:1262–1268

Plant Expression Systems for the Production of Vaccines

J.K.-C. Ma and N.D. Vine

1 Introduction

Plant biotechnology is a rapidly expanding area, and in the last 5–10 years it has become apparent that plant systems may be particularly valuable for the expression and production of recombinant molecules. A considerable effort has focused on genetic engineering for the improvement of plant characteristics and agricultural properties, but the technology has also been applied to the production of "high-value" products, namely pharmaceutical compounds and vaccines. A particular attraction of this approach is the potential for growing immunotherapeutic reagents on an agricultural scale, thereby significantly reducing the costs of production. However, as this review will attempt to demonstrate, there are many other advantages related to the use of plants. This chapter will review the major advances in plant genetic engineering for the production of products for both active and passive immunisation. It will discuss the advantages and disadvantages of the plant

Unit of Immunology, Department of Oral Medicine and Pathology, 28th Floor, Guy's Tower, UMDS Guy's Hospital, London Bridge, London SE1 9RT, UK

expression systems as well as strategies that have been devised to overcome some of the specific problems.

2 Expressing Recombinant Molecules in Plants

There are two strategies for production of vaccines in plants. Genetic transformation of the plant genome can be achieved relatively easily, using Agrobacterium T-DNA vectors, or by micro-projectile bombardment. This results in transfected plants that can express the recombinant protein either constitutively or in a tissue-specific manner. As the foreign gene is integrated into the plant genome, traditional breeding techniques can be used to generate transgenic seed stocks for easy and stable storage or distribution. Furthermore, sexual propagation between different transgenic plants can be used to accumulate multiple foreign genes in a single plant, which is a particular benefit in the production of either multi-component vaccines or multimeric proteins, such as immunoglobulins (Igs) (Table 1).

An alternative approach involves the use of viral vectors, in which plant viruses are genetically modified to encode the foreign genes. Plants are subsequently infected with the modified virus and, thus, act as viral culture vessels. By this means, very large quantities of recombinant proteins can be generated quite rapidly, using relatively small numbers of plants (Table 2). As described below, there are limitations to the use of both transgenic plants and genetically modified plant viral vectors. It is likely that both systems will have important applications for different vaccine candidates. What is clear, however, is that both contribute to the overall versatility of using plants for production of vaccines.

2.1 Transgenic Plants for the Production of Immunogenic Proteins

The first reported vaccine candidate to be expressed in transgenic plants was the cell surface-adhesion protein – spa A or streptococcal antigen I/II from *Streptococcus mutans* (Curtiss and Cardineau 1990). This is the major bacterial cause of tooth decay, and there is a considerable body of evidence that demonstrates protective immunity against dental caries following systemic or oral immunisation with this antigen. The *spa A* gene was introduced into tobacco by agrobacterium-mediated transformation, and the protein was expressed at levels of up to 0.02% total leaf protein. An interesting aspect of this work, apart from being somewhat in advance of the rest of the field, is that spa A is a large protein, comprising over 1500 amino acids, with an approximate Mr 185,000. Thus, transgenic plants can accommodate large gene inserts and, subsequently, express sizeable recombinant proteins.

Hepatitis B surface antigen (HBsAg) was the first commercial recombinant vaccine produced in yeast. Although this vaccine is effective and elicits a protective immune response, in common with most other vaccines, the expense is prohibitive

Table 1. Recombinant proteins expressed in transgenic plants

Recombinant protein	Plant host	Best level of expression	Immunogenicity	Protective immunity
Spa A of *S. mutans*	Tobacco	0.02%	ND	ND
Surface antigen of hepatitis B	Tobacco	0.01%	Yes	ND
	Potato	0.01%	ND	ND
M protein of Hepatitis B	Potato	0.001%	ND	ND
LT of *E. coli*	Tobacco	0.0005%	Yes	ND
	Potato	0.01%	Yes	ND
CT of *Vibrio cholerae*	Potato	0.3%	ND	ND
Capsid protein of Norwalk virus	Tobacco	0.23%	Yes	ND
	Potato	0.37%	?	ND
Rabies glycoprotein	Tomato	0.001%	ND	ND
Glutamic acid decarboxylase	Tobacco and potato	0.4%	Yes	Yes
VP1 of foot and mouth disease virus	Arabidoposis	ND	Yes	Yes
V_H domain	Tobacco	0.1–1	Yes	N/A
scFv	Tobacco	0.06–0.5	Yes	N/A
scFV-KDEL	Tobacco	6.8%	Yes	N/A
Fab	Tobacco	0.04	Yes	N/A
	Arabidoposis	1.3	Yes	N/A
IgG	Tobacco	1.3	Yes	ND
SIgA	Tobacco	5.0	Yes	Yes

N/A Not applicable; *ND* Not done.

Table 2. Viral vectors for expression of coat protein – antigen fusion proteins

Viral host	Antigenic insert		Viral yield (mg/g fresh plant tissue)	Immunogenicity demonstrated
Tobacco mosaic virus	Malarial	(4aa)	ND	–
	sporozoite	(4aa)3	0.4	±
	epitopes	(6aa)2	1.2	–
	H. influenza HA	8aa	2.0	–
		18aa	0.8	–
	HIV-1 gp120	13aa	1.6	–
	Murine zona pellucida	13aa	2.5	+
Cowpea mosaic virus	Foot and mouth	8aa	–	–
	disease virus VP1	19aa	ND	–
	Human rhinovirus – 14 VP1	14aa	1.2–1.5	+
	HIV-1 gp41	22aa	1.2–1.5	+
	Mink enteritis virus VP2	17aa	1.2	+
Tomato bushy-stunt virus	HIV-1 gp120	13aa	0.9	+
Alfalfa mosaic virus/TMV	Rabies virus glycoprotein	40aa	ND	+
	HIV-1 gp120	47aa	ND	+

ND Not done.

for use in many developing countries. In 1992, MASON et al. reported the expression of HBsAg in transgenic tobacco, which assembled into virus-like particles (VLPs) similar to those of serum and yeast derived HBsAg. This group went on to demonstrate that the plant-derived HBsAg was immunogenic in mice (THANAVALA et al. 1995). A crude preparation of plant HBsAg, in which the HBsAg was approximately 3% of the total protein, was administered parenterally on three occasions, and the resulting immune responses were compared with those following immunisation with a commercial hepatitis vaccine. After priming with plant HBsAg, it was shown that the mice developed T-cell proliferative responses to the yeast recombinant HbsAg, a synthetic peptide representing a partial analogue of the protective determinant in the *S* region of HBsAg, as well as an internal image anti-idiotype monoclonal antibody that mimics the same determinant. The antibody response following immunisation was lower in magnitude than that induced by yeast recombinant HBsAg in those mice immunised with plant HBsAg. However, the data supports the notion that HBsAg expressed in transgenic tobacco is immunogenic, and it was suggested that the reduction in antibody response was due to the impurity and low concentration of the plant HBsAg preparation.

Purification of proteins from tobacco is complicated by the presence of alkaloids and phenolic compounds, a problem that could be overcome by the use of other plant hosts. Indeed HBsAg has also been expressed in transgenic potatoes (DOMANSKY et al. 1995), in addition to a chimeric gene encoding the M protein of hepatitis B virus (EHSANI et al. 1997). The immunogenicity of these potato-derived antigens has not been reported, but it would be surprising if there was any significant loss related to expression in a different plant host.

The use of transgenic potatoes to produce immunogenic proteins has in fact been established, using the B subunit of the heat labile enterotoxin (LT) from *E. coli* (HAQ et al. 1995). Enterotoxic *E. coli* cause diarrhoea through the production of toxins, including LT, which comprises a toxic A subunit and a pentameric B-subunit structure responsible for binding to the membrane glycolipid (Gm1) ganglioside that is present on all cells, including gut epithelium. The B subunits link in a non-covalent manner, and antibodies that block the binding of the pentameric B-subunit structure to cells can prevent disease. LT-B was expressed at low levels, up to 0.01% of total protein in potato tubers (and approximately tenfold less in tobacco leaves), but the recombinant protein produced was at least partially assembled, as functional binding to Gm1 ganglioside was demonstrated.

More significantly, mice that were fed the raw transgenic potato expressing LT-B generated both specific serum IgG and mucosal IgA responses. However, as with the plant HBsAg, the magnitude of the antibody responses was lower than in mice immunised with purified bacterial LT-B by gavage. In the same study, a crude extract of LT-B expressed in tobacco was used to immunise mice by gavage and similar antibody results were determined. These antibodies had neutralising activity against *E. coli* LT in vitro. This demonstrates that the host plant species is not critical for recombinant protein expression. The authors suggest that the reduced antibody response observed in animals that received plant antigens might be due to plant contaminants that limit or interfere with antigen reactivity with lymphoid tissue.

Cholera toxin (CT) from *Vibrio cholerae* is a multimeric protein that is structurally, functionally and antigenically highly homologous to *E. coli* LT. The expression of the B subunit of CT has recently been reported in transgenic potato plants (Arakawa et al. 1997). The B subunits formed oligomers (possibly pentamers) that were biochemically and immunologically similar to bacterial CT-B, and bound specifically to the Gm1 ganglioside. Expression levels of up to 0.3% of total plant protein were achieved. The interest in expressing LT-B and CT-B in plants lies not only in their potential for developing vaccines against diarrhoeal diseases, but also in the fact that these two proteins can act as oral adjuvants, particularly for mucosal antibody responses. Alternatively, CT-B can promote systemic tolerance for co-administered antigens. Thus, these recombinant proteins are powerful tools for the design of orally delivered vaccines.

Two further demonstrations of the immunogenicity of plant-derived recombinant proteins have been reported. Mason et al. (1996) have expressed the capsid protein of Norwalk virus in tobacco and potato. Norwalk virus is a common cause of acute epidemic gastro-enteritis in humans. The capsid protein expressed in baculovirus self assembles into VLPs that are immunogenic. When produced in plants, the recombinant capsid protein also assembled into VLPs that were virtually indistinguishable from those observed previously. Immunisation by gavage with the plant antigen resulted, again, in serum and mucosal antibody responses. Surprisingly, however, of the 20 mice fed transgenic potato expressing the Norwalk virus protein, 11 developed a serum IgG response and only 1 was found to have a specific mucosal IgA response. The reason for this disappointing response is unclear and, although the author suggests that impurity and low antigen concentration might again be responsible, expression levels were 0.37%, which were significantly higher than that achieved for HBsAg.

For oral vaccination with edible, transgenic plant material, potato may not be ideal, as the cooking process might denature the recombinant protein and raw potato is rather unpalatable. McGarvey et al. (1995) have used tomato for expression of the rabies virus glycoprotein, with a view to developing an oral vaccine. Using the complete unmodified gene under the control of the 35S promoter of cauliflower mosaic virus, expression of glycoprotein was found in leaf and fruit tissue. The tomato-expressed glycoprotein occurred in two forms, both with a marginally faster migration rate on sodium dodecyl sulphate polyacrylamide gel electrophoresis (SDS-PAGE) than the native glycoprotein purified from denatured rabies virus. The differences might be due, in part, to plant-specific cleavage or differences in glycosylation. Plant glycans, as discussed later, are characteristically smaller than those of animal origin. The tomato-expressed glycoprotein was, however, recognised by specific anti-sera and a monoclonal antibody, suggesting preservation of important epitopes, although the levels of expression were low (0.001% soluble protein at best).

The studies described, so far, have all demonstrated the feasibility of expressing immunogenic proteins in transgenic plants. There are a few reports that go on to demonstrate efficacy in vivo, for example protection against disease challenge. Ma et al. (1997) reported the expression of glutamic acid decarboxylase (GAD), an auto-antigen associated with diabetes, in tobacco and potato. Relatively high levels of expression were achieved and experimental mice had their diet supplemented with transgenic plant tissue (either tobacco leaves or potato tuber), with the aim of inducing oral tolerance. After 4 weeks of feeding, T-cell proliferation in response to GAD was suppressed in mice fed with plant GAD compared with control plant-fed mice. Antibody responses in GAD-fed mice also appeared to be biased towards T-helper cell type 2 (Th2) rather than Th1 responses. Of most significance was the finding that in non-obese diabetic (NOD) mice that are genetically susceptible to diabetes, feeding transgenic plant tissue containing GAD resulted in prevention of disease in 10 of 12 animals, as opposed to 4 of 12 in the control mouse group fed control plant tissue. These results compare favourably with similar reports for prevention of diabetes using GAD in other systems.

The expression and characterisation of the VP1 structural protein of foot and mouth disease virus (FMDV) in plants has recently been described, along with evidence that immunisation with this plant antigen results in protective immunity in mice (Carrillo et al. 1998). FMDV is a highly prevalent problem in livestock and VP1 contains critical epitopes that are responsible for the induction of neutralising antibodies. In this study, VP1 was expressed in *Arabidopsis thaliana*. An entirely crude plant extract was used for intra-peritoneal inoculation, and all the immunised mice developed specific serum antibodies to intact FMDV particles, VP1 and to a synthetic peptide derived from the sequence of VP1. When challenged with a virulent strain of FMDV, all 14 of the immunised mice were protected against infection, whereas the 6 sham immunised mice and the 6 non-immunised control mice all became infected. These findings represent an important demonstration of protective immunity in animals from a transgenic plant product, and it is worthwhile noting that this approach will be equally as important in veterinary science as it is in human vaccinology.

2.2 Expression of Antibody Molecules for Passive Immunotherapy

The production of antibody in transgenic plants was first described in 1989, and demonstrated an important principle, that co-expression of two recombinant gene products could lead to a correctly folded and assembled multimeric molecule in plants that was functionally identical to its mammalian counterpart (Hiatt et al. 1989). Since then, a number of groups have also expressed other antibody molecules ranging from single chain molecules to multimeric secretory antibody in whole plants and in plant cell culture. In many cases, the aim has been to modify or improve plant performance, but there is also a clear potential for the exploitation of plants as bioreactors for large-scale production of antibodies. Economic production of kilogram quantities would open many new areas of use for antibodies for medical and veterinary purposes, in particular passive immunisation.

2.2.1 Antibody Fragments

A wide range of functional recombinant antibody fragments have been described in the literature, and most of the antibody fragments described in *E. coli* have also been produced in transgenic plants. These include a single-domain antibody (dAb) in tobacco (Benvenuto et al. 1991), single-chain Fv (scFv) molecules (Owen et al. 1992), Fab (De Neve et al. 1993) and $F(ab')_2$ production in tobacco and arabidopsis. The requirements for assembly of these molecules are quite undemanding; thus, they can be produced in *E. coli* and many other heterologous expression systems. They are small, which can be advantageous for some applications, but have a reduced binding avidity and cannot elicit secondary immune effector mechanisms. As processing through the endoplasmic reticulum (ER) is not essential, the antibody fragments can be accumulated intracellularly if required, or a signal sequence can be added to direct secretion into the extracellular space.

Originally, only low levels of expression levels were achieved in plants, similar to those described previously for antigen expression. Since then, various strategies have been devised to improve yields, such as the expression of genes for antibody fragments in fusion with the endoplasmic-reticulum retention signal KDEL, which can increase scFv yield tenfold (Schouten et al. 1996) or up to 4%–6.8% of total soluble protein (Fiedler et al. 1997). Another approach that has been successful is to target scFv for expression in seeds, where the antibody fragment can accumulate to 3%–4% of the total soluble seed protein (Fiedler et al. 1997).

2.2.2 Full-Length and Multimeric Antibodies

An important advantage of plants as a recombinant expression system is the ability to assemble full-length heavy chains with light chains to form full-length antibody efficiently (Hiatt et al. 1989). Full-length antibodies are not readily assembled in bacterial-expression systems, and bivalent antibody molecules can only be produced in *E. coli* by rather complex molecular engineering. Although the constant region of the Ig has no role in antigen recognition or binding, it has important effector functions and contains several important functional regions involved in glycosylation, complement activation, phagocyte binding, the hinge region and the site for association with joining (J) chain and secretory component (in α and μ chains). Importantly, it allows bivalent antigen binding with full flexibility of the antibody molecule at the hinge region, which may be important if aggregation is a major protective mechanism. In addition, binding antigen bivalently significantly affects the strength of the antigen–antibody interaction, by virtue of an increase in avidity.

Several groups have reported the production of full-length antibodies in plants. Antigen recognition and binding is a critical and sensitive test for correct assembly, as, in the vast majority of cases, individual light or heavy chains or misfolded Ig molecules are not functional. In mammalian plasma cells, the mechanism of this assembly is only partially understood. The Ig light and heavy chains are synthesised as precursor proteins, and signal sequences direct translocation into the lumen of the ER. Within the ER, there is cleavage of the signal peptides. Stress proteins, such as BiP/GRP78 and GRP94, and enzymes, such as protein disulphide isomerase (PDI), function as chaperones that bind to unassembled heavy and light chains and direct subsequent folding and assembly.

In plants, passage of Ig chains through the ER is also required, since, in the absence of a signal peptide related either to the light- or heavy-chain gene, assembly of antibody does not take place (Hiatt et al. 1989). However, both plant and non-plant signal sequences from a variety of sources are effective for correct targeting (During et al. 1990; Hein et al. 1991). Plant chaperones homologous to mammalian BiP, GRP94 and PDI have been described within the ER (Fontes et al. 1991; Denecke et al. 1991), and expression of Ig chains in plants is associated with increased BiP and PDI expression. Furthermore, BiP and PDI are associated with Ig chains in plants (Ma et al., unpublished data). Thus, it seems likely that there are broadly similar folding and assembly mechanisms for antibodies in mammals and plants.

Several IgG mAbs that may have therapeutic applications in humans or animals have been produced in transgenic plants by academic and commercial groups. For example, one group has produced an antibody that recognises the carcinoembryonic antigen (CEA) associated with human adenocarcinoma, and might be clinically useful for imaging, as well as treatment of these cancers and their metastases. Agracetus (now part of Monsanto) in Wisconsin, USA, is also producing, in corn, an "anti-cancer" antibody that is entering phase-I clinical trials, and in soybeans, human antibodies against herpes simplex virus type 2.

Guy's 13 is a murine IgG1 that binds to the adhesion protein of *Streptococcus mutans*, which is the primary cause of dental caries. The strategy used to produce this antibody in plants was to express each Ig chain separately in different plants and to introduce the two genes together in the progeny plant by cross pollination of the individual heavy- and light-chain expressing plants. This involves two generations of plants to generate an antibody-producing plant and, by this technique, the yield of recombinant antibody is consistently high; between 1% and 5% of total soluble plant protein (Hiatt et al. 1989; Ma et al. 1994). Other groups have used double-transformation techniques (De Neve et al. 1993) or cloned the light- and heavy-chain genes together in a single agrobacterium T-DNA vector (During et al. 1990; van Engelen et al. 1994), and this can save time and effort. Guy's 13 IgG was expressed in tobacco at high levels and is relatively easy to purify in large quantities. Functionally, there was no discernible difference between the antibody expressed in plants and that expressed in other systems (Ma et al. 1994).

The ability to accumulate genes in transgenic plants by successive crosses between individually transformed parental plants is a considerable advantage when attempting to construct multimeric protein complexes, such as secretory antibodies. Secretory (s) IgA is the predominant form of Ig found in mucosal secretions, such as those of the gastro-intestinal tract. Until recently, attempts to produce monoclonal sIgA have been frustrated by the complexity of this molecule, which consists of two basic Ig monomeric units (heavy and light chains) that are dimerised by a J chain and further associated with a fourth polypeptide, secretory component (SC)

(MESTECKY and MCGHEE 1987). These modifications are believed to enhance the activity of sIgA in the mucosal environment. Dimerisation by J chain increases the avidity of binding of the antibody and enhances the potential for bacterial aggregation, whilst the secretory component confers a degree of resistance against proteolysis – an important property for antibodies within the harsh environment of the gastro-intestinal tract.

The early work on passive immunotherapy against dental caries was performed with Guy's 13, which is an IgG monoclonal antibody. However, it is likely that, in the mucosal environment of the oral cavity, a secretory antibody would be preferable to IgG. Furthermore, there is some evidence that bacterial aggregation is an important protective mechanism in this model (MA et al. 1990), and so the increased valency and avidity of a secretory antibody might be an added advantage. The carboxyl-terminal domains of the Guy's 13 IgG antibody heavy chain were modified by replacing the Cγ3 domain with Cα2 and Cα3 domains of an IgA antibody, which are required for binding to J chain and SC (MA et al. 1994).

Four transgenic *Nicotiana tabacum* plants were generated to express either the Guy's 13 κ chain, the hybrid IgA–G antibody heavy chain described above, murine J chain or rabbit SC). A series of sexual crosses was performed between these plants and filial recombinants in order to generate plants in which all four protein chains were expressed simultaneously. In the final quadruple transgenic plant, three forms of antibody were detectable by Western-blot analysis of samples prepared under non-reducing conditions. These bands were approximately M_r 210K (the expected size of monomeric IgA–G), M_r 400 K (IgA–G dimerised with J chain) and M_r 470K (dimeric IgA–G associated with SC). The assembly was very efficient, with greater than 50% of the SC being associated with dimeric IgA–G, and the sIgA–G yield from fully expanded leaves was in excess of 5% of total soluble protein, or 200–500μg per gram of fresh weight material (MA et al. 1995).

Functional studies confirmed that the sIgA–G molecule bound specifically to its native antigen and by BIAcore analysis that the binding affinity of each antigen binding site was no different to that of the native IgG. However, the avidity and functional affinity of the entire molecule was greater, which helped to confirm that a dimeric, tetravalent antibody had been assembled. It was confirmed in vivo, that the secretory form of Guy's 13 did indeed survive longer in the mucosal environment (up to 3 days) compared with the IgG form (up to 24h). Finally, in a human trial, the plant secretory Guy's 13 antibody prevented oral colonisation by *Streptococcus mutans*, thereby demonstrating, for the first time, the therapeutic application in humans of a recombinant product derived from plants (MA et al. 1998).

2.3 Modified Plant Viruses for Expression of Vaccines in Plants

Whereas transgenic plants have the advantage of stable integration of the foreign gene into the plant nuclear chromosome, the use of genetically modified plant viruses offers an alternative, more rapid means of generating extremely high levels of recombinant proteins. In general, two approaches have been used: first, by foreign gene transcription, in which the foreign gene is expressed as a soluble protein and, second, by engineering of viral coat proteins (cp) in fusion with antigenic peptides or proteins, whilst allowing the continuing assembly and formation of infectious virus particles that display antigen on their surface. Several plant viruses have been used, most successfully tobacco mosaic virus (TMV) and cowpea mosaic virus (CPMV). This field is presently dominated by two biotechnology companies: Biosource Technologies Inc., California, USA and Axis Genetics, Cambridge, UK.

2.3.1 Tobacco Mosaic Virus

TMV is a well-characterised RNA virus with a broad host range. After infection, the amount of recoverable virus is extremely high and can reach up to 50% of the dry plant weight. Thus, TMV RNA is a good candidate vector for expression of foreign genes in plants. Early attempts met with limited success, however, sometimes due to unstable constructs or because the genetic engineering of the cp interfered with virus particle formation and systemic long-distance viral spread through the plant. One successful approach to overcome this was to engineer a hybrid virus vector that included a heterologous cp gene from a second virus, odontogrossum ringspot virus (ORSV), and used the TMV and ORSV promoters to direct synthesis of the sub-genomic RNAs for the foreign gene and the ORSV cp gene, respectively (Donson et al. 1991). Following infection of tobacco, soluble foreign gene product accumulated to levels of at least 2% of the total soluble protein (Kumagai et al. 1993).

Since then, this technology has been developed further using similar viral vectors with dual sub-genomic promoters for the expression of human α-galactosidase, γ-interferon and single chain Fv antibodies. An advantage of this system is the speed with which foreign gene products can be made. DNA sequences are cloned into the viral vector, transcribed into infectious RNA, passaged through a laboratory plant that acts as a packaging host to prepare chimeric viruses, which are then used for infection of plants in the field. The foreign protein accumulates to high levels and can be harvested within days, the entire process occupying only a few weeks (Della-Cioppa and Grill 1996). Biosource Technologies has been testing this system in field trials since 1991 and currently use a 15-acre tobacco processing plant in Kentucky. Their overall estimates for therapeutic protein production suggest that yields of approximately 18kg/acre of pre-GMP purified protein or 1kg/acre of GMP pharmaceutical-grade protein could be achieved. Although as yet unreported, it should be possible to express mulitmeric proteins using this system too. It is unlikely that the size of the protein will present a problem, as, so far, proteins up to 78–80kDa have been expressed. As the recombinant proteins are targeted through the secretory pathway of the plant cell, glycosylation would also be expected.

An alternative, more widely used and described system for producing antigens using viral vectors involves the fusion of peptides or polypeptides to the cp that normally envelopes the virus. This principle was first demonstrated in 1986, when a TMV cp fusion with an eight-amino acid polio virus capsid peptide was engineered (Haynes et al. 1986). The modified protein was expressed in *E. coli* and, due to the self-assembling properties of TMV cp, formed into VLPs that displayed the polio virus epitope. These subsequently induced neutralising antibodies in experimental rats. One advantage of this approach is that the foreign proteins are expressed in multiple copies and in particulate form on a large carrier – a combination that can be highly immunogenic.

TMV cp can be manipulated in several ways to express fusion proteins without affecting assembly or the ability of the virus to spread and infect the plant systemically. Thus, foreign peptides can be expressed as amino- or carboxyl-terminal fusions or inserted at a structurally unimportant site within the cp. Constructs can also be designed that have a leaky termination codon, separating the cp from the DNA encoding foreign peptide. These viral vectors produce both native and recombinant cp resulting in functional virus that contains a proportion of cp displaying the foreign peptide.

Turpen et al. (1995) described the expression of malarial epitopes in TMV. One or three tandem copies of the sporozoite B-cell epitope AGDR was inserted into a surface loop of the TMV cp. A third construct was made in which two copies of the epitope QGPGAP were inserted at the carboxyl terminus of TMV cp, downstream of a leaky stop codon. Following infection of plants, recombinant viruses expressing cp with three copies of AGDR or two copies of QGPGAP were recovered in high yields: 0.4–1.2mg/g fresh weight of plant material. Both reacted to monoclonal antibodies specific for the relevant epitopes in enzyme-linked immunosorbent assay (ELISA) and Western blot, which suggests that the epitopes were displayed at the surface of the viral particles. The authors also indicated preliminary results that suggest immunogenicity to the AGDR peptide in experimental mice.

Two peptides (18 amino acids and 8 amino acids) from *H. influenzae* haemagglutinin and a 13 amino acid peptide from the human immunodeficiency virus (HIV) envelope protein have also been expressed as carboxyl-terminal fusions of TMV cp (Sugiyama et al. 1995). In each case, the foreign DNA was cloned downstream of the cp gene using a leaky stop codon. Virus particles were purified from infected plants and these were recognised in Western blot by antisera specific for the relevant peptides, demonstrating correct expression. The fusion proteins were susceptible to trypsin digestion (although native TMV cp was not), and positive precipitin reactions were also demonstrated to suggest that the foreign peptides were located on the viral surface.

TMV cp has also been modified by the insertion of a 13 amino acid sequence from the murine zona pellucida ZP3 protein that is associated with antibody-mediated contraception (Fitchen et al. 1995). Modified VLPs containing the hybrid cp were purified from infected plant tissue and used in immunisation studies in mice. The two strains of mice generated an antibody response to both the ZP3 epitopes and the carrier protein. The circulating anti-ZP3 antibodies localised to the zona pellucida in vivo, but although there was some evidence of ovarian pathology, this had no impact on the fertility of the treated mice. Further work may be needed to define a more effective epitope, but this study demonstrates the feasibility of using modified TMV to produce immunogenic peptides for use as vaccines.

2.3.2 Cowpea Mosaic Virus

CPMV is a positive-strand RNA virus. Viral particles consist of 60 copies each of two protein subunits, large (L) and small (S), which are arranged with icosohedral symmetry. Usha et al. (1993) reported that a peptide from FMDV could be inserted into an eight amino acid surface loop of the Scp, resulting in chimeric viral particles displaying 60 copies each of the peptide. However, the genetic modification was found to be unstable, homologous recombination resulting in reversion to wild-type viral sequence.

Porta et al. (1994) modified the viral vector by selecting a different site for insertion of foreign sequences, whilst ensuring that no CPMV-specific sequences were deleted. By this strategy, a 19-amino acid peptide from VP1 of FMDV, a 14-amino acid peptide from VP1 of human rhinovirus-14 and a 22-amino acid peptide from gp41 of HIV-1 (the "Kennedy" epitope) were all successfully and stably expressed by modified CPMV. The HRV-14 and HIV modified viruses were harvested from infected plants and, in both cases, immunisation of mice with the purified viral particles resulted in antisera specific for the expressed foreign peptide. Anti-HIV-1-neutralising antibodies were elicited in three different strains of mice using doses as low as 1μg virus (equivalent to 17ng HIV-1 gp41 peptide) (McLain et al. 1996).

The use of the CPMV expression system has been extended to protective vaccination in animals. Dalsgaard et al. (1997) have described the display of a 17 amino acid linear epitope from the VP2 capsid protein of mink enteritis virus (MEV). Chimeric virus particles were propagated in black-eyed bean plants (*Vigna unguiculata*), resulting in yields of 1–1.2mg/g of plant (one plant gave approximately 10g fresh material). The virus particles were mixed with a saponin adjuvant and adsorbed to aluminium hydroxide gel, and 100μg or 1mg (equivalent to 2μg and 20μg of peptide, respectively) were administered as a single subcutaneous dose to groups of six mink. Antibody responses to MEV VP2 epitope were elicited in a dose-dependent manner and protection against infection was observed with reduced viral shedding. Significantly, 11 of 12 immunised mink were protected against clinical disease. These results meet the potency requirements for inactivated MEV vaccine, as defined by the United States' Food and Drug Association (FDA). Anti-CPMV antibodies were also detected in 9 of 12 animals that received the recombinant viral vaccine. This may prove to be problematic if repeated immunisations are required or if CPMV is used to deliver several vaccines sequentially.

The CPMV expression system is not entirely predictable and will need further refining. At least one insert (an FMDV epitope) resulted in virus that failed to spread systemically through the plant, thereby compromising viral infectivity and yield. Indeed the virus particles appeared to remain adherent to plant cell membranes. This, however, might also have been due to the presence of an RGD sequence within the FDMV epitope, which is a consensus cell attachment sequence (PORTA et al. 1994).

2.3.3 Other Viruses

TMV and CPMV have received the majority of attention in the literature, but there are several other viral vector candidates and these are reviewed by PORTA and LOMONOSSOFF (1996).

Tomato bushy stunt virus (TBSV) has been used to display a 13 amino acid V3 loop peptide from HIV1 gp120 (JOELSON et al. 1997). The foreign peptide was fused to the carboxyl terminus of the cp and, following infection and purification from tobacco plants, the structure of the virus particle resulted in external display of 180 copies of the antigen per virus. The epitope was recognised by a specific monoclonal antibody as well as by sera from HIV-1-positive patients. Immunisation of mice with the chimeric TBSV particles resulted in a specific primary antibody response to the peptide, but also a strong antibody response to TBSV. Somewhat surprisingly, a second immunisation failed to boost either antibody response.

An important limitation to the engineering of most viral cps is the size restriction imposed on the antigenic insert, due to potential interference with virus particle assembly. Thus, only peptides up to 25 amino acids have been introduced successfully into TMV cp, and 30 amino acids in CPMV cp. The cp of alfalfa mosaic virus (AlMV) is quite flexible and can form particles of different sizes and shapes depending on the length of the encapsulated RNA. YUSIBOV et al. (1997) have used it as part of a chimeric TMV construct, and inserted two antigens into the AlMV cp reading frame, under the control of a TMV cp subgenomic promoter within a vector derived from TMV. The amino terminus of the cp is at the surface of the viral particles and does not interfere with viral assembly. They used a synthetic 40 amino acid sequence, derived from Rabies virus glycoprotein and a 47 amino acid sequence from the V3 loop of HIV-1 MN isolate. Both epitopes were expressed by recombinant virus, and immunised mice generated antisera that gave 90% neutralisation of rabies virus and up to 80% neutralisation of an HIV-1 MN isolate.

Another virus vector that stands out by offering the potential to express large fusion proteins is potato virus X (PVX). The outer coat of this virus is composed of several thousands of copies of the cp, thus allowing very large numbers of antigenic molecules to be displayed per viral particle. CRUZ et al. (1996) have described the fusion between a 27kDa marker protein with the amino terminus of the 25kDa PVX cp. Viral assembly still took place and, although the chimeric viral particles were over twice the diameter of normal wild-type virions, local and systemic viral spread and infection in the host plants was still possible. One disadvantage of this system may be related to purification of the recombinant antigen, as the modified virus can be prone to precipitation and, subsequently, be resistant to resolubilisation. However this virus system greatly broadens the possibilities for foreign protein production in plants and has exciting potential in the further development of plant derived vaccines.

3 Glycosylation of Recombinant Proteins in Plants

Protein modification by glycosylation is found in all higher eukaryotes and plant proteins contain *N*-linked as well as *O*-linked glycans. Variations between the glycans associated with native proteins and recombinant forms may complicate immunotherapy, whatever the heterologous expression system, and this is not a problem that is specific to recombinant proteins produced in plants. However, it is important to understand the differences between plant and mammalian glycans in order to evaluate their relative importance.

It had previously been demonstrated that the *N*-linked core, high-mannose type glycans have identical structures in plants, mammals and other organisms (Sturm et al. 1987; Faye et al. 1989), which are subsequently modified to complex glycans via a number of steps. Native complex glycans in plant proteins can be quite heterogeneous, but they tend to be smaller than mammalian complex glycans and differ in the terminal sugar residues. For example, a xylose residue-linked β(1,2) to the β-linked mannose residue of the glycan core, and/or an α(1,3)-fucose residue in place of an α(1,6)-fucose linked to the proximal glucosamine, are frequently found in plants, but not in mammals (Sturm et al. 1987). However, *N*-acetyl neuraminic acid (NANA), which is a prevalent terminal residue in mammals has not been identified in plants (nor for that matter in insect cells or yeast).

A structural comparison of the glycans associated with Guy's 13 IgG expressed either in plants or in murine hybridoma cells has recently been performed (Lerouge et al., accepted for publication). The results demonstrated that the same glycosylation sites were utilised in both systems and that, compared with the murine antibody, the glycans on the plant antibody were more heterogeneous. In addition to high-mannose type glycans, approximately two-thirds of the plant antibodies had β(1,2)-xylose and α(1,3)-fucose, as described previously. The differences in glycosylation patterns of plant antibodies have no effect on antigen binding or specificity. However, these types of plant glycans associated with plant glycoproteins can be quite immunogenic in humans, although it remains to be demonstrated whether plant glycans presented by mammalian proteins are equally immunogenic.

In the recent human study of oral administration of plant secretory antibody, no evidence for an immune response to the plant recombinant glycoprotein was detected after six applications of antibody (Ma et al. 1998). Nevertheless, for systemic administration, it may be necessary to remove the complex glycans, or to alter the heavy-chain sequence to remove the sites for *N*-linked glycosylation. An alternative, more elegant approach could also be adopted, by the use of mutant plants that lack enzymes involved in the complex glycosylation pathway (von Schaewen et al. 1993).

4 Overall Advantages in Expressing Vaccines in Plants

There are a number of considerations to be taken into account in the development of any recombinant vaccine. These include fidelity of the antigen, in terms of antigenicity and immunogenicity which, in turn, depends on folding, structure and glycosylation; antigen stability; ease of purification; the potential for scale-up production to produce sufficient quantities; and cost and safety issues. The strength of bioengineering in plants is that there are significant advantages over other expression systems with respect to many of these issues. With regard to protein folding and structure, small peptides, polypeptides and even complex proteins can be expressed in plants that are fully assembled and functional. For larger molecules, this is associated with the presence of ER-resident chaperones that are homologous with those involved in protein assembly in mammalian cells. Targeting recombinant proteins for secretion through the ER and Golgi apparatus is achieved using either native or plant leader sequences, and this also ensures that *N*-glycosylation take place. In plants, glycosylation differs from mammals in the complex glycans, but for the recombinant proteins expressed so far, this has not led to any loss of structure or function.

The storage of genes and gene products in plants can be very stable. Transgenic plants can be conveniently self-fertilised to produce stable, true breeding lines, propagated by conventional horticultural techniques and stored and distributed as seeds. The expressed recombinant protein can be targeted to stable environments within the plant, for example the extracellular apoplastic space. Alternatively, tissue-specific promoters can be used to direct expression in storage organs such as seeds or tubers. Extraction and purification from these sites is generally simple.

One of the most obvious benefits of plants is the potential for scale-up production, in which virtually limitless amounts of recombinant protein could be grown at minimal cost. Plants are easy to grow; unlike bacteria or animal cells their cultivation is straightforward and does not require specialist media or equipment or involve toxic chemicals. Various estimates have been made of the commercial advantages of expressing IgG antibodies in plants. Agracetus developed a strain of corn that would allow the production of 1.5Kg of pharmaceutical-quality antibodies per acre of corn and The Scripps Research Institute have estimated that antibodies could be produced at approximately US$1/g. The use of plants also avoids many of the potential safety issues associated with contaminating mammalian viruses, as well as ethical considerations involving the use of animals.

There are currently two approaches for introducing foreign genes into plant cells, both of which are versatile and can utilise a number of plant hosts. The use of viral-based vectors results in the rapid production of very large quantities of recombinant product, but at present appears to be less suitable for larger complex molecules, such as full-length antibodies; for this, the transgenic approach is favoured. However, a persistent problem in the expression of vaccine candidates in transgenic plants has been the disappointingly low levels of expression, although several approaches are being adopted to overcome this technical problem. The exception is antibodies for which levels of 1–5% total protein are achieved consistently. The reasons for this are still unclear, but may be due, in part, to the stability conferred by assembly of a multimeric protein. It seems likely that, even at current levels, sufficient antibody could be "grown" in plants, for most medical applications, on only a few acres of land and at minimal cost relative to alternative methods of production.

References

Arakawa T, Chong DK, Merritt JL, Langridge WH (1997) Expression of cholera toxin B subunit oligomers in transgenic potato plants. Transgenic Res 6:403–413

Benvenuto E, Ordas RJ, Tavazza R, Ancora G, Biocca S, Cattaneo A, Galeffi P (1991) 'Phytoantibodies': a general vector for the expression of immunoglobulin domains in transgenic plants. Plant Mol Biol 17:865–874

Carrillo C, Wigdorovitz A, Oliveros JC, Zamorano PI, Gomez N, Salinas J, Escribano JM, Borca MV (1998) Protective immune response to foot-and-mouth disease virus with VP1 expressed in transgenic plants. J Virol 72:1688–1690

Cruz SS, Chapman S, Roberts AG, Roberts IM, Prior DA (1996) Assembly and movement of a plant virus carrying a green fluorescent protein overcoat. Proc Natl Acad Sci USA 93:6286–6290

Curtiss R, Cardineau GA (1990) World Intellectual Property Organization PCT/US89/03799

Dalsgaard K, Uttenthal A, Jones TD, Xu F, Merryweather A, Hamilton WD, Boshuizen RS, Kamstrup S, Lomonossoff GP, Vela C, Casal JI, Meloen RH, Rodgers PB (1997) Plant-derived vaccine protects target animals against a viral disease. Nat Biotechnol 15:248–252

De Neve M, De Loose M, Jacobs A, Van Houdt H, Kaluza B, Weidle U, Depicker A (1993) Assembly of an antibody and its derived antibody fragment in Nicotiana and Arabidopsis. Transgenic Res 2:227–237

Della-Cioppa G, Grill LK (1996) Production of novel compounds in higher plants by transfection with RNA viral vectors. Ann N Y Acad Sci 792:57–61

Denecke J, Goldman MH, Demolder J, Seurinck J, Botterman J (1991) The tobacco luminal binding protein is encoded by a multigene family [published erratum appears in Plant Cell (1991) 3:1251]. Plant Cell 3:1025–1035

Domansky N, Ehsani P, Salmanian A-H, Medvedeva T (1995) Organ specific expression of hepatitis B surface antigen in potato. Biotechnol Lett 17:863–866

Donson J, Kearney CM, Hilf ME, Dawson WO (1991) Systemic expression of a bacterial gene by a tobacco mosaic virus-based vector. Proc Natl Acad Sci USA 88:7204–7208

During K, Hippe S, Kreuzaler F, Schell J (1990) Synthesis and self assembly of a functional antibody in transgenic Nicotiana tabacum. Plant Mol Biol 15:281–293

Ehsani P, Khabiri A, Domansky NN (1997) Polypeptides of hepatitis B surface antigen produced in transgenic potato. Gene 190:107–111

Faye L, Johnson KD, Sturm A, Chrispeels MJ (1989) Structure, biosynthesis and function of asparagine-linked glycans on plant glycoproteins. Physiol Plant 75:309–314

Fiedler U, Phillips J, Artsaenko O, Conrad U (1997) Optimization of scFv antibody production in transgenic plants. Immunotechnology 3:205–216

Fitchen J, Beachy RN, Hein MB (1995) Plant virus expressing hybrid coat protein with added murine epitope elicits autoantibody response. Vaccine 13:1051–1057

Fontes EBP, Shank BB, Wrobel RL, Moose SP, O'Brian GR, Wurtzel ET, Boston RS (1991) Characterization of an immunoglobulin binding-protein homolog in the maize floury-2 endosperm mutant. Plant Cell 3:483–496

Haq TA, Mason HS, Clements JD, Arntzen CJ (1995) Oral immunization with a recombinant bacterial antigen produced in transgenic plants (comments). Science 268:714–716

Haynes JR, Cunningham J, von Seefried A, Lennick M, Garvin RT, Shen S (1986) Development of a genetically engineered, candidate polio vaccine employing the self-assembling properties of the tobacco mosaic virus coat protein. Biotechnology 4:637–641

Hein MB, Tang Y, McLeod DA, Janda KD, Hiatt AC (1991) Evaluation of immunoglobulins from plant cells. Biotechnol Prog 7:455–461

Hiatt AC, Cafferkey R, Bowdish K (1989) Production of antibodies in transgenic plants. Nature 342:76–78

Joelson T, Akerblom L, Oxelfelt P, Strandberg B, Morris TJ (1997) Presentation of a foreign peptide on the surface of tomato bushy stunt virus. J Gen Virol 78:1213–1217

Kumagai MH, Turpen TH, Weinzettl N, Della-Cioppa G, Turpen AM, Donson J, et al. (1993) Rapid, high-level expression of biologically active a-tricosanthin in transfected plants by an RNA viral vector. Proc Natl Acad Sci USA 90:427–430

Ma JK-C, Hunjan M, Smith R, Kelly C, Lehner T (1990) An investigation into the mechanism of protection by local passive immunisation with monoclonal antibodies against Streptococcus mutans. Infect Immun 58:3407–3414

Ma JK-C, Lehner T, Stabila P, Fux CI, Hiatt A (1994) Assembly of monoclonal antibodies with IgG1 and IgA heavy chain domains in transgenic tobacco plants. Eur J Immunol 24:131–138

Ma JK-C, Hiatt A, Hein MB, Vine N, Wang F, Stabila P, van Dolleweerd C, Mostov K, Lehner T (1995) Generation and assembly of secretory antibodies in plants. Science 268:716–719

Ma SW, Zhao DL, Yin ZQ, Mukherjee R, Singh B, Qin HY, Stiller CR (1997) Transgenic plants expressing autoantigens fed to mice to induce oral immune tolerance. Nat Med 3:793–796

Ma JK-C, Hikmat BY, Wycoff K, Vine N, Chargelegue D, Yu L, Hein MB, Lehner T (1998) Characterization of a recombinant plant monoclonal secretory antibody and preventive immunotherapy in humans. Nat Med 4:(in press)

Mason HS, Lam DM, Arntzen CJ (1992) Expression of hepatitis B surface antigen in transgenic plants. Proc Natl Acad Sci USA 89:11745–11749

Mason HS, Ball JM, Shi JJ, Jiang X, Estes MK, Arntzen CJ (1996) Expression of Norwalk virus capsid protein in transgenic tobacco and potato and its oral immunogenicity in mice. Proc Natl Acad Sci USA 93:5335–5340

McGarvey PB, Hammond J, Dienelt MM, Hooper C, Fu Z-F, Dietzschold B, Koprowski H, Michaels FH (1995) Expression of the rabies virus glycoprotein in transgenic tomatoes. Biotechnology 13:1484–1487

McLain L, Durrani Z, Wisniewski LA, Porta C, Lomonossoff GP., Dimmock NJ (1996) Stimulation of neutralizing antibodies to human immunodeficiency virus type 1 in three strains of mice immunized with a 22 amino acid peptide of gp41 expressed on the surface of a plant virus. Vaccine 14:799–810

Mestecky J, McGhee JR (1987) Immunoglobulin A (IgA): molecular and cellular interactions involved in IgA biosynthesis and immune response. Adv Immunol 40:153–245

Owen M, Gandecha A, Cockburn B, Whitelam G (1992) Synthesis of a functional anti-phytochrome single chain Fv protein in transgenic tobacco. Biotech 10:790–794

Porta C, Spall VE, Loveland J, Johnson JE, Barker PJ (1994) Development of cowpea mosaic virus as a high-yielding system for the presentation of foreign peptides. Virology 202:949–955

Porta C, Lomonossoff GP (1996) Use of viral replicons for the expression of genes in plants (review) [73 refs]. Mol Biotechnol 5:209–221

Schouten A, Roosien J, van Engelen FA, de Jong GAM, Borst-Vrenssen AWM, Zilverentant JF, Bosch D, Stiekema WJ, Gommers FJ, Schots A, Bakker J (1996) The C-terminal KDEL sequence increases the expression level of a single-chain antibody designed to be targeted to both the cytosol and the secretory pathway in transgenic tobacco. Plant Mol Biol 30:781–793

Sturm A, Van Kuik JA, Vliegenthart JF, Chrispeels MJ (1987) Structure, position, and biosynthesis of the high mannose and the complex oligosaccharide side chains of the bean storage protein phaseolin. J Biol Chem 262:13392–13403

Sugiyama Y, Hamamoto H, Takemoto S, Watanabe Y, Okada Y (1995) Systemic production of foreign peptides on the particle surface of tobacco mosaic virus. FEBS Lett 359:247–250

Thanavala Y, Yang YF, Lyons P, Mason HS, Arntzen C (1995) Immunogenicity of transgenic plant-derived hepatitis B surface antigen. Proc Natl Acad Sci USA 92:3358–3361

Turpen TH, Reinl SJ, Charoenvit Y, Hoffman SL, Fallarme V, Grill LK (1995) Malarial epitopes expressed on the surface of recombinant tobacco mosaic virus. Biotechnology 13:53–57

Usha R, Rohll JB, Spall VE, Shanks M, Maule AJ, Johnson JE (1993) Expression of an animal virus antigenic site on the surface of a plant virus particle. Virology 197:366–374 van Engelen FA, Schouten A, Molthoff JW, Roosien J, Salinas J, Dirkse W, Schots A, Bakker J, Gommers FJ, Jongsma MA, et al. (1994) Coordinate expression of antibody subunit genes yields high levels of functional antibodies in roots of transgenic tobacco. Plant Mol Biol 26:1701–1710

von Schaewen A, Sturm A, O'Neill J, Chrispeels MJ (1993) Isolation of a mutant arabidopsis plant that lacks N-acetyl glucosaminyl transferase I and is unable to synthesise Golgi-modified complex N-linked glycans. Plant Physiol 102:1109–1118

Yusibov V, Modelska A, Steplewski K, Agadjanyan M, Hooper DC, Koprowski H (1997) Antigens produced in plants by infection with chimeric plant viruses immunize against rabies virus and HIV-1. Proc Natl Acad Sci USA 94:5784–5788

Subject Index

Printing: Saladruck, Berlin
Binding: Buchbinderei Saladruck, Berlin

Current Topics in Microbiology and Immunology

Volumes published since 1989 (and still available)

Vol. 196: **Koprowski, Hilary; Maeda, Hiroshi (Eds.):** The Role of Nitric Oxide in Physiology and Pathophysiology. 1995. 21 figs. IX, 90 pp. ISBN 3-540-58214-2

Vol. 197: **Meyer, Peter (Ed.):** Gene Silencing in Higher Plants and Related Phenomena in Other Eukaryotes. 1995. 17 figs. IX, 232 pp. ISBN 3-540-58236-3

Vol. 198: **Griffiths, Gillian M.; Tschopp, Jürg (Eds.):** Pathways for Cytolysis. 1995. 45 figs. IX, 224 pp. ISBN 3-540-58725-X

Vol. 199/I: **Doerfler, Walter; Böhm, Petra (Eds.):** The Molecular Repertoire of Adenoviruses I. 1995. 51 figs. XIII, 280 pp. ISBN 3-540-58828-0

Vol. 199/II: **Doerfler, Walter; Böhm, Petra (Eds.):** The Molecular Repertoire of Adenoviruses II. 1995. 36 figs. XIII, 278 pp. ISBN 3-540-58829-9

Vol. 199/III: **Doerfler, Walter; Böhm, Petra (Eds.):** The Molecular Repertoire of Adenoviruses III. 1995. 51 figs. XIII, 310 pp. ISBN 3-540-58987-2

Vol. 200: **Kroemer, Guido; Martinez-A., Carlos (Eds.):** Apoptosis in Immunology. 1995. 14 figs. XI, 242 pp. ISBN 3-540-58756-X

Vol. 201: **Kosco-Vilbois, Marie H. (Ed.):** An Antigen Depository of the Immune System: Follicular Dendritic Cells. 1995. 39 figs. IX, 209 pp. ISBN 3-540-59013-7

Vol. 202: **Oldstone, Michael B. A.; Vitković, Ljubiša (Eds.):** HIV and Dementia. 1995. 40 figs. XIII, 279 pp. ISBN 3-540-59117-6

Vol. 203: **Sarnow, Peter (Ed.):** Cap-Independent Translation. 1995. 31 figs. XI, 183 pp. ISBN 3-540-59121-4

Vol. 204: **Saedler, Heinz; Gierl, Alfons (Eds.):** Transposable Elements. 1995. 42 figs. IX, 234 pp. ISBN 3-540-59342-X

Vol. 205: **Littman, Dan R. (Ed.):** The CD4 Molecule. 1995. 29 figs. XIII, 182 pp. ISBN 3-540-59344-6

Vol. 206: **Chisari, Francis V.; Oldstone, Michael B. A. (Eds.):** Transgenic Models of Human Viral and Immunological Disease. 1995. 53 figs. XI, 345 pp. ISBN 3-540-59341-1

Vol. 207: **Prusiner, Stanley B. (Ed.):** Prions Prions Prions. 1995. 42 figs. VII, 163 pp. ISBN 3-540-59343-8

Vol. 208: **Farnham, Peggy J. (Ed.):** Transcriptional Control of Cell Growth. 1995. 17 figs. IX, 141 pp. ISBN 3-540-60113-9

Vol. 209: **Miller, Virginia L. (Ed.):** Bacterial Invasiveness. 1996. 16 figs. IX, 115 pp. ISBN 3-540-60065-5

Vol. 210: **Potter, Michael; Rose, Noel R. (Eds.):** Immunology of Silicones. 1996. 136 figs. XX, 430 pp. ISBN 3-540-60272-0

Vol. 211: **Wolff, Linda; Perkins, Archibald S. (Eds.):** Molecular Aspects of Myeloid Stem Cell Development. 1996. 98 figs. XIV, 298 pp. ISBN 3-540-60414-6

Vol. 212: **Vainio, Olli; Imhof, Beat A. (Eds.):** Immunology and Developmental Biology of the Chicken. 1996. 43 figs. IX, 281 pp. ISBN 3-540-60585-1

Vol. 213/I: **Günthert, Ursula; Birchmeier, Walter (Eds.):** Attempts to Understand Metastasis Formation I. 1996. 35 figs. XV, 293 pp. ISBN 3-540-60680-7

Vol. 213/II: **Günthert, Ursula; Birchmeier, Walter (Eds.):** Attempts to Understand Metastasis Formation II. 1996. 33 figs. XV, 288 pp. ISBN 3-540-60681-5

Vol. 213/III: **Günthert, Ursula; Schlag, Peter M.; Birchmeler, Walter (Eds.):** Attempts to Understand Metastasis Formation III. 1996. 14 figs. XV, 262 pp. ISBN 3-540-60682-3

Vol. 214: **Kräusslich, Hans-Georg (Ed.):** Morphogenesis and Maturation of Retroviruses. 1996. 34 figs. XI, 344 pp. ISBN 3-540-60928-8

Vol. 215: **Shinnick, Thomas M. (Ed.):** Tuberculosis. 1996. 46 figs. XI, 307 pp. ISBN 3-540-60985-7

Vol. 216: **Rietschel, Ernst Th.; Wagner, Hermann (Eds.):** Pathology of Septic Shock. 1996. 34 figs. X, 321 pp. ISBN 3-540-61026-X

Vol. 217: **Jessberger, Rolf; Lieber, Michael R. (Eds.):** Molecular Analysis of DNA Rearrangements in the Immune System. 1996. 43 figs. IX, 224 pp. ISBN 3-540-61037-5

Vol. 218: **Berns, Kenneth I.; Giraud, Catherine (Eds.):** Adeno-Associated Virus (AAV) Vectors in Gene Therapy. 1996. 38 figs. IX,173 pp. ISBN 3-540-61076-6

Vol. 219: **Gross, Uwe (Ed.):** Toxoplasma gondii. 1996. 31 figs. XI, 274 pp. ISBN 3-540-61300-5

Vol. 220: **Rauscher, Frank J. III; Vogt, Peter K. (Eds.):** Chromosomal Translocations and Oncogenic Transcription Factors. 1997. 28 figs. XI, 166 pp. ISBN 3-540-61402-8

Vol. 221: **Kastan, Michael B. (Ed.):** Genetic Instability and Tumorigenesis. 1997. 12 figs.VII, 180 pp. ISBN 3-540-61518-0

Vol. 222: **Olding, Lars B. (Ed.):** Reproductive Immunology. 1997. 17 figs. XII, 219 pp. ISBN 3-540-61888-0

Vol. 223: **Tracy, S.; Chapman, N. M.; Mahy, B. W. J. (Eds.):** The Coxsackie B Viruses. 1997. 37 figs. VIII, 336 pp. ISBN 3-540-62390-6

Vol. 224: **Potter, Michael; Melchers, Fritz (Eds.):** C-Myc in B-Cell Neoplasia. 1997. 94 figs. XII, 291 pp. ISBN 3-540-62892-4

Vol. 225: **Vogt, Peter K.; Mahan, Michael J. (Eds.):** Bacterial Infection: Close Encounters at the Host Pathogen Interface. 1998. 15 figs. IX, 169 pp. ISBN 3-540-63260-3

Vol. 226: **Koprowski, Hilary; Weiner, David B. (Eds.):** DNA Vaccination/Genetic Vaccination. 1998. 31 figs. XVIII, 198 pp. ISBN 3-540-63392-8

Vol. 227: **Vogt, Peter K.; Reed, Steven I. (Eds.):** Cyclin Dependent Kinase (CDK) Inhibitors. 1998. 15 figs. XII, 169 pp. ISBN 3-540-63429-0

Vol. 228: **Pawson, Anthony I. (Ed.):** Protein Modules in Signal Transduction. 1998. 42 figs. IX, 368 pp. ISBN 3-540-63396-0

Vol. 229: **Kelsoe, Garnett; Flajnik, Martin (Eds.):** Somatic Diversification of Immune Responses. 1998. 38 figs. IX, 221 pp. ISBN 3-540-63608-0

Vol. 230: **Kärre, Klas; Colonna, Marco (Eds.):** Specificity, Function, and Development of NK Cells. 1998. 22 figs. IX, 248 pp. ISBN 3-540-63941-1

Vol. 231: **Holzmann, Bernhard; Wagner, Hermann (Eds.):** Leukocyte Integrins in the Immune System and Malignant Disease. 1998. 40 figs. XIII, 189 pp. ISBN 3-540-63609-9

Vol. 232: **Whitton, J. Lindsay (Ed.):** Antigen Presentation. 1998. 11 figs. IX, 244 pp. ISBN 3-540-63813-X

Vol. 233/I: **Tyler, Kenneth L.; Oldstone, Michael B. A. (Eds.):** Reoviruses I. 1998. 29 figs. XVIII, 223 pp. ISBN 3-540-63946-2

Vol. 233/II: **Tyler, Kenneth L.; Oldstone, Michael B. A. (Eds.):** Reoviruses II. 1998. 45 figs. XVI, 187 pp. ISBN 3-540-63947-0

Vol. 234: **Frankel, Arthur E. (Ed.):** Clinical Applications of Immunotoxins. 1999. 16 figs. IX, 122 pp. ISBN 3-540-64097-5

Vol. 235: **Klenk, Hans-Dieter (Ed.):** Marburg and Ebola Viruses. 1999. 34 figs. XI, 225 pp. ISBN 3-540-64729-5